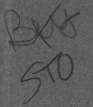

INDUSTRIAL AND ENGINEERING MATERIALS

Henry R. Clauser

Materials Consultant
Formerly Editor and Publisher, *Materials Engineering*

**McGRAW-HILL
BOOK COMPANY**

New York
St. Louis
Dallas
San Francisco
Auckland
Düsseldorf
Johannesburg
Kuala Lumpur
London
Mexico
Montreal
New Delhi
Panama
Paris
São Paulo
Singapore
Sydney
Tokyo
Toronto

Library of Congress Cataloging in Publication Data

Clauser, H R
  Industrial and engineering materials.

  Includes bibliographies.
  1. Materials. I. Title.
TA403.C54   620.1'1   74–22478
ISBN 0–07–011285–1

INDUSTRIAL AND ENGINEERING MATERIALS

Copyright © 1975 by McGraw-Hill, Inc. All rights reserved. Printed in the United States of America. No part of this publication may be reproduced, stored in a retrieval system, or transmitted, in any form or by any means, electronic, mechanical, photocopying, recording, or otherwise, without the prior written permission of the publisher.

1 2 3 4 5 6 7 8 9 0 DODO 7 8 3 2 1 0 9 8 7 6 5

The editors for this book were Donald E. Hepler and Alice V. Manning, the designer was Paddy Bareham, and the production supervisor was Laurence Charnow. It was set in Melior by Monotype Composition Company, Inc.
Printed and bound by R. R. Donnelley & Sons Company.

CONTENTS

PREFACE	vii
1 INTRODUCTION	1
2 MATERIALS APPLICATION PRINCIPLES	7
Materials Application Analysis	7
Materials Selection	13
3 THE NATURE OF MATERIALS	23
Microstructures	24
Macrostructures	32
4 PROPERTY DEFINITIONS	37
Mechanical Properties	37
Thermal Properties	60
Chemical Properties	63
Electrical and Magnetic Properties	67
High-Energy Radiation Effects	69
Other Properties	70
5 METALLIC MATERIALS	75
Structure of Metals	76
Metal Deformation	84

Heat Treatments	88
Metal-Processing Methods	93

6 FERROUS METALS — 103

Composition and Structure	103
Steel	114
Cast Irons	139

7 NONFERROUS METALS — 155

Aluminum	156
Magnesium	166
Titanium	169
Beryllium	173
Copper	174
Zinc	184
Low-Melting Metals	186
Nickel and Cobalt	189
Refractory Metals	193
Precious Metals	197

8 PLASTICS MATERIALS — 203

Nature and Major Characteristics	204
Plastics Processing	210

9 THERMOPLASTICS AND THERMOSETS — 221

Thermoplastics	221
Thermosetting Plastics	250
Fiber-Reinforced Plastics	260

10 ELASTOMERS — 273

Structure and Properties	274
Types of Elastomers	278

11 WOOD AND PAPER — 297

Nature and Structure	297
Kinds and Grades	299

	Properties	304
	Modified Woods	310
	Plywoods	316
	Paper	320
12	**FIBERS AND TEXTILES**	**329**
	Fibers	330
	Textiles	331
	Properties of Fibers and Textiles	336
13	**CERAMIC MATERIALS**	**345**
	Composition and Structure	346
	Processing	347
	Properties and Characteristics	349
	Technical and Industrial Ceramics	352
	Glass	361
	Carbon and Graphite	370
	Mica	377
	Asbestos	379
14	**COMPOSITE MATERIALS**	**381**
	Fiber Composites	387
	Particulate Composites	393
	Laminar Composites	398
	Flake Composites	404
	Filled Composites	406
15	**FINISHES AND COATINGS**	**409**
	Organic Coatings	411
	Metallic Coatings	419
	Conversion Finishes	424
	Ceramic Coatings	426
	INDEX	**431**

PREFACE

Materials are all around us all the time. We use them almost daily no matter what our occupation or profession. And there is hardly a technical or industrial course where subject matter in one way or another does not relate to materials.

Because of their pervasive importance, a knowledge of the nature, properties, and processing of *all* types of materials is essential today. Traditionally, courses and texts were confined to only single-materials groups or families. However, the old boundaries, such as between metals and plastics, are disappearing as manufacturing and building industries now consider all materials in the design of their products. Industries develop or select materials with the right combination of properties to meet performance and economic requirements, and they are unconcerned whether the chosen material is a metal, a plastic, or a green cheese.

The purpose of this book, therefore, is to provide the student with a detailed overview of the whole field of industrial and engineering materials. It is designed for use in first courses given in engineering colleges, community and junior colleges, and technical institutes. Although not highly technical, it assumes a basic knowledge of chemistry.

In recent years, general texts on materials have strongly stressed materials science and have given only cursory coverage to the engineering or application aspects of materials. While in-depth knowledge of materials science is important to materials specialists, the great majority of students during their industrial or professional careers will be concerned more often with selection and application of materials. For this reason, the content of this book has a practical focus. Although the nature, composition, and structure of materials are covered, major emphasis is on application properties, processing, and use of materials.

The first four chapters set forth the principles and methodology of materials selection and application, explain the types of properties used to evaluate materials, and describe broadly the basic nature and structure of all materials. After this foundation is laid, subsequent chapters survey and compare the principal materials families. These families include ferrous and nonferrous metals, polymeric materials (plastics, rubber, wood, fibers, and textiles), ceramics, glass, graphite, and composite materials. Finishes are also discussed.

Coverage of each major group of materials includes an explanation of nature and structure, a description of performance properties, and a discussion of how each is processed and fabricated into products. In presenting this information and data, emphasis is placed on viewing the whole spectrum of materials as a continuum of properties, to give students the broad perspective needed to properly select and use materials in the products they will be called on to plan, design, produce, and maintain.

I am indebted to many individuals and organizations for their generous help in providing advice and information. I am particularly grateful to my former colleagues on the staff of *Materials Engineering* as sources of information, and to my wife for her able editorial and secretarial assistance. Special thanks also go to the faculty and students at Stout State University, with whose help the course was developed that became the basis for this book.

Henry R. Clauser

INTRODUCTION

We live in a world of man-made materials. Steel, aluminum, plastics, ceramics, copper, glass, paints, and all the others are the concrete substance of our ideas, our designs, our product plans and blueprints.

Of course, materials have always been vital to human civilization. From history books we have all learned, at one time or another, that three of humanity's earliest eras are called the Stone Age, the Bronze Age, and the Iron Age, because the civilization of each was almost entirely dependent on the material after which the era was named. But now, in the twentieth century, materials—not just one, but many—have become the most important single factor on which the advance of technology and industry depends. Our progress in space, in electronics, and in atomic energy is directly linked to the solution of crucial materials problems. Even in many of the less glamorous manufacturing fields, materials are of major importance in the planning, design, and manufacture of products. Whether it is a rocket nose cone that must withstand the tremendous heat of reentry into the atmosphere, or a washing machine, the use of proper materials is indispensable to the success of the product.

Let's take a closer look at this materials world which is so vital to both our everyday products as well as advanced technology. Until the beginning of this century, the world of materials was relatively small. It was composed of only the few common materials with which we all are familiar: iron, copper, lead, wood, glass, ceramics, and rubber. But then steel and aluminum were produced commercially, and the first commercial plastics were developed. In the last 50 years, the number of new metals, alloys, plastics, rubber, and ceramics has increased exponentially (Fig. 1-1). Today there are several hundred times as many different materials as in 1900, and it is estimated that there are between 50,000 and 70,000 different compositions and grades available now. For example, in 1900, less than 100 different materials were used in automobiles. Today's car has at least 4,000 different materials in it. And one large manufacturing corporation uses 14,000 different materials in the wide variety of industrial and consumer products it produces.

Despite the amazing variety of modern-day materials, the search still goes on for better materials to meet new and critical service requirements, not only in an advanced technology area such as aerospace, but in the manufacture of industrial and consumer products as well. Also, better materials are needed to meet higher quality and reliability standards, epitomized by long-term product warranties and the demand for safer products. Finally, the search goes on constantly for new and better materials to lower costs. Every company is constantly reevaluating its materials usage because at least 40 percent of the manufactured cost is in materials.

1-1 *Materials Activities*

The huge world of materials activity can be divided into two hemispheres. One involves production and the other application of materials (Fig. 1-2). In the production hemisphere are all the activities

Fig. 1-1 Chart shows rapid pace at which new materials have been developed since 1900 (L. H. van Vlack, University of Michigan).

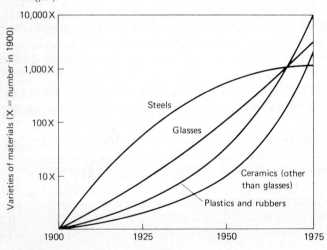

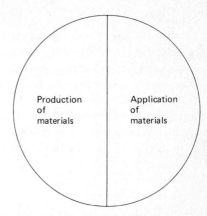

Fig. 1-2 The world of materials activity divided into production and use.

concerned with the making of materials for industrial use. These include mining and refining the raw materials from which usable materials are made, and the production of basic materials forms. For example, steelmaking plants produce hundreds of different steels in many different sizes, forms, and shapes: I beams, wires, rods, and sheets. Similarly, aluminum plants produce the basic forms of many different aluminum alloys. And chemical and rubber plants produce the basic forms of plastics and rubber and glass.

The other hemisphere involves the utilization of materials—that is, engineering and processing them into products. This activity includes (1) design and development, (2) processing and fabrication, and (3) research and development.

In the design stage, where products are conceived and planned to meet specific needs and requirements, materials become the concrete embodiment of conceptual models and designs. In the materials processing stage, the product is produced, often by a series of complex manufacturing operations. Also, the material is shaped and/or assembled into a usable form. Both stages involve decisions about the kinds of materials used and the gross amount needed to yield the finished item.

Supporting and serving these two functions of design and processing, are materials and process (or manufacturing) research and development. Here knowledge is acquired about the character, properties, and behavior of materials in order to develop new materials and processes and to improve existing ones. This knowledge is also applied to achieve more efficient use and processing of materials.

1-2 *Engineering or Industrial Materials*
This book is concerned with the utilization of materials in manufactured parts and products. To distinguish these materials from other

types of substances, such as chemicals and fuels, we will refer to them as either engineering or industrial materials.

Another broad category of materials is construction or building materials. Most industrial materials are used in building products and structures, and some, such as steel and wood, are mainly used in the building and construction field. The emphasis here, however, will be on the manufactured products area. Also, materials used almost exclusively in building and construction applications, such as cement, concrete, and the like, will not be covered.

One other distinction needs to be made. In this book we will be concerned almost entirely with matter in the solid state. While liquids and gases are used in manufactured products (for example, Freon gas in refrigerators), they are predominantly thought of and used as chemicals.

Solids differ from liquids and gases in that solids possess both a definite shape and volume. Also, compared to the other two states of matter, solids have useful mechanical properties such as strength, hardness, and rigidity. At the microlevel, solids are characterized by a greater regularity of structure than gases and liquids. Crystalline materials, such as metals and ceramics, represent the true microstructure of the solid state, where molecules are arranged in regular and repeatable patterns. The microstructure of polymers like plastics, although more amorphous than that of metals, is made up of molecular chains composed of regular repetitive units.

1-3 *Classification of Materials*
Historically, and in the broadest sense, solid materials are divided into two major categories: organic and inorganic. The name organic evolved originally from the strict distinction made between living matter, or matter derived from a living organism, and all other compounds considered to be of mineral origin. Today, organic materials are defined as compounds in which carbon is the major element in combination with hydrogen, oxygen, nitrogen, sulfur, and phosphorus, in decreasing order of occurrence. Because hydrogen is most often present with carbon, organic materials are also frequently referred to as hydrocarbons. Inorganic materials now include all substances other than hydrocarbons and their derivatives.

In industry, it is more common to divide materials into three broad categories: metals, polymers, and ceramics. Metals are inorganic substances composed of metallic elements (iron, aluminum, and so on). In the periodic table of elements, they are found in the left and middle sections. They are characterized chiefly by close-packed crystal structures and by a small number of valence electrons. Polymers are primarily noncrystalline hydrocarbon substances, and are composed of large molecular chains whose major element is the nonmetallic element, carbon. Ceramics are crystalline substances composed of compounds of metallic and nonmetallic elements. These three major categories of materials are not all-inclusive. Some materials, such as silicone polymers, have silicon, often considered a metal, as the principal element. Some substances, such as semiconductors, can be classified as metal or ceramic. And graphite, a form of carbon, does not fit into any of the three classes.

Review Questions

1. Name the technical areas into which materials activity can be divided.
2. Give three reasons why effective materials application is important in the development of today's products.
3. Explain the difference between production and application of materials.
4. Give three ways in which solid materials differ from liquids and gases.
5. Identify the following as organic or inorganic materials: metals, wood, textiles, glass, rubber, newspaper, leather, and plastics.
6. Define a hydrocarbon.
7. What are the three broad classes of industrial materials?
8. Name three industrial materials that are also used as building or constructional materials.

Bibliography

"The Challenge of the Materials Explosion," *Materials Engineering*, November 1968.
Clauser, H.R. (ed.): "Conservation in Materials Utilization," Report of Federation of Materials Societies for the National Commission on Materials Policy, November 1973.
Hempel, C. A., and G. G. Hawley (eds.): *The Encyclopedia of Chemistry*, 3d ed., Van Nostrand Reinhold Co., New York, 1973.
Smith, C. S.: "Materials and the Development of Civilization and Science," *Science*, May 14, 1965.

MATERIALS APPLICATION PRINCIPLES

A materials application can be considered as a functional system. As shown in Fig. 2-1, the material is a functional block (or black box). The input is the application's imposed conditions, and the output is the performance of the material.

Materials Application Analysis

2-1 *Performance Requirements*
A material in a given product has two sets of performance requirements. One is related to the functions performed by the part or product. The other is related to the part as a physical object (Fig. 2-2).

Functional Performance. A product or a part is planned, designed, and produced to meet a set of functional objectives or requirements. For example, an automobile engine is designed to deliver a certain horsepower or torque over a certain period of time without failing. And, within the engine, each of the piston rods transmits the energy of combustion in the cylinder to the crankshaft for the engine's life.

The material of which a part is composed must be capable of embodying or performing a part's function without failure. In the piston rod, for example, the material of which the rod is composed must be capable of withstanding the load imposed on it as the rod transmits the energy of combustion in the cylinder to the crankshaft. Therefore material functional requirements are derived from the functions the

8 Industrial and Engineering Materials

Fig. 2-1 Materials application as a functional system.

product has been designed to perform. While it is not always possible to assign quantitative values to these functional requirements, they must be related as precisely as possible to specified values of the most closely applicable mechanical, physical, electrical, or thermal properties.

The material must not only perform the desired functions specified by the design, it must remain stable in order to perform the functions over a period of time. All materials—metal or nonmetal—change with time. For example, many metals corrode in time under certain conditions, and short- and long-term creep or flow has an important bearing on the life of plastic products. The evaluation of materials performance over long periods of time is difficult and only a few standard tests have been devised. Usually, evaluation must rely on past experience or extrapolation of short-time, simulated service tests.

A material in a given application must also be reliable. Simply stated, reliability is the degree of probability that a product, and the material of which it is made, will remain stable enough to function in service for the intended life of the product without failure. Materials reliability is difficult to measure because the level of reliability is dependent not only on its inherent nature and properties, but also on its production and processing history. Specified reliability also depends on the status of a material's technology. That is, relatively new and nonstandard materials will tend to be rated as having lower reliability than established standard materials.

A material must also safely perform its functions. Safety has several

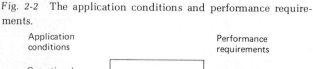

Fig. 2-2 The application conditions and performance requirements.

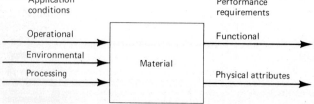

aspects. One is related directly to reliability. That is, where failure of a product could be catastrophic—such as in airplanes and high-pressure systems—the reliability level of the material must be much higher than for noncritical products. Another aspect of safety has to do with the nature of the material itself. Safety glass is a well-known example of a material designed for fail-safe performance. And, finally, there are special applications where some materials may not be used because under certain conditions their behavior can be hazardous. For example, materials that give off sparks when struck are safety hazards in a coal mine.

Physical Attributes. In addition to functional requirements, products have physical attributes such as configuration, size, weight, and appearance. Physical attributes sometimes also serve functional requirements. For instance, the functioning of a gyroscope or a flywheel is directly related to the weight of the materials used.

2-2 *Application Conditions*
Performance requirements proscribe the performance output of materials used in an application. On the input side are the conditions imposed on the materials during their life cycle in that application. They, of course, are closely related to the performance requirements and are often determined by them.

The application input conditions are of three kinds: operational, environmental, and processing.

Operating Conditions. These are related to the functioning of the part and will vary from ones that are few and simple to a highly complex set of interacting forces. For example, in container applications, the operating conditions may be simply the pressure of the contents. In the case of automobile piston rods and bearings, the conditions include a complex of exploding gas, impact, and abrasion, and so on. In electrical products, such as relays, for example, the materials are exposed to electrical currents as well as to structural stresses, impact, and abrasion.

Environmental Conditions. The environment in which a product operates strongly influences service performance. Humidity, water, or chemicals can cause deterioration and subsequent failure of materials. Other environmental conditions, such as high or low temperature, alter or adversely affect the service performance of most materials. Environments are also sometimes beneficial. For example, the oxide which forms on many metals when exposed to air improves their corrosion resistance. In ablative rocket nose cones, the heat environment during reentry reacts with and changes the material into a more heat-resistant substance for its short operational lifetime.

Processing Conditions. These are chiefly set by the physical attributes required of the part. Thus shape, form or configuration, and the physical tolerances, largely determine what processing and fabrication will be done on the material to meet the physical requirements of the part.

2-3 *Properties*

The performance requirements and application conditions just discussed in a general way can be reduced to more specific terms. All three types of application conditions are energy forms of one or more kinds (Fig. 2-3):

Mechanical: loads or stresses.

Thermal: heat or cold.

Chemical: atmosphere, water, chemicals.

Electrical: power, current.

Radiation: light, ultraviolet, nuclear.

(Although electricity is a form of electromagnetic radiation, it is listed separately because of its importance.)

Again viewing a material application as a functional system, we find that the input energies interact with or operate on the material, and the net responses are the outputs of the system. The outputs, which are directly related to the functional performance and the physical attributes of materials in service, are of two major kinds: energy changes and state changes in the materials.

Fig. 2-3 Relation of materials properties to energy input conditions and resulting energy and state changes.

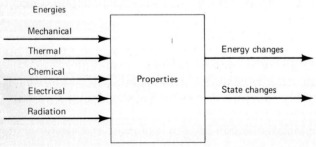

Energy changes can be either qualitative or quantitative, or both. Qualitative changes are those in which the input energy is transformed from one type of energy to another. Quantitative changes are those in which the output level or quantity of energy is different from that of the input. For example, when an electric current is imposed on a piece of material, at least some of the electrical energy is converted to heat (qualitative). At the same time there will be a voltage drop from the input to the output side of the material (quantitative).

State changes that occur when a material is exposed to energy are evident at micro- and macrolevels. At the microlevel, the chemical and/or structural nature of the material can be altered. These internal responses can then produce a number of gross state changes at the macrolevel of performance. Principal among them, from a practical performance standpoint, are changes in configuration, size, density, and appearance; deflection and displacement; and fracture and deterioration.

In materials application practice, materials are evaluated for intended use in terms of engineering or performance properties, which are derived from the behavior or performance of the material when subjected to input energies. Therefore, when viewing a material as a functional system, we can say that a *property* is an expression of some relationship between the application conditions and the functional performance of the material. There are a number of possible types of such relationships (see Chap. 4, where property characterization is discussed). For example, properties can be expressed as the energy input required to bring about a certain state change, as in the case of tensile strength; or as the energy output resulting from a given energy input, as in the case of thermal transfer properties; or as the ratio between energy input and a resulting state change, as in the case of modulus of elasticity.

2-4 Constraints

In every material application there are external factors that place limits on what materials can be used. Termed *constraints,* they are listed and discussed briefly below.

1. Existing facilities: The type of processing equipment available in the company or plant will often restrict the choice to only those materials that can be processed by the existing equipment. However, buying a part or component made outside can remove this constraint.
2. Compatibility: Whenever more than one material is involved in an

application, compatibility becomes a constraint. That is, materials operating together in a given service environment should not cause adverse or damaging reactions. In thermal environments, for example, thermal expansion of all the materials in a part has to be similar to avoid stress buildup. In a water or high-humidity environment, materials that will be in physical contact must be chosen carefully to avoid galvanic corrosion. In established products or designs, color and appearance match may be important. Also, the joinability of new materials to existing materials in a product design must be considered.
3. Marketability: In the case of consumer products, this constraint can seriously limit materials choice. For example, in some items a feeling of solidity or lightness is desirable even though it has nothing to do with product performance. However, even this constraint can sometimes be circumvented. An example is the rapid increase in the use of plastics as imitation wood for furniture.
4. Availability: Obviously a material must be readily available, and available in large enough quantity, for the intended application. In times of materials scarcity, this constraint becomes significant. And, in the future, with the projected scarcity of many material resources, this constraint will assume increasing importance.
5. Disposability and recyclability: These are the newest of the constraints, and increasingly important factors in materials selection.

2-5 *Economic Factors*

Cost, perhaps more often than any other constraint, is the controlling factor in a given materials application problem. For, in every application, there is a cost beyond which one cannot go that prescribes the limit that can be paid for a material to meet the application requirements. If it becomes apparent that this limit will be exceeded, the design will be changed to alter materials requirements. This fact of limiting cost is as true in the aerospace field as in consumer-products fields. The only difference is that the limiting cost in aerospace systems is considerably higher than for consumer products.

The total original cost of a material for a given application is made up of two components: the cost of the materials and the cost of processing the materials into the finished part or product.

Materials Cost. The common way of expressing the cost of material itself is by unit weight. However, since the density of materials varies widely, the price per unit volume is just as important, and, depending on the application, it can be more significant as the comparison criterion. For example, the ratio of the cost of aluminum to hot-rolled carbon steel on a weight basis is about 4 or 5 to 1. But, on a volume basis, they are nearly equal in cost. In the case of a number of plastics, on a cost per pound basis, they are five or more times as costly as carbon steel. Yet, on a volume basis, some are lower in cost than steel.

Even price per unit volume can be misleading because it may not be possible from a design standpoint to substitute one material for another on an equal volume basis. Actually, what is needed is at least a rough design of the part as it might be made for each material under consideration. Then the cost for the amount of one material needed for a particular job may be compared with a fair degree of accuracy with the cost of others in appropriate forms and quantities.

Processing Costs. The other factor making up the total cost of material is the processing cost. Differences exist in the cost of processing different materials at each stage in their fabrication and assembly. Even a choice between two grades of plain carbon steel must often be made on the basis of differences in processing costs.

What are some of the production differences that affect the total cost of a material? One material may be more costly because it is more difficult to machine or weld. Another may be more difficult to form without special processing. Another may require more finishing operations. In the case of some materials, special safety precautions or special processing atmospheres are required. These and many other considerations must be taken into account when evaluating the total cost of a material.

Materials Selection

One of the most important requisites to the development and manufacture of satisfactory products at minimum cost is to make a sound, economic choice of materials.

2-6 *The Selection Process*

The materials selection process involves the following major operations:

1. Analysis of the materials application problem. This requires a study of the performance requirements, including functional performance, physical attributes, and application conditions (Secs. 2-1 and 2-2).
2. Translation of the materials application requirements to materials

property values (Sec. 2-3). In some cases, this is relatively easy, as, for example, in parts where unidirectional stresses are involved. Here mechanical-strength properties, such as yield and compressive strength, can be directly derived from the measured applied loads encountered in the application. However, for many service conditions there is no direct or simple correspondence between the condition or requirement and a measurable materials property. Thus the translation problem may be quite complex and depend on predictions based on simulated service tests or on property tests closely related to product service.
3. Selection of candidate materials. Once the required properties are clearly specified, the rest of the selection process involves the search for the material (or materials) that best meets those properties. The starting point for selection of candidate materials would appear to be the entire universe of materials. Of course, a survey of the many thousands of available materials seldom, if ever, is practical, nor is it necessary. Past experience and a simple, cursory survey eliminate whole classes of materials, so that in reality the starting point usually involves only a fraction of all possible choices.

The principal objectives, then, are to narrow the selection of candidate materials to a manageable number for subsequent detailed evaluation and, at the same time, to make certain that no important possible solution is omitted.

In choosing candidate materials, any one or more of a number of criteria can be used. Past experience and materials presently being used are often guides or starting points. Another method is to base selection on the most important or most critical requirement. Different approaches to the solution of the materials problems, as distinct from simply choosing candidate materials for evaluation, should be considered. For example, composite or combination materials, or exploiting a mechanism such as ablation, are two examples of many that can be considered.
4. Evaluation of the candidate materials. The objective of the evaluation step is to weigh the candidate materials against the specified properties to find the one best suited for the application. In principle, this step is a continuation of the previous one in that it is essentially an elimination or screening operation.

2-7 *Evaluation Phases*
Although no universally used formal procedure has been developed for the evaluation operation, it can be divided into three phases: (1) screening, (2) selection, and (3) design data phases.

In the screening phase, one starts with a large number of candidate materials and proceeds to narrow them down to a select few that look promising. In the selection phase, detailed evaluation of the candidate materials takes place. Properties and characteristics of the various materials are related to performance requirements to finally arrive at the optimum material for the application. In the design-data phase, the pertinent properties of the selected material are determined in detail and depth to obtain statistically reliable measures of the material's performance under the specific operational, environmental, and processing conditions. To carry out this three-phase process, screening criteria are established for each phase.

Screening Criteria. The first screening criteria are usually of the go–no-go kind. Here, obvious or self-evident constraints figure importantly. For example, existing processing facilities or compatibility are important first considerations. When designating some properties as screening criteria, an absolute lower or upper limit is established. Any materials going beyond the limit are unacceptable.

Selection Criteria. Selection criteria are those properties and characteristics that are not absolute or firmly fixed and whose values can be changed, within limits, during a tradeoff process. They can be referred to as the "wants" and "don't-wants," "desirables" and "undesirables" as contrasted to the go–no-go screening criteria.

In practice there is no one best way to perform the selection phase. The particular approach used will, among other things, depend on the nature of the application, the company, and the engineering organization. The selection phase may begin with the most critical property and then proceed to those of lesser importance; or, where most of the requirements are of equal importance, the candidates may be compared on the basis of all the pertinent properties. In still other applications, rating systems or analytical techniques, such as weighted indices or failure analysis, can be used to advantage (see Sec. 2-8).

Design-Data Properties. The design-data properties and characteristics are those of the selected material in its finished, fabricated state. They must be known with sufficient confidence to permit the design and fabrication of a product that will function with a specified reliability. Establishing which are the final design properties depends, of course, greatly on the cooperative efforts of the designer and materials specialist. Figure 2-4 illustrates how the evaluation screening phases would apply in selecting a material for the pilot canopy on a supersonic military aircraft.

16 Industrial and Engineering Materials

Operational or Design Environment	Material Characteristics		
	Screening (Mat'l A, B, C...Z)	Selection (Mat'l A, B, C, D)	Design Data (Mat'l A)
Thermal: Steady 275°F for 3 hrs/flt Max 420°F for 5 min/flt Min −65°F Gradient: 275° ext to 100° int Defogging 180°F Natural Weathering: Sun Humidity Erosion: Wind, Rain, Sand	Luminous transmittance Color stability Tensile strength Craze resistance	Tensile strength/density Compression strength/density Craze resistance limits Modulus of elasticity Crack propagation characteristics	Tensile strength vs. temp Compression strength Edge joints—tension

Pressure:
Internal
  Steady 8 psi
  Max   20 psi (burst)
External
  Steady Gradient
  Max   50 psi for 10 min underwater
	Crack propagation resistance
	Notch fatigue
		Thermal expansion
		Specific heat
		Thermal conductivity
		Creep
		Dimensional stability
			Creep
			Stress rupture

Optics:
Transmission—Per Spec
  MIL-P-25690
Deviation—Critical Area—<1'
          Other Area—<3'
Special Requirements:
Ejection through canopy
	Measure of ability to eject through canopy	Fabricability tests
		Joinability (bonding)
		Formability
		Trade-off factors:
		  Availability
		  Fabricability
		  Weight, cost
		  Shape of canopy
			Fabrication limits
			Joining
			Stretching

The above discussion may have given the impression that the materials selection process always proceeds directly and step by step along the problem-solving path. This, however, is seldom the case. The process often operates in an iterative manner. That is, in collecting, analyzing, and evaluating information in any given stage, new insights may be gained or new problems may be uncovered that require a repetition of earlier steps.

2-8 Systematic Selection Methods

As we have just seen, the materials application process is divided into two main phases: defining the problem and establishing property requirements, and searching for materials that best meet the requirements. In the past, engineers relied largely on cut-and-try or on successive approximation methods. However, in recent years, systematic methods have been developed. Some of these are failure analysis, cost versus property indices, and weighted property indices.

Weighted Property Indices. One of the most potentially useful methods of evaluating complicated combinations of materials is the weighted index. Briefly, it is a method of evaluating materials in which each parameter is assigned a certain weight, depending on its importance, then individual property weights of each material are summed up to give a comparative materials performance figure. An example of the weighted index system is that developed by the NASA Committee on Materials Research for Supersonic Transports (see bibliography at the end of this chapter). Although the SST program has been discontinued, this weighted index system should be applicable to other products.

A key step in the weighted index is the determination of performance requirements and properties that are important for the product. In the case of the SST, for example, properties such as strength, stiffness, toughness, corrosion resistance, joinability, and cost are some of the major requirements. The next step separates the requirements into three screening groups: (1) go–no-go, (2) nondiscriminating, and (3) discriminating parameters.

Go–no-go parameters are the constraints. They involve the certain fixed minimum value which a material must meet. As shown in Fig. 2-5, in the case of the SST these include such properties as corrosion

Fig. 2-4 (on facing page) Evaluation phases in selecting a material for pilot canopy on a supersonic military aircraft. (Adapted from NMAB Report No. 246.)

resistance, weldability, and brazability. These are classified as go–no-go parameters because the material is either satisfactory or unsatisfactory. Any merit in excess of a minimum level is of no special advantage, nor could it be used to make up for a material's deficiency in meeting another requirement. Thus go–no-go requirements do not lend themselves to compromise or relative rating or tradeoff.

Nondiscriminating parameters, such as availability and producibility, are also constraints—requirements that must be met if a material is to be considered at all. Like the go–no-go group, they represent parameters that do not allow comparison or quantitative discrimination. Thus, if a material is not readily available, or cannot be formed into the shape desired, it should not be considered.

Discriminating parameters are those to which quantitative values can be assigned. It is on the basis of these parameters that tradeoffs can be made in terms of the relative importance of parameters. As shown in Fig. 2-5, typical discriminating properties are strength, toughness, and cost.

Once the various parameters are established, an overall rating of the candidate materials is arrived at by means of a rating table. The table only lists those materials that are available, producible, or formable for the application. In the case of the go–go-no screening parameters, the materials are rated as S (satisfactory) or U (unsatisfactory). If any material is rated U, it is given no further consideration.

Each of the discriminating parameters is assigned a weighting factor, depending on its importance. For example, in the case of the SST, strength has a weighting factor of 5, indicating that strength is a factor of great importance. On the other hand, cost is relatively unimportant and is given a weighting factor of 1. For other applications, such as, for example, automobile trim, these factors could be reversed, with cost having a factor of 5, and strength a factor of 1.

Each of the pertinent properties of the candidate materials is then assigned a rating depending on how closely it meets the requirement. The ratings range from 1 for the poorest to 5 for the best. These ratings in turn are multiplied by the weighting factor for each parameter. As shown in the righthand column of Fig. 2-5, the final rating number for each candidate is obtained by taking the sum of the relative rating numbers and dividing it by the sum of the weighting factors used. If the rating process is right in concept and execution, then the material with the highest material rating number is the optimum material.

Failure Analysis. Failure analysis is a method of materials selection based on predicting and anticipating all the ways that a product could

Material	Go–No-Go[3] Screening	Relative Rating Number ([2]Rating Number × [1]Weighting Factor)									Material Rating Number
Alloy and Condition	Corrosion / Weldability / Brazability	Strength (5)	Toughness (5)	Stiffness (5)	Stability (5)	Fatigue (4)	As Welded Strength (4)	Thermal Stress (3)	Cost (1)		$\dfrac{\Sigma \text{ Rel Rating No.}}{\Sigma \text{ Sigma Rating Factors}}$

[1] Weighting factor (range = 1 poorest to 5 best)
[2] Range = 1 poorest to 5 best
[3] Code – S = Satisfactory
  U = Unsatisfactory

Fig. 2-5 Example of weighted index rating chart.

fail and then selecting materials so that failure does not occur. That is, it is a systematic approach to the measurement, control, and improvement of reliability. Obviously, the theory is based on the fact that materials are the lowest common denominator in the failure of a product or equipment. That is, product failure can be analyzed by studying how materials are applied, the stresses they will be subject to, and the effect of process defects in weakening materials. Thus all failure theory and generalizations are drawn almost exclusively at the materials level.

From this viewpoint, failure analysis can be considered a "backward" method of materials selection. That is, one thinks primarily in terms of the product in service and from that point works backward and anticipates every stage in the history of the product that might ultimately cause failure. Consequently, every detail of the functional requirements and operating environment has to be analyzed, as well as the complete processing, fabrication, raw materials, and vendor history of the product.

Several specific techniques have been developed to carry out failure analysis. They include environment profiles, fabrication and process flow diagrams, and failure models. *Environment profiles* are designed to provide a complete description of the environment to which the product will be exposed. These profiles include such obvious requirements as temperature, time, and chemical environment. They also include all of the less obvious things that can lead to failure. For example, the engineer should be wary of such things as the effects of various lubricants on materials, the effects of incidental vibrations, and the possibility of stray electrical conditions that might cause galvanic corrosion.

Fabrication and process flow diagrams indicate the condition of materials and detail the step-by-step controls through the processing phase. These flow diagrams are made not only for controlling in-plant processing, but for controlling vendor processing as well.

Failure models describe all the possible types of failures that can occur. The failures are described in terms of the causes of failure, such as a functional failure of materials, failure resulting from inability to withstand a particular environment, or failure caused by improper manufacturing or processing. The purpose of these steps is to document the various failure mechanisms and the conditions that produce them. Once the mechanisms are known, it is possible to make statistical analyses and mathematical modes describing the failure phenomena. Once the failure mechanisms are analyzed, controls are established on the materials in the form of suitable procurement documents, vendor specifications, and materials and processes specifications.

Review Questions

1. When analyzing a materials application problem, what are the materials requirements that must be considered? (List four.)
2. Give an example for each of the following application conditions to which materials can be subjected:
   (a) Operational.
   (b) Environmental.
3. Name three environmental conditions that must be considered when selecting the material for a refrigerator shelf.

4. Name three forms of energy to which a refrigerator shelf is subjected.
5. When viewing a material in a given application as a functional system, how is a material property expressed?
6. Name two constraints that will become increasingly important considerations in the future when selecting materials.
7. Name three service conditions that can affect the stability or service life of a material.
8. Explain two kinds of safety that might be considered when selecting a material for a product.
9. What are two major components of the total material cost in a given application?
10. The price of a material is usually given in terms of unit weight. Name two other ways of expressing material cost.
11. After a materials application problem is analyzed, what major steps are involved in the materials selection process?
12. In the materials selection process, what is the purpose or objective of the screening phase? the selection phase? the design-data phase?
13. In the weighted property indices method, what is the difference between a discriminating and a nondiscriminating parameter?

Bibliography

An Approach for Systematic Evaluation of Materials for Structural Applications, National Materials Advisory Report (NMAB-246), February 1970.

Clauser, H. R., R. J. Fabian, and J. A. Mock: "How Materials are Selected," *Materials Engineering,* July 1965.

Gabrovic, L. J., J. V. Hackworth, and H. M. Lampert: "The Physics of Failure Applied to Materials Selection," Paper no. 64-MD-44, American Society of Mechanical Engineers, June 1964.

Raring, R. H.: "SST Materials," *Astronautics & Aeronautics,* September 1964, p. 26.

THE NATURE OF MATERIALS

The question of the nature of matter or materials has always preoccupied and perplexed people, particularly scientists and philosophers. Before Democritus proposed the original atomistic theory during the fifth century B.C., Thales of Miletus declared that water was the fundamental substance from which all matter was derived. Over the centuries since the first crude theories were proposed, our ideas of matter have seldom remained fixed for long. The last 400 years have seen a variety of theories come and go.

The idea that permeates all modern thinking on the nature of materials is the concept of structure. This is in contrast to the major attention that has almost always been given to the fundamental units or the "bricks" of which materials are composed. That is, modern theory says that what makes one material differ from another is not only the "stuff" of which it is made, but, just as importantly, the form or pattern and organization of the basic units. And, perhaps even more importantly, it is this form and structure that determines the properties of engineering materials. So, by studying material structures, we can predict behavior and properties and ultimately tailor the structures to our needs.

We will now take a broad overview of the nature and structure of materials at two levels: first, at the microlevel of molecules, crystals, and phases of what we will term the monolithic materials; and then at the macrolevel of composite materials. Here we will describe only the broad outlines of the structural hierarchy of materials. In later

Microstructures

3-1 Atoms and Bonding

In studying microstructures, we will consider atoms the basic structural units of materials. We will not delve deeply into the atom nor will we consider subatomic particles. Atoms are composed of a nucleus surrounded by spherical layers of moving electrons (Fig. 3-1). Atoms differ from one another in the number of electrons spinning around the nucleus. Up until recent years and the discovery of atomic energy, it was generally thought that there were 92 different atoms (or elements) ranging from hydrogen with one electron to uranium with 92. Today, however, there are more than 100 known elements.

It is the total number of electrons and the number of layers or shells of electrons that determine most of the properties of that material. The electrons in the outermost ring particularly influence the properties that are of most engineering interest. The reason for their importance is that all the atoms composing a solid material are bonded to each other by means of these outermost valence electrons.

For example, let us consider two materials, both composed of carbon atoms. Graphite consists of layers made up of carbon atoms arranged in flat hexagonal rings (Fig. 3-2a) that resemble old-fashioned tile walls or floors. Each of the hexagonal "tiles" is formed by carbon atoms located at each of the six points of the hexagon. Therefore each carbon atom is bonded to three others. Graphite is black and relatively soft. It is frequently used as a lubricant because the planes of atoms can fairly readily slide over each other. Diamonds are also composed of carbon atoms. But, unlike graphite, diamonds are extremely hard, and

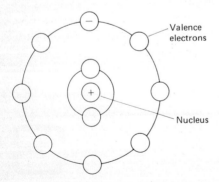

Fig. 3-1 Atoms are composed of a nucleus around which electrons revolve.

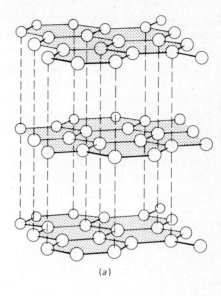

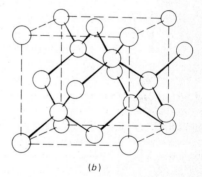

Fig. 3-2 Structure of two forms of carbon: (a) graphite, (b) diamond.

colorless. Also, diamonds are nonconductors, whereas graphite conducts electricity.

Why these great differences between two materials that are made of the same kind of atom? The answer lies in the difference in *structure*. In the diamond, instead of the carbon atoms being arranged in layers, they have a three-dimensional structure so that each carbon atom is bonded to *four* other carbon atoms, which are all in different planes (Fig. 3-2b). As a result, the atoms are bonded to each other in a much stronger fashion than in the graphite structure.

The bonding mechanism between atoms derives from the "desire" of atoms to remain in or revert to a stable condition. And, in accordance with basic natural laws, this stability depends on an atom's ability to maintain eight electrons in its outer ring. An atom not having eight electrons in its outer layer can achieve stability by (1) gaining an electron—ionic bond, (2) sharing electrons—covalent bond, or (3) losing electrons—metallic bond (Fig. 3-3).

In ionic bonding, one of two different atoms receives an electron from the outer layer of the other atom. The atom losing the electron becomes a positively charged atom, known as an ion. The receiving atom is negatively charged; consequently, the two atoms are attracted to each other, thus forming a bond. Ionic-bonded crystals, commonly found in ceramics, are quite strong but very brittle.

In covalent bonding, commonly found in polymeric materials, especially in plastics and rubber, atoms share negatively charged elec-

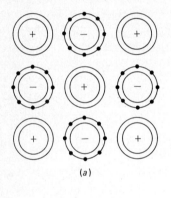
(a)

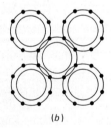
(b)

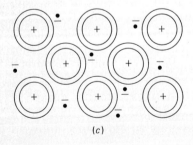
(c)

Fig. 3-3 Bonding mechanisms: (a) ionic, (b) covalent, (c) metallic.

trons, and thus become strongly linked together. The atoms may be of the same kind, such as carbon atoms in diamonds, or they may be different, as in many plastics, where carbon and hydrogen atoms are linked with a covalent bond.

The types of bonds just described are called primary bonds. There are also weaker secondary bonds called van der Waals forces. These forces provide bonding between molecules as a result of electrical imbalances that can occur in covalently bonded molecules.

Metallic bonding, as the name implies, occurs in metals. Because the atoms of metallic materials have only a few electrons in their outer shells, the valence electrons are detached relatively easily. The "free" electrons form an electron cloud in the metal structure and the atoms become ions which are positively charged. The cloud has a negative charge as opposed to the positive charge of the ions. Thus the attracting forces between the unattached electrons and the ions serve to bond the atoms together.

There are three broad types of atomic patterns or structures in solid materials produced by these bonding mechanisms: the molecular structures of polymers, the crystal structures of metals and ceramics, and the amorphous structure of glass.

3-2 *Molecular Structures*

Common materials with molecular structures are plastics, rubber, paper, and wood. The basic structural unit in molecular materials is the monomer molecule (sometimes called a mer) (Fig. 3-4a). A monomer is composed of a limited number of atoms strongly bonded together, usually with a covalent bond. In the case of plastics, monomers, with some exceptions, are composed of two carbon atoms with attached ribs of other atoms, such as hydrogen, fluorine, or chlorine.

The next level in molecular structure is achieved by joining the monomers end-to-end to make long, continuous, chainlike molecules (Fig. 3-4b). These chains, called polymers and sometimes macromolecules, are literally giant molecules composed of hundreds or thousands of monomer units. When a large number of these molecular chains are mixed together, a plastic material or resin is formed (Fig. 3-4c). A plastic therefore consists of millions of these giant polymer molecules or chains.

The process by which monomers are joined together to form the chainlike polymer molecules is known as polymerization. This is a chemical reaction brought about by application of heat and/or pressure, or by the use of catalysts. When a polymer molecule is built up of a single kind of monomer, such as polyethylene from ethylene or poly-

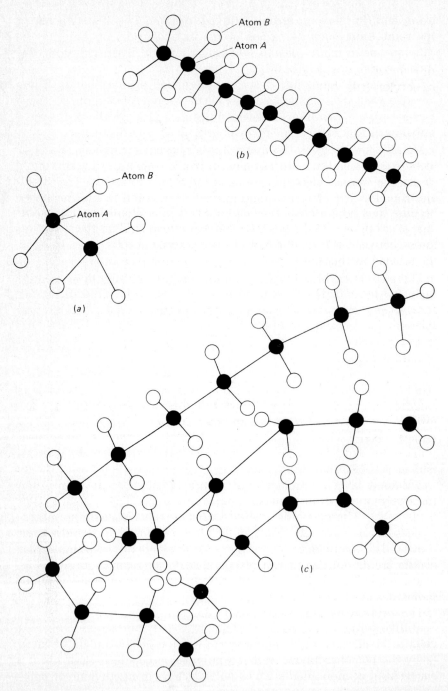

Fig. 3-4 Basic structure of polymers: (a) monomer or mer; (b) polymer or molecular chain; (c) plastic resin.

vinyl chloride from vinyl chloride, it is called a homopolymer. When the polymerized material is constructed from two or three different monomers, it is known as a copolymer or a terpolymer. An example of a terpolymer is the commonly known ABS plastic, which consists of acrylonitrile, butadiene, and styrene monomers.

The length of a molecular chain, or the number of monomer units in the chain, and the kind and strength of bonds between the chain molecules also determine the general type of polymeric material. For example, thermoplastic resins consist of long molecules that may have side chains or groups that are not attached to other molecules, that is, that are not cross-linked. Hence bonding between molecular chains is quite weak, and they can be repeatedly softened and hardened by heating and cooling. Heating will soften the plastic so that it can be formed. Cooling hardens it into the final desired shape. In thermosetting resins, cross-links are formed between portions of the long molecules during polymerization, and therefore the intermolecular bonding is strong. Thus, once polymerized or cured, the plastic cannot be softened by heating.

Natural and synthetic rubbers are somewhere between the thermoplastics and thermosets. They have molecular chains held together by a few primary bonds or cross-links. Heat causes rubbers to become elastic without breaking down.

3-3 *Crystal Structures*

We have now seen that in polymeric materials there is a repetition of the pattern of atoms along the length of the polymer chain. In the case of crystals, the atom pattern is repeated in three dimensions.

The three-dimensional building blocks of which crystals are constructed are known as unit cells. A unit cell is constructed of atoms regularly arranged in intersecting plane surfaces. There are seven main unit-cell patterns, but the most common ones are those with a cubic or hexagonal pattern, since most metals are of these types.

Cubic unit cells are of two major kinds (Fig. 3-5). In the body-centered cubic, there is an atom at each corner of the cube and another in the center of the cube. Iron, at room temperature, has a body-centered structure. In the face-centered structure, in addition to an atom at each corner of the cube, one is located at the center of each face. A typical face-centered metal is copper.

A crystal, then, is simply a lattice structure built up of unit cells. The unit-cell pattern is repeated indefinitely in three dimensions as shown schematically in Fig. 3-6. In principle, any piece of a crystalline material should be made up of one crystal. However, except in our most advanced materials technologies, metals and ceramics are usually polycrystals. That is, they are composed of many crystals having

30 Industrial and Engineering Materials

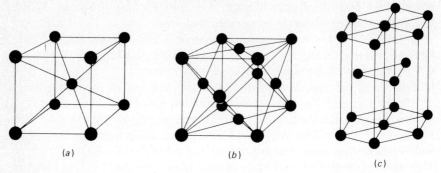

Fig. 3-5 Common kinds of unit cells: (a) body-centered cubic, (b) face-centered cubic, (c) hexagonal.

Fig. 3-6 The crystal lattice structure.

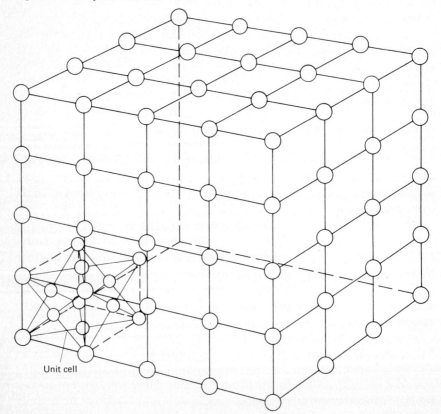

dissimilar orientations. The borders or boundaries between adjacent crystals are called grain boundaries.

If the structure of all the crystals making up a material is of the same type, it is a single-phase polycrystalline material. Pure metals and a few alloys are of this type. Most metal alloys and ceramics, however, are multiphase or polyphase in that they are composed of two or more different crystal structures (Fig. 3-7). Steels, for example, are composed of crystals of iron, iron carbide, and often one or more other alloying constituents.

The crystals making up an industrial material are seldom all perfect. Although the flaws are usually less than 1 percent of a crystal structure, this small imperfect fraction has a major influence on the properties of the material.

There are a number of kinds of crystal imperfections. Some of the most common ones are foreign or substitute atoms and lattice vacancies and dislocations (see Sec. 5-3). In many cases, foreign atoms are intentionally added to a material in order to provide certain properties. For example, zinc atoms are added to copper to produce brass, which, in general, is stronger and harder than pure copper.

3-4 *Amorphous Structures*
The third major type of material structure—the amorphous or noncrystalline structure—differs from the other two in that it does not have a repetitive pattern. (Amorphous literally means without form.)

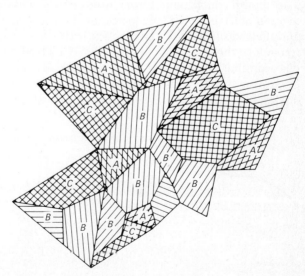

Fig. 3-7 Multiphase crystal structure. *A*, *B*, and *C* represent different crystal structures.

Glass Structure

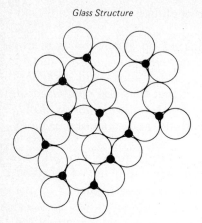

Fig. 3-8 The amorphous, noncrystalline structure of glass.

Amorphous materials include gases, liquids, and glasses. We are concerned here only with glass, which is often thought of as a rigid liquid. Actually, glass structures do exhibit what is called short-range order. In a typical simple glass, a small atom is bonded to three surrounding larger atoms. As a result, a continuous structure of strongly bonded atoms is developed (Fig. 3-8).

Macrostructures

The major constituent forms used in the structuring of composite macromaterials are fibers, particles, laminas or layers, flakes, fillers, and matrices (Fig. 3-9). The matrix is the "body" constituent. It serves to enclose the composite and to give it its bulk form. The fibers, particles, laminas, flakes, and fillers are the "structural" constituents that determine the character of the internal structure of the composite. They are generally, but not always, the "additive" phase.

Fig. 3-9 The major structural constituents of composites.

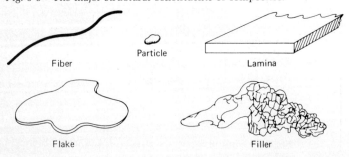

3-5 *Types of Composite Structures*

All composites can be divided into five general types on the basis of the structural constituents just described (Fig. 3-10):

1. Laminar composites, composed of layer or laminar constituents.
2. Particulate composites, composed of particles with or without a matrix.
3. Fiber composites, composed of fibers with or without a matrix.
4. Flake composites, composed of flat flakes with or without a matrix.
5. Filled (or skeletal) composites, composed of a continuous skeletal matrix filled by a second material.

Perhaps the most typical composite is that composed of a structural constituent embedded in a matrix. However, many composites have no matrix, but rather are composed of one (or more) constituent form(s) consisting of two or more different materials. Sandwiches and laminates, for example, are composed entirely of layers. The layers taken together give the composite its form.

Because the different constituents are intermixed or combined, there is always a region contiguous to them. These regions can take two forms, that of an interface, a surface forming the common boundary of the constituents (Fig. 3-11), or of an interphase. The interface surface is in some ways analogous to the grain boundaries in monolithic materials.

The interphase region is a distinct added phase. Examples are the coating on the glass fibers in reinforced plastics and the adhesive that bonds together the layers of a laminate. When such an interphase is present, there are two interfaces—one between each surface of the interphase and its adjoining constituent.

Fig. 3-10 Five types of composites based on structure.

Fiber composite

Particulate composite

Laminar composite

Flake composite

Filled composite

34 Industrial and Engineering Materials

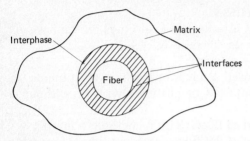

Fig. 3-11 Typical form of composites, including an interphase and interfaces.

In still other composites, the surfaces of the dissimilar constituents interact to produce an interphase. For example, in some cermets, bonding between the particles and matrix is achieved as a result of a small amount of solubility at the surface of one or both of these constituents (see Sec. 14-9).

3-6 Distribution of Constituents
The constituents making up a composite can be distributed in one of two general ways. Perhaps most commonly, the constituents are present in a regular and repetitive pattern. They have a relatively uniform cross section both in material and structure, and uniform density. Composites such as the matrix-particle and some matrix-fiber types, in which the structural constituent is evenly distributed throughout the matrix, are of this homogeneous type.

The second possible distribution is a variable, nonrepetitive pattern of constituents—nonrepetitive in either their internal form or their material. Materials of this type can be termed graded or gradient composites (Fig. 3-12). Laminated materials, which are composed of several different layers, are of this type. Also, filament-wound composites can be designed with variable fiber distribution.

Fig. 3-12 Gradient, or variable density, composites.

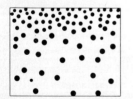
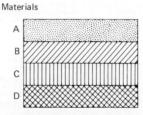

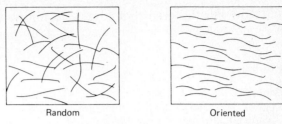

Fig. 3-13   Directional distribution of structural constituents: random and oriented.

Finally, in both the homogeneous and gradient composites, the constituents can either be arranged in one or more definite directions or they can be randomly oriented (Fig. 3-13).

Review Questions

1. What are valence electrons?
2. Explain why two different materials, although composed of the same kind of atoms, may have different properties.
3. Explain the difference between covalent, ionic, and metallic bonding.
4. Indicate which kind of bonding is found in
    (a) Metals.
    (b) Plastics.
    (c) Ceramics.
    (d) Rubber.
5. What is the basic structural unit of
    (a) Plastics.
    (b) Rubber.
    (c) Metal.
    (d) Wood.
6. What is a major difference in structure between crystalline and polymeric materials?
7. Name three structural levels of
    (a) Polymeric materials.
    (b) Crystalline materials.
8. How does the structure of glass differ from that of polymers and metals?
9. Name the five structural constituents used in composite materials.
10. What is the function of the matrix in a composite material?

11. How are the grain boundaries in microstructures and the interphases (or interfaces) in macrostructures similar?

Bibliography

Bonigan, S. (ed.): *The Binding Force,* Walker and Co., New York, 1966.

Clauser, H.: "The Concept and Nature of Composites," *Materials Engineering,* September 1963.

Holden A., and P. Singer: *Crystals and Crystal Growing,* Anchor Books, Doubleday & Co., Inc., Garden City, N.Y., 1960.

Jastrzebski, Z. D.: *Nature and Properties of Engineering Materials,* John Wiley & Sons, Inc., New York, 1959.

Kaufman, M.: *Giant Molecules,* Doubleday & Co., Inc., Garden City, N.Y., 1968.

Toulmin, S., and J. Goodfield: *The Architecture of Matter,* Harper & Row, New York, 1962.

Van Vlack, L.: *Elements of Materials Science,* Addison-Wesley Publishing Co., Inc., Reading, Mass., 1969.

PROPERTY DEFINITIONS

The term *property* is often used loosely and in several different ways to characterize materials. In this book it will be used broadly to refer to materials application behavior or performance. It also will be used more precisely, as explained in Sec. 2-3, to refer to relationships between application or service conditions imposed on a material and the response of that material. Service conditions are expressed in terms of energy inputs to the material, and the response is expressed in terms of energy outputs and/or state changes. State changes can be "internal," such as compositional and structural, and/or physical, such as size change, deformation, fracture, or deterioration.

The properties used in the evaluation and application of materials number in the hundreds. Many are highly specialized and have only limited scope. We will cover only the most widely applicable properties, most of which have been accepted as "standard" properties when evaluating and selecting materials.

Based on the kinds of energy inputs to materials, the properties we will cover can be divided into the following broad groups: mechanical, chemical, thermal, electrical and magnetic, and radiation.

Mechanical Properties

In practical engineering terms, the mechanical properties of a material, such as strength, hardness, and elasticity or stiffness, are measures of a material's ability to carry or resist the application of mechanical forces or energy. If we consider the material as a property system, the input is the applied mechanical energy or force, which is usually expressed as load or stress. The output is the net change that the material undergoes as a result of the input. This change can be deformation and/or fracture or rupture (Fig. 4-1). The term strain is used to represent deformation and dimensional change and is considered to be one kind of deformation.

Mechanical properties are expressed in three ways: (1) in terms of the input—load, stress, or energy—for a specified output (for example, tensile strength and impact strength); (2) in terms of output for a specified input and/or set of conditions (for example, elongation or ductility); and (3) in terms of a ratio between input and output (for example, modulus of elasticity).

4-1 Stress

The mechanical force or energy that causes or produces deformation or fracture in a material is called stress. Strictly speaking, stress refers to the internal interatomic forces that react to an externally applied force. However, because it is difficult and not practical to measure or calculate these internal reaction forces, stress is commonly expressed in terms of the externally applied force.

There are three basic stress modes (Fig. 4-2): Tensile stress results when applied forces tend to increase the length and decrease the cross-sectional area of the bar (Fig. 4-2a). When the applied forces tend to shorten the length and increase the cross-sectional area, we have compressive stress (b). Shear stress results from opposing applied forces tending to cause one part of the material to slip or slide with respect to the other part (c). In practice, the stresses resulting from applied forces are combinations of these three modes. For example, when a bending load is applied to a piece of material as shown in Fig. 4-2d, half of the material contains compressive stresses, while the other half is stressed in tension.

Fig. 4-1 A model of mechanical properties.

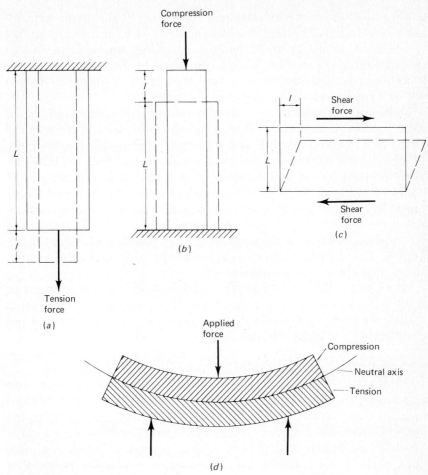

Fig. 4-2 Three kinds of applied stress: (a) tension, (b) compression, (c) shear, and (d) combination of tension and compression stresses in beam bending.

4-2 Strain and Deformation

The deformation of a material subjected to mechanical energy forces is referred to as strain. All materials—without exception—deform when a mechanical force, or load, is applied. The amount of deformation, of course, depends on the magnitude of the force and the kind and dimensions of the material. Often the deformation is so small that it is not visible to the naked eye, and, in some cases, it may not even be measurable by conventional instruments.

This deformation of a material under load, whether visible or not, is a result of the inner infinitesimal movement of the atoms held together by various bonding mechanisms (Sec. 3-1). In crystalline materials such as metals, the interatomic bonding forces, which resist the applied load, are very strong compared to those in plastics, for example. Consequently, a heavy load that causes a plastic to deform will not cause noticeable change in a metal.

There are two basic types of deformation or strain. One kind, called elastic strain, is reversible. That is, when the stresses that cause the atoms in a material to move and shift their position are removed, the atoms return almost to their original positions, and thus the bulk material assumes its original dimensions. A rubber band returning to its unflexed size after being stretched is an obvious illustration of elastic strain. In contrast, plastic strain is essentially permanent deformation. When the stresses are released, the atomic structure does not return to its original state, and, consequently, the material will remain deformed. The deformation or strain remaining after load release is referred to as the set, or permanent set.

There is a third kind of strain, viscoelasticity, which is found in many polymers, such as plastics. Here, after an applied stress that is within the material's elastic range is removed, the material will not completely recover its original size.

4-3 Stress-Strain Relationships

Conventionally, stress is measured in terms of applied force per unit area. In the English system, for structural materials stress is expressed in pounds per square inch (psi) or kips per square inch (ksi). In the metric system, kilograms per square centimeter are the common units. Strain is customarily defined as the change per unit of length in a linear dimension of the material body. It is often expressed in terms of millionths.

In accordance with the above definitions, stress and strain are calculated as follows. Referring to Fig. 4-3, when a load or force is applied to a bar of length L having a cross section A, the bar will deform—elongate a certain amount, l. The stress on the bar then is

$$S = \frac{F}{A}$$

where S = stress, psi
F = load or force, pounds
A = cross-sectional area normal to stress, in.^{2}

Strain, then, is

$$e = \frac{l}{L}$$

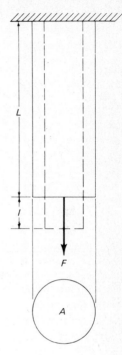

Fig. 4-3   Method of calculating stress and strain.

where e = strain
       L = original length of material
       l = amount of extension

The stress-strain diagram is perhaps the most widely used means of charting stress and strain and their relationships. As shown in Fig. 4-4, it is a plot of stress on the vertical axis versus strain on the horizontal axis. The elastic and plastic deformation regions are also indicated. Curves of this type are valuable in studying many of the mechanical properties we will discuss in this chapter.

4-4   Fracture

An end condition that can result from application of mechanical forces to a material is a complete separation, or fracture, of the material into two or more pieces. Fracture is described in several different ways. The terms *ductile* and *brittle* characterize the material behavior prior to the fracture. A ductile fracture is one that occurs after the material has undergone appreciable plastic strain before breaking. In a brittle fracture, there is little or no plastic deformation.

The terms *shear* and *cleavage* describe the crystallographic nature of the fracture. Shear fractures result from excessive plastic deformation caused by slippage of atomic planes, whereas cleavage fractures

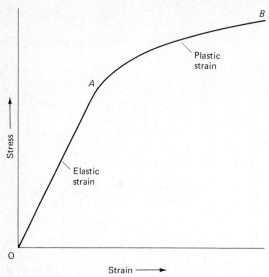

Fig. 4-4 Stress-strain curve showing elastic (OA) and plastic (AB) region.

are caused by separation across cleavage planes under tensile stresses. A brittle material therefore usually fractures by the cleavage mode.

The terms *fibrous* and *granular*, which generally are applicable only to metal fractures, are themselves descriptive of the appearance of the fractured surface. As a general rule, a fibrous fracture is characteristic of the shear mode, and a granular fracture indicates the cleavage mode.

4-5 Failure

Failure, simply defined as unacceptable deformation or fracture, is a relative term. The criterion or measure of failure depends on the materials application. For example, in many products, a certain amount of deformation—elastic or plastic—is tolerable and, in some cases, desirable (for example, elastic deformation in fishing poles). Therefore any definition of failure is arbitrary and frequently involves subjective judgment.

There are three basic ways in which a material can fail when under an applied mechanical force. One way is by fracture—the separation or pulling apart of the material (Sec. 4-4). Another is by buckling or collapsing, as in the case of an overloaded column. The third is deforming by plastic deformation past the point that is desirable in the given application.

4-6 *Static Strength*

The static strength of a material can be defined in a number of ways. In all cases, though, static strength is expressed in terms of an applied stress that is related to a specific kind or magnitude of deformation—either strain or fracture.

Tensile Strength. To describe various kinds of static tensile strength, we will use the stress-strain diagram to follow the behavior of a mild carbon steel as it is slowly loaded in tension (Fig. 4-5). When a load is applied to the test specimen, the material, up to a certain point A, exhibits elastic strain, which is proportional to the stress in accordance with Hooke's law. This point is called the proportional limit and is defined as the maximum stress at which stress is proportional to elastic strain.

As the stress increases, the curve begins to deviate from a straight line. For many materials, elastic strain continues briefly above the proportional limit to some point B. This point, designated the elastic limit, is defined as the maximum stress that can be applied to a material without producing permanent deformation (set). Hence, if the load is removed at any point in the curve before stress B is reached (stage I in the figure), the specimen will return to its original length.

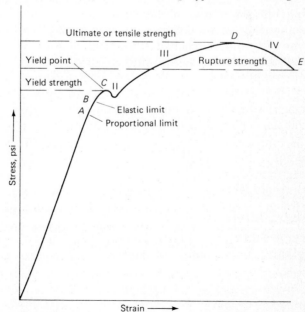

Fig. 4-5 Stress-strain curve showing types of static strength.

After this point is passed, the material exhibits plastic deformation. The nature of initial plastic strain varies with materials. A number of metals and alloys, such as low-carbon steel, for example, suddenly begin to deform sharply without an increase, or sometimes even with a decrease, in stress, at point C. This strain is sometimes referred to as plastic yielding (stage II in Fig. 4-5). The point at which this sudden plastic yielding begins is known as the yield point, and the stress at this point is defined as yield strength—one of the most widely used measures of materials strength. Because it represents the elastic strength of a material, the yield point is for practical purposes the dividing line between elastic and plastic behavior. Also, in the case of metals, it usually represents the highest useful strength level.

Many materials (for example, aluminum and copper) do not have a sharply defined yield point. Therefore an arbitrary yield point on which to base the yield strength of such materials has been established. It is defined as the stress corresponding to a definite amount of permanent set, usually 0.10 to 0.20 percent of the original length (gage length) of the test specimen (Fig. 4-6).

After the material passes the yield point and enters the plastic deformation range (stage III in Fig. 4-5), the stress either remains essentially constant or rises only a relatively small amount to point D, which marks the ultimate strength. Often termed tensile strength, this is the maximum stress a material sustains prior to fracture, and it is the highest point on the stress-strain curve. Because of the large amount of plastic deformation that takes place in many materials before this point is reached, ultimate strength is seldom realized or used in practice. Nevertheless, ultimate strength is widely used as a criterion of strength, especially in general comparisons of materials.

The last point, E, on the stress-strain curve is the stress at which a material breaks or fractures and represents rupture strength. For some metals, such as low-carbon steel and other ductile metals, the rupture strength is lower than the ultimate strength. This is because, beyond point D, the specimen's cross-sectional area decreases sharply, resulting in a decreasing resisting area. Other metals, particularly many in the cold-worked or hardened condition, fracture at the ultimate-strength level without an appreciable reduction of cross-sectional area (Fig. 4-7c). This is also true of brittle materials, such as ceramics, which exhibit little or no plastic deformation (Fig. 4-7a). In the case of many plastics, final fracture of the specimen requires an increase in stress (Fig. 4-7d).

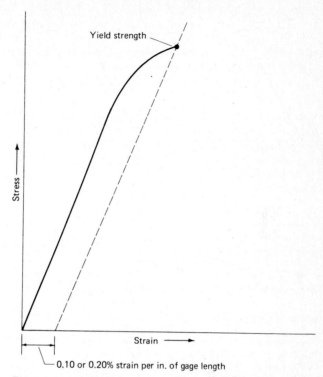

Fig. 4-6 Determination of yield strength for materials without a definite yield point.

Compressive Strength. In principle, compressive strength is the opposite of tensile strength. The material first goes through the elastic strain range and then enters the plastic deformation range. With brittle materials that fracture under compression, a definite compressive-strength value can be obtained. In the case of ductile materials that do not rupture, but rather "bulge," compressive strength is a value representing a stress at which the specimen has distorted to a degree regarded as an effective failure. Compression stress-strain curves have the same types of critical or property points as described under tensile strength—elastic limit, proportional limit, and yield strength.

Shear and Torsion Strengths. What is termed direct shear strength or ultimate shearing strength is defined as the stress at which fracture occurs under application of loading in the shear mode. Torsion

46   Industrial and Engineering Materials

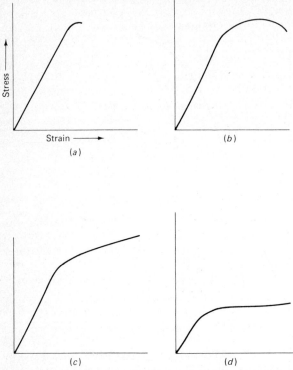

Fig. 4-7 Different shapes of stress-strain curves for various types of materials: (a) Brittle material, (b) ductile metals like low-carbon steel, (c) hardened metals, (d) plastics.

strength is also an expression of shear strength since the stressing is in the shear mode when a piece of material is twisted. Torsion strength, which is sometimes referred to as modulus of rupture in torsion, is the maximum torsional stress that a material can sustain without rupture.

Flexural Strength. Referred to also as bending strength, flexural strength is the ability of a material to resist combined tensile and compressive stresses developed under a bending load (Fig. 4-2d). It is expressed in terms of the maximum stress a material can sustain without fracture under loading parallel to the longitudinal axis of the test specimen.

4-7 *Elastic Modulus and Stiffness*
The modulus of elasticity, which is indicative of behavior in the elastic range, is one of the most fundamental and important mechanical properties of materials. It is the ratio of stress to elastic strain:

$$E = \frac{S}{e}$$

where E = modulus of elasticity, psi
S = stress, psi
e = strain, in./in.

This relationship for a majority of engineering materials is linear as indicated in Fig. 4-5—that is, stress is closely proportional to elastic strain. Therefore if, for example, we double the load, the amount of elastic deformation, such as bending or stretching, will also double. Steels, for example, are almost perfectly elastic throughout the entire elastic range. But concrete, cast iron, and some nonferrous metals exhibit the direct stress-strain proportionality in only a small range. For materials such as these, special definitions of the modulus of elasticity have been established, based on the slope of lines drawn through certain points on the stress-strain curve. These are known as the tangent modulus and the secant modulus.

It should be evident that the modulus of elasticity is a measure of stiffness, that is, the higher the modulus, the lower the elastic deformation under a given load. Consequently, modulus is an extremely important and useful property in the design of products subject to mechanical loads and forces.

There are three moduli, giving the stiffness under the three basic modes of stress. The most common—the tensile modulus—is referred to as the modulus of elasticity or as Young's modulus. Under simple shear, stiffness is termed either shearing modulus of elasticity or modulus of rigidity. And stiffness under compression stress is called bulk modulus or coefficient of compressibility.

4-8 *Plastic Flow*

As previously discussed, when stress on a material passes beyond the elastic limit, plastic strain, or plastic flow, resulting in permanent deformation occurs. The properties related to this plastic flow behavior are measures of the amount a material will plastically deform before or at some specified end point.

Ductility. Plastic flow behavior in metals at normal temperatures is termed ductility. It is usually measured in terms of the percentage of elongation and reduction of area in the tension test specimen or the ability of a specimen to withstand a cold-bend test.

Elongation is calculated by dividing the total permanent increase in the length of the test specimen by the original length and multiplying by 100. Reduction of area is calculated by dividing the difference between the original and the smallest final cross-sectional area by the original cross-sectional area and multiplying by 100. Elongation is the value most widely used as an indicator of ductility.

Ductility is an important property for two reasons. First, it is a measure of the extent to which a metal will plastically deform before fracture. Ductile metals such as aluminum and mild steel will undergo considerable plastic flow before rupture compared to brittle materials like ceramics, which fracture rather abruptly after only a small amount of deformation. Second, ductility is one measure of the ease of forming a metal with such shaping operations as extrusion, stamping, drawing, and forging.

Creep. Most materials, when loaded at a certain minimum stress over a period of time, exhibit plastic flow. Depending on temperature, stresses even below the elastic limit can cause some permanent deformation. This behavior, termed creep, has no standard definition. It is most generally defined as time-dependent strain occurring under stress. Sometimes it is more specifically defined as total strain occurring under constant load. And sometimes it refers to only that part of the total time-dependent strain that occurs in the plastic range. In this book, creep or creep strain is defined as the total strain occurring under a constant load over a specified time period.

The amount of creep depends on the material, the stress level, temperature, and time (Fig. 4-8). Given suitable conditions and sufficient time, creep in a material proceeds through three stages as shown in Fig. 4-9. When a constant load is applied, an initial strain OA immediately occurs. Depending on the material and conditions, this

Fig. 4-8 Effect of time, stress level, and temperature on creep strain.

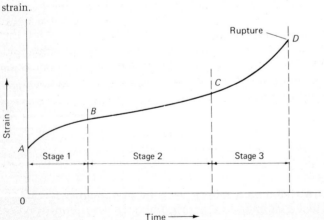

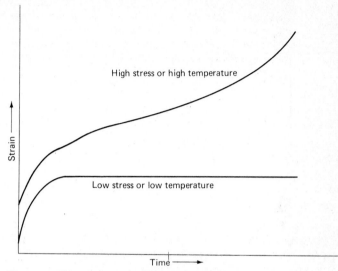

Fig. 4-9. Three stages of creep.

strain will roughly relate to the modulus of elasticity or will be a combination of elastic and plastic strain. Then, in stage 1, called primary creep, the rate decreases over a relatively short period of time, shown as AB. In stage 2, the creep strain, called steady-state or secondary creep, continues at an approximately constant rate BC. And, in stage 3, termed tertiary creep, the rate can markedly increase until rupture.

In the case of polymeric materials, creep can occur even at normal temperatures and relatively low stresses when a constant load is present for an extended period of time. In most structural metals and ceramics, creep at normal temperatures is small and insignificant. However, at elevated temperatures, creep in metals is appreciable and therefore an important consideration.

Because creep involves three variables—time, temperature, and stress—tests are rather complex. Also, since actual service applications often involve extended periods of time, data from short-time creep tests are often extrapolated for selection and design purposes. Creep behavior is expressed in several different ways (Fig. 4-10). For metals, one way is to plot the second-stage creep rate for a number of different temperatures in percent per specified number of hours versus stress (Fig. 4-10a). Another common method is to plot stress versus temperature for specified second stage creep rates (Fig. 4-10b). The stress for a specified rate of strain at a constant temperature, read from a plot such as this, is often termed creep strength. Thus, in Fig. 4-10b, stress x

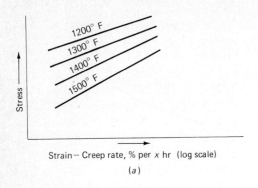

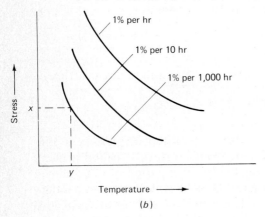

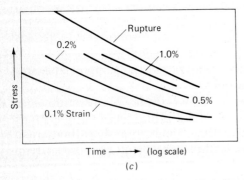

Fig. 4-10 Methods of expressing creep behavior: (a) stress vs. creep rate for various temperature levels, (b) stress vs. temperature for various creep rates, (c) stress vs. time for various quantities of creep strain and for rupture.

is the creep strength at y temperature for a strain rate of 1 percent/1,000 hr. A common strain rate for which creep strength is specified is 1 percent/10,000 hr. Creep strength can also be defined as the constant stress that will cause a specified quantity of creep in a given time period at a constant temperature (Fig. 4-10c). It is obtained from a plot in which stress versus time is plotted for various amounts of strain at a constant temperature. The curve labeled rupture is often referred to as rupture strength.

In plastics engineering practice, a commonly used measure of creep is known as creep (apparent) modulus. It is calculated by dividing the creep strain into the initial applied stress. In this case, the creep strain is the total plastic strain. From a plot, like that in Fig. 4-11, the creep modulus for a given time period can be determined.

There is no standard, established creep test for plastics, although the American Society for Testing and Materials test ASTM D 674 is identified as "a recommended practice for creep tests." The tests described measure creep by applying a constant load to test bars placed in a constant temperature chamber. Creep under tension stress and under bending load is determined in this way.

4-9 *Toughness*

The ability of a material to absorb energy without rupturing is known as toughness. Related to both strength and ductility properties, it is often defined as the total area under the stress-strain curve, as plotted from tensile test data (Fig. 4-12). Known as modulus of toughness, it represents the amount of work per unit volume of a material required to produce rupture under static loading. Using this criterion of toughness, the difference between tough ductile materials and low-toughness brittle materials is shown by the stress-strain curves. Ductile materials display considerable plastic flow before fracture (Fig. 4-12a), whereas brittle materials do not (Fig. 4-12b). The appearance of fractured metal surfaces generally shows these differences in toughness and ductility. Fractured ductile metals have fibrous surfaces indicative of a predominantly shear fracture compared to the granular appearance of brittle fractures, which are in the cleavage mode.

Toughness is not an easy property to measure accurately. It is affected by a number of different factors, including the speed at which

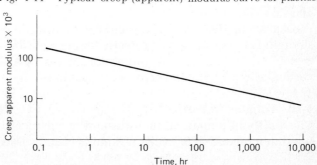

Fig. 4-11 Typical creep-(apparent)-modulus curve for plastics.

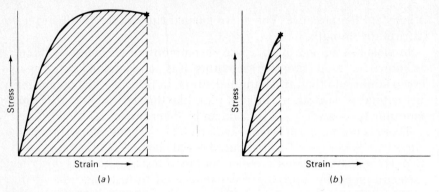

Fig. 4-12   Modulus of toughness as measured by area under stress-strain curve: (a) ductile material, (b) brittle material.

a load or force is applied, and the material's so-called notch sensitivity. A common method of expressing toughness is in terms of impact strength, or impact toughness, which is a measure of a material's ability to absorb the energy of a dynamic impact force. Two tests, the Charpy and the Izod impact tests, are widely used to determine impact strength. In both tests, impact strength is determined by measuring the energy in foot-pounds required for a weighted swinging pendulum to fracture a standard test specimen. The specimens are rectangular, and, depending on material and other considerations, are notched in the shape of a V, a keyhole, or a U. In the Charpy test, the specimen is supported as a simple beam. In the Izod test, it is mounted as a cantilevered beam.

Toughness is also significantly affected by temperature. This is shown in Fig. 4-13. There is one narrow temperature region in which impact strength changes sharply. On the low side, the material exhibits a brittle-fracture mode, while on the high side, fracture is of the ductile type. With a number of materials, including many steels, this region (or temperature), termed the transition temperature, occurs near room temperature. Polymeric materials exhibit brittle behavior below the glass transition temperature (see Sec. 8-5).

Because of the relative inaccuracy of impact test values, impact strength is not usually used directly for design. Nevertheless, it and the transition temperature are important guides in evaluating materials, particularly for low-temperature applications.

4-10   *Fatigue Properties*

The ability of a material to withstand mechanical loads or forces applied in a repetitive or cyclic fashion is closely associated with tough-

Property Definitions 53

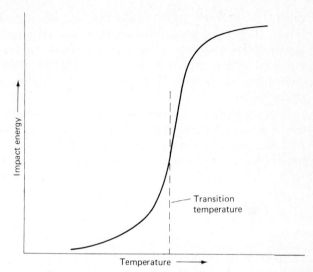

Fig. 4-13 Transition temperature for metals indicates change from ductile- to brittle-type fracture.

ness. In contrast to effects of static loading on toughness, cyclic repetition of stresses may decrease the capacity of a material to deform plastically and thus precipitate a fatigue fracture at loads below the nominal ultimate strength, and often below the yield strength. This type of fracture is one of the most common causes of failure in structural and dynamic machine parts. Often, over the service life of automotive and airplane parts, motor and turbine parts, switching and relay devices, and many similar parts, fluctuating stresses may be repeated millions or even billions of times.

The exact nature of fatigue failure is not fully understood. However, it is generally recognized as proceeding in two stages. The fracture is initiated by stress concentrations at a local crack or a surface defect. Under repetitive applied stress, the material loses its ability to flow plastically. Then, in the second stage, as the local crack begins to propagate, cross-sectional area of the material is reduced until stress per unit area exceeds static strength, and fracture occurs.

The stress at which a material fails under repeated loading conditions is referred to as fatigue strength. For many materials there is a limiting stress below which the stress may be applied repeatedly for an indefinitely large number of times without causing fracture. This stress level is called the endurance limit.

Fatigue properties are affected by both the nature and magnitude of fluctuating stress. While in service, a great many different kinds of repetitive stresses are encountered, for testing and evaluation purposes, they have been reduced to a few standard types. Figure 4-14 shows three of these general cyclic stress modes: fluctuating stress, repeated stress, and reversed stress. Unless otherwise specified, the endurance limit usually represents a completely reversed bending stress mode.

A number of tests have been devised to determine fatigue properties. Perhaps the simplest and most widely used one employs a completely reversed flexural loading on rotating beam specimens. A number of identical specimens are subjected to a repeated stress cycle, each with a different maximum stress. For each specimen, the cycles to rup-

Fig. 4-14 Some types of cyclic stresses: (a) fluctuating, (b) repeated, (c) reversed.

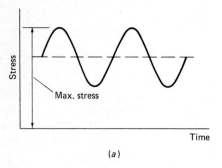

(a)

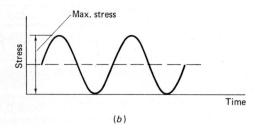

(b)

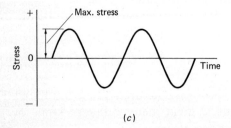

(c)

ture are plotted against the maximum stress to obtain what is known as an SN diagram or curve.

Figure 4-15 shows the two typical kinds of SN curves. Curve A is typical of most ferrous and some nonferrous metals. The endurance limit is the stress level at which the curve flattens out. Curve B is characteristic of many polymers and nonferrous metals. Since there is no flattening out of the curve, these materials have no endurance limit.

Fig. 4-15 Typical fatigue curves: (a) for most ferrous metals and some nonferrous metals, and (b) for many nonferrous metals and polymers without an endurance limit.

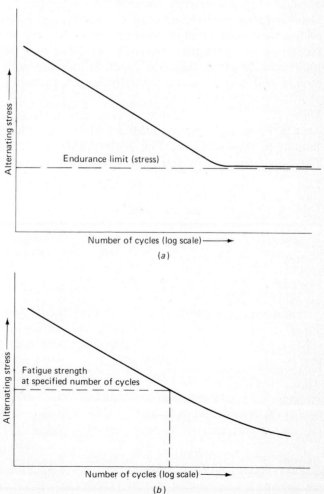

Therefore fatigue behavior can only be specified in terms of fatigue strength at given stress levels.

With many metals there is a rough relationship between the endurance limit and the ultimate strength. For example, for wrought steels, the endurance limit is about half the strength; for copper alloys, endurance limits range from 25 to 50 percent of the ultimate strength.

4-11 Hardness

Although hardness is one of the properties most widely used to distinguish between various kinds of materials, there is no single, universally applicable measure or definition of it. There are, however, a number of characteristics associated with hardness. These include resistance to indentation, resistance to scratching, resistance to wear and abrasion, machinability, bearing qualities, and resilience. Because hardness, like strength properties, is related to interatomic and intermolecular bonding, hardness tends to increase with strength. Hence, metals and ceramics are generally harder than polymeric materials.

Mohs' Hardness. The oldest and simplest method of indicating hardness is the scratch test, associated with the so-called Mohs' scale, well known to mineralogists. The scale (Table 4-1) consists of hardness numbers from 1 to 10 representing increasing hardness. Each number

Table 4-1 *Mohs' Hardness Scale*

Standard Mineral	Mohs No.	Rockwell No.	Brinell No.	Common Object	Typical Materials
Diamond	10				
Corundum or sapphire	9			Spark plug	Ceramics
Topaz	8	C70–80		Cutting tip	Tungsten carbide
		C65–70	745–800	Saw blade	Tool steels
Quartz	7	C58–65		Hand tools	Hardened steel
Feldspar	6				Glass
Apstite	5			Knife blade	
Fluorite	4	C20–30	230–280	"Tin" can	Annealed steel
		B70–90		Penny	Copper
Calcite	3	H50–80	20–45	Storm window	Aluminum
Gypsum	2		10–40	Fingernail	Plastics
Talc	1	H20–30		Solder	Lead

is associated with a reference mineral that will scratch all minerals lower on the scale. Therefore the Mohs' number of a material indicates that it can be scratched by the reference minerals with higher numbers. Obviously a crude test, it can be used only to give a rough, comparative approximation of a material's hardness.

The most widely used methods of evaluating hardness are based on resistance to indentation. In all these methods, a precisely defined indenter is used to produce a surface penetration. The resulting hardness value is then determined on the basis of the size or depth of the indentation and the applied load.

Brinell Hardness. In the Brinell test, a small steel ball (10 mm in diameter) is pressed into the surface under loads of either 3,000 kg, for relatively hard metals, or 500 kg, for softer materials. The standard application time is 30 sec. The diameter of the indentation is then measured. The Brinell hardness number (Bhn) is based on the relationship expressed in the following formula:

$$\text{Bhn} = \frac{\text{load (or pressure) on ball}}{\text{area of indentation}}$$

or

$$\text{Bhn} = \frac{F}{(\pi D/2)(D - \sqrt{D^2 - d^2})}$$

where Bhn = Brinell no., force/unit area
F = load
D = diameter of indenter ball
d = average diameter of indentation

The Brinell test is not generally suitable for hard materials because of the relative softness of the indenter ball, although diamond indenters are available. Nor is it recommended for testing thin specimens where the indentation could be greater than the specimen thickness.

Rockwell Hardness. The Rockwell test uses various loads and indenters depending on the material and other conditions. It differs from the Brinell test in that the loads and indenters are smaller. Rockwell hardness numbers are arbitrary values. The numbers are read directly from the instrument dial and are related inversely to the depth of indentation. Hence the smaller the depth, the higher the number and hardness. The test procedure involves applying a minor load of 10 kg followed by the major load, which can be from 60 to 100 kg, depending on the indenter and its size, as well as the material being tested. Steel balls and diamond cones are used for indenters.

The Rockwell test is applicable to a wide range of hardnesses and is used for testing materials with hardnesses that cannot be measured by the Brinell test. Also, because it is simple, direct reading, and rapid, the test is widely used in production work. Table 4-2 lists the various Rockwell scales and the materials for which they are usually applicable.

Vickers Hardness. The Vickers test is similar to the Brinell in that the hardness value is based on the ratio of applied force to indentation area. However, in this method, the indenter is a diamond in the shape of a square pyramid. The applied load can range from 5 to 120 kg and depends on material hardness and permissible size of impression area. The formula used to calculate the *Vickers hardness number* (DPH) is:

$$DPH = \frac{1.8544\,F}{D^2}$$

where DPH = Vickers hardness no.
F = load
D = average diagonal of indentation

Table 4-2 *Rockwell Hardness Scales*

Scale Symbol	Applicable To
A	Thin steel, shallow case-hardened steel
B	Soft steels, aluminum alloys, copper alloys, malleable iron
C	Steel, hard cast iron, deep case-hardened steel, pearlitic malleable
D	Thin steel, medium case-hardened steel
E	Aluminum alloys, magnesium alloys, bearing metals, cast iron
F	Thin and soft sheet metals, annealed copper alloys
G	Malleable iron, phosphor bronze, beryllium, copper
H	Aluminum, lead, zinc
K,L,M,P,R,S,V	Soft, thin materials such as bearing metals and plastics

The Vickers method is especially suitable for thin and very hard materials.

Other. There are several tests and devices specially designed to determine hardness over a very small area. These involve the use of a so-called Knoop indenter or a Vickers indenter, both of which are diamonds of a size and shape to produce a very small impression that is measured with the aid of a microscope.

The Durometer hardness tester is applicable primarily to polymeric materials. A conical steel indenter is loaded to penetrate the surface. The hardness number, known as *Durometer hardness,* is a measure of the depth of indentation. The values are read directly from a dial and range from 0 to 100, representing penetration depths from 0 to 0.100 in. Thus the higher the number, the higher the Durometer hardness. The Rockwell test is also widely used for indicating hardness of plastics. However, since elastic recovery is involved in testing these materials, it is usually not valid to compare hardness of various kinds of plastics only on the basis of Rockwell data.

Hardness values based on material resilience are determined by measuring the rebound of a standard object dropped from a specified height onto the surface of the specimen. Known as Scleroscope hardness, the rebound-height scale is graduated arbitrarily from 0 to 140. A heat-treated high-carbon steel, for example, has a scale reading of

Table 4-3 *Approximate Equivalent Hardness Numbers for Steel*

Brinell No. Standard	Rockwell No. B Scale	Rockwell No. C Scale	Vickers No.	Shore Scleroscope
110–150	65–81	—	116–156	15–23
151–200	81.2–93.6	—	157–211	23–31
201–250	93.8	14–24	212–264	31–37
251–300	—	25–32	265–318	37–45
301–350	—	33–37	319–371	45–51
351–400	—	38–43	372–424	51–58
401–450	—	43–47	425–478	58–64
451–500	—	48–51	479–533	65–68
501–550	—	52–54	534–586	68–73
551–600	—	54–57	587–639	73–77
601–650	—	57–59	640–694	77–81
651–700	—	60–61	695–775	81–86

about 100. Because the hardness numbers are arbitrary, only numbers on similar types of material are comparable.

4-12 *Wear Resistance*
Wear resistance is the general term referring to a material's ability to resist the action of abrasive or other forces on the surface. Wear is manifested by a loss of surface material, either in a regular or irregular form, and this loss, in turn, results in dimensional changes. Because of the many variables involved in wear, and its complexity, there is no one quantitative wear criterion or test. Most evaluations are based on past experience and tests of materials in service.

However, some mechanical properties are rough indicators of a material's probable wear resistance. Hardness is perhaps most often considered. Abrasive wear is usually a result of scratching action. And, since a material can be scratched only by a harder substance, it would appear that the harder a material, the greater the wear resistance. While this correlation is valid in many applications, it will not hold in others—for example, in the case of brittle materials where broken surface fragments will act as abrasive particles.

Galling, or adhesive wear, is different from abrasion in that the wear is caused by the surfaces of two materials tending to adhere to each other. Such adhesion causes pieces of the bearing surfaces to be torn out. Relatively soft metals and those with low yield strength are most susceptible to galling.

Thermal Properties

Materials in service are almost always subject to some thermal energy that may come from the service environment or that may arise from or be a result of the material's operational function. Also, many processing methods involve application of heat to either form the material hot or in a molten condition. Heat entering or generated by a material frequently causes instability and internal as well as bulk structural or state changes that alter mechanical and other properties. In addition, heat-treating materials to deliberately change properties is based on input and/or withdrawal of heat energy from the material.

4-13 *Temperature and Heat*
The concepts of temperature and heat, which are intuitively familiar because they are so much a part of our physical environment, are not one and the same as is often supposed. Heat is thermal energy produced in a material by either supplying heat directly to it or by ap-

plying some other form of energy to it, such as mechanical work or an electric current. In both cases, the internal energy of the material is increased, and one manifestation or output is a temperature increase. Hence temperature can be thought of as a measure of the degree or level of thermal activity.

Thermal energy is commonly expressed in British thermal units (Btu) or as calories. One Btu is the quantity of thermal energy required to increase the temperature of one pound of water one degree Fahrenheit at the water's greatest density, which is 39°F. One calorie is the energy needed to increase the temperature of one gram one degree Celsius, at 15°C.

By definition, the quantity of heat required to raise a material's temperature one degree (at 39°F) is its heat capacity. In English units, it is expressed as Btu/lb/°F; in the metric system, it is cal/g/°C. Specific heat is defined as the ratio of the heat capacity of a material to that of water and is the same in the English and metric systems.

4-14 *Heat Resistance*

Heat resistance is a general term referring to the ability of a material and its properties to remain stable with changes in temperature. The temperatures at which significant changes occur in the structure and/or behavior of a material are termed transition or transformation points or levels. Of the many such points, those defined below are especially significant from an application standpoint.

An important requirement is that a material's melting or softening point be above the service temperature. The melting point is the temperature at which a material changes from a solid to a liquid state. Crystalline materials have a sharp melting point. Many polymeric materials and glasses, however, do not have a definite melting point. They become soft over a temperature range or may decompose appreciably before or without melting. The softening point (Vicat) of plastics is defined as the temperature at which a flat-ended needle of 1-mm^2 circular or square cross section penetrates the test sample to a 1 mm depth.

The heat distortion point in plastics indicates the temperature at which the plastic begins to soften and yield under a load. Under a given set of test conditions (ASTM D 648), it is the temperature at which the total deflection is 0.10 in.

The maximum service temperature is a widely used indicator of the heat resistance of plastic materials. Although not very precise, it indicates in a general way the maximum temperature at which the material can be used continuously over a relatively long period of time (days or weeks) and essentially retain the room-temperature properties required in a given application.

Another aspect of a material's heat resistance is the effect of heat on mechanical, electrical, and chemical properties. Depending on the material and the temperatures involved, property changes can be quite large, as we noted in our discussion of mechanical properties. And, finally, the heat resistance of a material also depends on thermal stresses generated within the material (see Sec. 4-16).

4-15 Thermal Conductivity and Emissivity

Heat energy flows through a material from higher to lower temperature regions. The rate at which the transfer takes place is termed the coefficient of thermal conductivity (K). It is defined as the quantity of heat (Btu) that flows in a unit of time (1 hr) through a unit area of plate (1 ft^2) of unit thickness (1 ft) having a unit difference (gradient) of temperature (1°F) between faces. In general, in a given material, thermal conductivity varies with temperature. In metals, K usually decreases with rising temperatures, whereas in other materials, it usually increases.

Thermal conductivity varies widely among different kinds of materials and depends on structure, state, and purity. In some applications, such as heat-sink devices, high thermal conductivity is desirable. In others, such as in heat insulation, a low K factor is sought. This is best achieved with porous materials because of the low conductivity of air compared to solids.

Heat transfer also takes place in materials by means of radiation. The rate with which a solid emits thermal radiation depends on the material's surface and the temperature. This type of heat transfer is termed emissivity and is defined as the ratio of the heat emitted by a material to that radiated by an ideal black body at the same temperature. An ideal black body is a material that would absorb all heat radiation impinging on it without any reflection or transmission.

4-16 Thermal Stresses and Expansion

When heated, most solids expand; and, conversely, when cooled, they contract. Although dimensional change takes place in all dimensions of a solid, the property, called the coefficient of linear expansion, is defined as the increase in length per unit length per degree rise in temperature [in./in./°F or cm/cm/°C].

The thermal expansion coefficient of a material usually changes with temperature, normally increasing at higher temperatures. It is also related to a material's specific heat and melting point. In general, plastics and elastomers, with relatively low softening points, have coefficients of expansion several times higher than those of metals. Dimensional changes also occur when thermal energy produces structural phase changes in a material. This is common in the heat treatment of metals.

Thermally induced dimensional changes are usually quite small—generally not more than around 5 percent. However, these volume changes create stresses within the material that can lead to thermal damage. When the internal stresses exceed the ultimate strength, and the material is restrained in some way, the material can crack. This is particularly true in the case of nonductile materials that cannot readily adjust internally to thermal stresses.

Chemical Properties

Environments in which materials operate always have a chemical aspect of some kind. These environments can be liquids, such as chemicals or water; gaseous, such as oxygen; other solids; or any combination of them. The reaction of a chemical environment with materials, termed corrosion, results in either a loss of material or a deterioration of the material or both. It can also affect and modify other properties, such as wear resistance and fatigue strength.

Evaluating, measuring, and predicting corrosion is difficult because the reactions of corrosive media on materials are influenced by a great many interacting variables. These include the exact chemical composition and concentration of the media, the degree of exposure (total or partial and constant or cyclic), the length of time of exposure, the temperature, and the nature of the media's movement. There are, however, a few standard procedures for evaluating or measuring the effects of corrosive attack. Perhaps the most common is to measure the weight change over a period of time, usually in inches per year. Another method is to measure the depth of pitting. In some laboratory tests, consumption of oxygen or amount of hydrogen evolved in the corrosion process is determined.

Corrosion is an exceedingly complex phenomenon, and an all-encompassing explanation of the process has not yet been developed. In application work, corrosive attack is usually divided into two major types: chemical and electrochemical. Chemical corrosion, the simpler of the two, occurs when a material is in direct contact with a liquid in which it is soluble to some degree. Corrosive attack in polymeric materials and ceramics is generally of this type. For example, in varying degrees, polymers are prone to attack by organic solvents.

4-17 Electrochemical Corrosion

Electrochemical corrosion is by far the more prevalent type of attack, especially in metallic materials. Figure 4-16 is a simplified representation of the electrochemical corrosion process. The liquid environment is in contact with the material and acts as an electrolyte, thus conducting a current between areas of different electric potentials. The two areas may be dissimilar metals or different parts of the same material. Current flow removes material from the anodic zone, which has a greater potential compared to the cathodic zone. No corrosion occurs in the cathodic area. However, depending on the nature of the corrosive media, current flow causes either evolution of hydrogen or absorption of oxygen. In the oxygen-absorption process, a corrosion product is deposited at the cathodic area. For example, in the case of steel, the corrosion product is the familiar rust formed on the surface, which becomes the cathode while the base metal acts as the anode.

There are three ways in which coupling can take place between anodic and cathodic areas to produce corrosion.

Composition Couples. This form involves areas of different material structure or composition. The most common case exists where two different metals are in contact in a liquid or in the presence of moisture, such as when steel screws or bolts are used as fasteners for steel or tinplate. The galvanic series (Table 4-4) is the common method of showing the tendencies of various metals to form galvanic couples or

Fig. 4-16 Schematic representation of electrochemical corrosion.

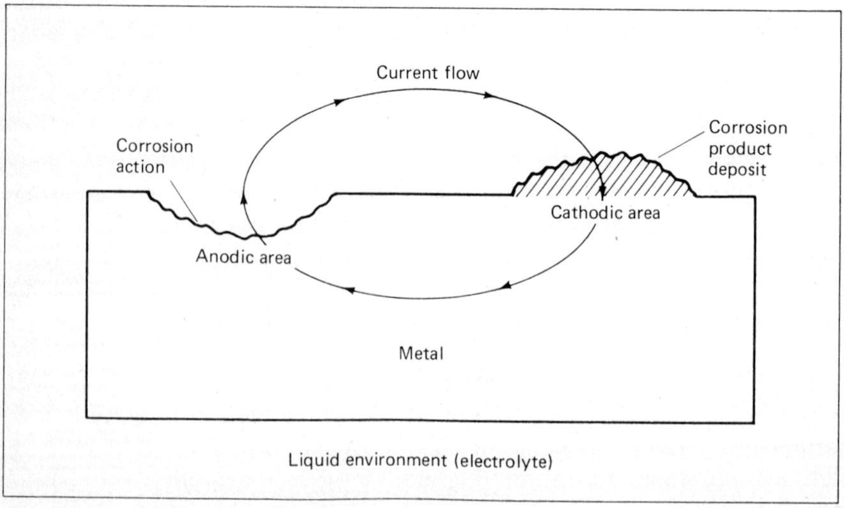

cells. The further apart the metals are, the greater will be their tendency to corrode when they are in contact with each other in the presence of a liquid. The metals in the top portion of the series behave anodically with respect to those below them. For example, when aluminum and copper are together and touching in water, the aluminum will suffer material loss whereas the copper will remain undamaged.

A common example of the action of composition coupling in a material is dezincification in some brass alloys which consist of separate phases of copper and zinc. Because of the fairly wide separation of copper and zinc in the galvanic series, an electric cell between the two develops when the alloy is in a liquid environment. This causes the zinc to corrode, leaving the alloy porous and weak.

Composition coupling can be used as a means of protecting metals in service. Known as cathodic protection, this method involves con-

Table 4-4 *Galvanic Series of Metals*

Anodic, Least Noble, Corroded End

Magnesium
Magnesium alloys
Zinc
Aluminum alloys (low strength)
Cadmium
Steel or iron
Cast iron
Stainless steels (active)
Lead
Tin
Nickel (active)
Brasses
Copper
Bronzes
Copper-nickel alloys
Nickel (passive)
Silver
Titanium
Graphite
Gold
Platinum

Cathodic, Most Noble, Protected End

necting a metal high in the galvanic series (magnesium or zinc, for example) to the metal to be protected. This creates a galvanic couple, and the added metal then functions as a sacrificial anode.

Stress Couples. This form results when certain areas of a material or part contain high internal stresses compared to adjacent areas. Stress couples can occur in weldments and cold-worked metals with high residual stresses. The stressed areas always are anodic and therefore suffer the corrosive attack. In intergranular corrosion, the grain boundaries, which are sites of high stress and/or different composition, become anodic in relation to adjacent crystals and therefore can suffer damage in a corrosive environment. Corrosion fatigue, as the name implies, is caused by a combination of corrosion and fatigue stresses. Corrosion products and pits become starting points for fatigue cracks.

Concentration Couples. This form results when the concentration of the corrosive media differs from one region to another. It frequently occurs in crevices where the media is relatively high in metal-ion concentration or low in oxygen, making these sites anodic with respect to adjacent areas.

4-18 *Oxidation*

Many materials, including most metals and polymeric materials, combine with oxygen in the environment. In metals oxidation often occurs rapidly on exposure to air until a protective oxide film or scale is formed on the surface. The rate of the oxidation process thereafter depends on the protective value of the formed film. In the case of carbon steel, for example, a rather porous oxide film provides little protection against further oxidation. In contrast, the oxide film formed on aluminum is dense and highly protective. In general, metals oxidize at a decreasing rate with time, but at an increasing rate with a rise in temperature.

Most plastics and rubbers oxidize in the presence of oxygen. In rubbers, which are especially susceptible to oxidation, the process is termed *aging*. The reaction of oxygen with the rubber initially reduces elasticity and increases hardness. As aging proceeds, the rubber begins to degrade, and eventually it completely loses its strength. In varying degrees, usually to a lesser extent and over longer periods of time, the aging process occurs in most other polymers. The process depends also on other factors, such as temperature, heat, and the nature of the atmosphere.

4-19 *Water Absorption*

Water absorption is a property of concern chiefly with polymeric materials, many of which have a tendency to absorb water. The result is a volume and weight increase. Also, water absorption can cause warping and swelling, and a deterioration of mechanical and electrical properties. The water absorption of a plastic is determined by measuring the amount of water a standard test sample absorbs when immersed for 24 hours. The water absorption is usually expressed as a percentage of the original weight.

Electrical and Magnetic Properties

Electrical properties involve the behavior of materials under an electric current and, principally, the ability of materials to transmit electrical energy. Magnetic properties involve the behavior of materials in an external electromagnetic field.

4-20 *Electrical Conductivity*

The ability of a material to transmit, or conduct, electric current is termed electrical conductivity. All materials can be classified into three broad groups in accordance with their conductivity: conductors, insulators (or dielectrics), and semiconductors. In this book, we will be concerned principally with the first two.

The reciprocal of electrical conductivity, electrical resistivity, refers to a material's resistance to the flow of an electric current. Electrical conductivity is usually expressed in terms of electrical resistivity, which is defined as the electrical resistance of a material per unit length and unit cross-sectional area. The values of electrical resistivity of metals are expressed in ohm-centimeters or ohm-inches. For dielectric materials such as plastics, volume resistivity is used. It also is usually given in ohm-centimeters. The electrical conductivity of metals is often expressed as a percent of the electrical conductivity of copper, which is arbitrarily set at 100. The resistivity of the International Annealed Copper Standard (IACS) is 1.7241 microhm-cm at 68°F.

4-21 *Insulation and Dielectric Properties*

Whereas electrical conductivity and resistivity are used as measures of the ease of transmittance of an electric current, the ability of a material to prevent the transfer of electrical energy, or an electric charge, is indicated by its dielectric strength. That is, dielectric strength is a measure of insulating qualities, or, more specifically, the ability of a material to resist breakdown when subjected to electrical stress. The breakdown electric stress is expressed as voltage per unit thickness of material, ordinarily in volts per mil. Since resistance to electrical voltage decreases with exposure time, and varies with many other factors, the relatively short time tests used to determine dielectric strength do not indicate the stress the material can withstand over a long time period.

Dielectric constant is a measure of the relative capacitance—that is, the capacity for storing electrical energy—of a material. It is expressed as a dimensionless constant, and is related to the capacitance of the material in a vacuum. Low values of dielectric constant are desirable when a material is used as an insulator, and high values when it is used as a capacitor.

Arc resistance is also related to the electrical insulating qualities. It is a measure of resistance of the surface to breakdown under electrical stress. It is expressed in terms of time in seconds required to cause a material to fail under an electric arc applied intermittently with increasing severity as prescribed by a standard test procedure.

Loss factor, also related to electrical insulation, is a measure of the electrical power loss when a material is subjected to alternating current. Conventionally designated DK, it is given in seconds per cycle, or ohm-farads.

4-22 Magnetic Properties

The magnetic properties of a material relate to its ability to be magnetized by an external electromagnetic field. The external field, or magnetizing force, is designated by the letter H, and is expressed in oersteds. The magnitude, or strength, of induced magnetic force or current, also referred to as magnetic flux density, is designated by the letter B. Magnetic flux density is expressed in gausses.

Permeability (μ) is defined as the ability of a material to be magnetized—more specifically, the ability to carry magnetic lines of force as compared to air or vacuum, which have a permeability of 1. All materials can be classified into three groups in accordance with their behavior in a magnetic field. Ferromagnetic materials such as iron, nickel, and cobalt have high permeability. Materials with permeabilities less than 1 are called diamagnetic; those with permeabilities slightly higher than 1 are called paramagnetic.

When an external magnetizing field is increased or decreased, there is a lag in changes in magnetic flux within a material. This lag is known as hysteresis. To evaluate the hysteresis characteristics of a magnetic material, a curve called the hysteresis loop is used. It consists of a plot of the changes in magnetic induction resulting from variations of the magnetic field.

High-Energy Radiation Effects

All materials are affected by high-energy electromagnetic radiation. However, when small doses are involved the effects are not usually noticeable. Organic materials are generally more susceptible to change or damage than are metals and ceramics.

4-23 *Radiation Types*
Radiation effects can be short-lived or permanent. The primary concern in evaluating materials is with the permanent changes produced by radiation. The permanent effects can be either beneficial, as when polymer branching in certain plastics is produced by neutron bombardment, or damaging, as the lowering of mechanical properties when metals are used in a high-energy radiation environment.

Three types of radiation must be considered:

1. Gamma radiation affects properties by interacting with a material's electrons, resulting in an ionization process. In the case of organic materials it usually causes softening and crumbling. However, gamma radiation has little effect on crystalline materials—metals and ceramics.
2. Fast neutrons, which are uncharged nuclear particles, move at high velocities and have a considerable amount of kinetic energy. When they collide with atoms in a material, they transfer some of their energy to the atoms, which in turn strike and displace other atoms, thus causing structural irregularities, such as vacancies, interstitials, and impurity atoms. The end result is a change of properties in the material. For example, fast neutrons tend to increase yield strength and decrease ductility in a number of metals.
3. Thermal neutrons, which move at velocities equal to the kinetic motion of molecules, induce radioactivity in materials. Otherwise they have little direct effect on properties. One exception is the cracking of glass vacuum tubes caused by absorption of thermal neutrons by boron in the glass.

4-24 *Measurements*
Radiation effects are rather difficult to measure quantitatively. They are often characterized in terms of cross section, threshold energy, and damage threshold. The cross section value of a material indicates the

probability of collisions between impinging radiation particles and atoms in the lattice of the material. Threshold energy is the minimum energy of particles in the radiation beam that will effectively displace atoms in the material. Damage threshold refers to the radiation exposure required to change at least one property of the material. When a property is changed by 25 percent, the material is said to have reached 25 percent damage.

Other Properties

An attempt has been made in the preceding sections to cover what we believe are the most pertinent and widely applicable properties in materials application. Because the lesser used and more specialized properties number in the hundreds, space is not available to cover them. However, we will now briefly define or describe a few of these properties that will be mentioned in other parts of this book. Most of them are related to light radiation and sound waves.

1. *Opacity or hiding power:* This is the degree to which a material or coating obstructs the transmittance of visible light.
2. *Transmittance:* This is the ratio of light transmitted by a material to the light incident on it.
3. *Sound absorption coefficient:* This is a measure of the effectiveness of a material in absorbing sound energy.
4. *Color:* This is that aspect of the appearance of a material that depends on the spectral composition of the light reaching the retina of the eye, and on the light's temporal and spatial distribution.
5. *Specular gloss:* This is the ratio of (1) the light reflected by a material at an angle corresponding to the mirror image of the incident beam to (2) the incident light. *Sheen* is 85-degree specular gloss of nonmetallic materials.
6. *Porosity:* This is the relative extent or volume of open pores in a material or the ratio of pore volume to overall volume of a material in percent.

Review Questions

1. Name the three basic modes of applied stress, and state the kind of deformation each tends to produce.
2. Name and define the two basic types of deformation that can result when mechanical energy is applied to a material.
3. Define viscoelasticity. In what materials does it commonly occur?

4. A load of 2,000 lb is hung from the end of a square bar 0.5 in. thick and 5 in. long. The bar increases in length to 5.2 in. after the load is applied. Calculate the stress and the strain.
5. Explain three ways in which a material can fail.
6. Explain the difference between ductile and brittle fracture.
7. Refer to the accompanying stress-strain diagram to give the value (stress) for each of the following properties:
   (a) Proportional limit.
   (b) Yield point and yield strength.
   (c) Ultimate tensile strength.
   (d) Rupture strength.

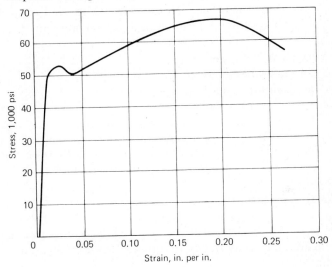

8. What property does the ratio of stress to strain represent?
9. Referring to the stress-strain diagram in Question 7, the elastic strain at a stress level of 40,000 psi is 0.01 in. per in. Calculate the modulus of elasticity.
10. Material A has a modulus of elasticity of 30,000,000 psi compared to 5,000,000 psi for material B. If the same load is applied in tension to both materials, which material will stretch or elongate more? How many times more?
11. Ductility is a term used to refer to what type of deformation?
12. Two ways of specifying or measuring ductility are in terms of percent elongation and percent reduction of area. Give the formula for calculating each.
13. Although there is no standard definition for creep, what are the three variables that influence creep in a material? What are the

three stages of creep? In what terms or units is creep strength expressed?
14. What is the difference between modulus of toughness and impact strength?
15. What information does the transition temperature give about a material's behavior under impact stress?
16. Fatigue strength is an important property to consider when selecting materials for what kind of service? What is the difference between fatigue strength and endurance limit?
17. Name three characteristics or qualities of materials associated with hardness.
18. Give three differences between the Brinell and the Rockwell tests.
19. Which hardness test or tests would you recommend for the following:
    (a) Very hard materials.
    (b) Thin materials.
    (c) Plastics and rubber.
    (d) Relatively soft metals.
20. Name two different kinds of wearing action.
21. Explain the difference between temperature and heat.
22. List four transition, or transformation, points that are indicative of heat resistance, and name the family of materials to which each applies.
23. Name two ways in which heat transfer occurs in or from a material body.
24. Why is the coefficient of linear expansion an important property to consider in applications involving two or more different materials?
25. What are the two essential "elements" that must be present for electrochemical corrosion to occur?
26. Indicate which metal will corrode when the following metals are in contact in the presence of water:
    (a) Zinc and copper.
    (b) Bronze and steel.
    (c) Aluminum and silver.
27. Most materials are susceptible to oxidation. What is one major reason why metals differ widely in the rate at which they deteriorate by oxidation?

28. What electrical property or properties are important in the following materials applications?
    (a) Electric transmission line insulators (two properties).
    (b) Electric light switch "button".
    (c) Residential wiring (two properties).
29. Name three properties related to the effect of light on materials.

Bibliography

Bowman, R. E.: "How Radiation Affects Engineering Materials," *Materials Engineering*, July 1960.

Campbell, J. B.: "Mechanical Properties and Tests," *Materials Engineering*, July 1954.

Davis, H. E., G. E. Troxell, and C. T. Wiskocil: *The Testing and Inspection of Engineering Materials*, McGraw-Hill Book Co., New York, 1964.

Fabian, R. J., and L. H. Seabright: "Corrosion," *Materials Engineering*, January 1963.

Koves, Gabor: *Materials for Structural and Mechanical Functions*, Hayden Book Co., New York, 1970.

Shigley, J. E.: *Machine Design*, McGraw-Hill Book Co., New York, 1956.

METALLIC MATERIALS

Of the 100 elements available to us, about three-quarters can be classified as metals. And, about half of these are of at least some industrial or commercial importance. While pure metallic elements have a wide range of properties, they have quite limited commercial use. Metal alloys, which are combinations of two or more elements, are far more versatile and for this reason are the form in which most metals are produced and used by industry. The number of metal alloys is tremendous, variously estimated to range between 35,000 and 40,000 different standard and trademarked compositions and grades. Allowing for close duplication of compositions among various producers, the total number of metal alloys probably runs over 10,000.

Although the word *metal* by strict definition is limited to the pure metal elements, common usage gives it wider scope to include alloys. In this book, the word metal will be used in this wider sense to refer to all metallic materials.

Some of the major characteristics of metals as a family, distinguishing them from other materials, particularly polymers, are:

1. They have a crystalline structure.
2. They are conductors of electricity and have relatively high heat conductance.
3. They have a combination of relatively high strength and toughness.

4. Almost all of them are serviceable at or above the boiling point of water and many resist considerably higher temperatures.
5. They are relatively stable over long periods of time.

There are two broad families of metallic materials—ferrous and nonferrous. Ferrous alloys are those in which the base or primary metal is iron. They range from cast irons and carbon steels with over 50 percent iron to high alloys that contain a variety of other elements adding up to 50 percent of the total composition. All other metallic materials automatically fall into the nonferrous category, which can be arbitrarily divided into the following:

Light metals: aluminum, magnesium, titanium, beryllium.
Heavy metals: copper, zinc, lead, tin, and so on.
Refractory metals: tungsten, nickel, molybdenum, chromium, and so on.
Precious metals: gold, silver, and platinum-group metals.

Metals are also often identified as to the method used to produce the forms in which they are used. When a metal has been formed or shaped in the solid, plastic state, it is referred to as wrought. Metal shapes that have been produced by pouring liquid metal into a mold are referred to as cast.

Structure of Metals

Metallic materials are crystalline solids. As stated in Sec. 3-3, individual crystals are composed of unit cells repeated in a regular pattern to form a three-dimensional crystal-lattice structure. A piece of metal is an aggregate of many thousands of interlocking crystals (grains) immersed in a cloud of negative valence electrons detached from the crystals' atoms. These loose electrons serve to hold the crystal structures together because of their electrostatic attraction to the positively charged metal atoms (ions). The bonding forces, being large because of the close-packed nature of metallic crystal structures, account for the generally good mechanical properties of metals. Also, the electron cloud makes most metals good conductors of heat and electricity.

5-1 *Unit Cells and Atom Packing*
As discussed in Sec. 3-3, three principal types of unit cells or lattices are found in metals (see Fig. 3-5). They are body-centered cubic, as in

iron and tungsten; face-centered cubic, as in aluminum and copper; and close-packed hexagonal, as in zinc and magnesium. The nature of the unit cell in a given metal determines the relative position of adjacent atomic planes in the crystal lattice as well as the way atoms are packed in the atomic planes. In both face-centered cubic and close-packed hexagonal crystals, the planes are close packed, with each atom in the plane surrounded by and in close contact with six other atoms.

The difference between the two structures is in the positioning of adjacent atomic planes (Fig. 5-1). In face-centered structures, atoms in one plane are positioned over pockets formed by groups of three atoms in the adjacent plane. In hexagonal structures, atoms in adjacent planes fit into pockets like those in the face-centered structures. However, the atoms in every second plane lie directly above each other. In body-centered crystal structures, the atom in the center of the unit cell results in a more open packing of the atoms in the lattice. Consequently, spacing between atoms in the atomic planes is greater.

The practical significance of these differences in crystal structure is reflected in various properties. For example, plastic deformation, which is closely related to ductility, is possible because close-packed atomic planes are able to slide over one another, with the atoms coming to rest in adjacent pockets between atoms (Fig. 5-2). Metals with face-centered and body-centered cubic crystal structures are relatively ductile because of the large number of possible planes where such slipping can occur. In contrast, hexagonal structured metals are generally brittle because they have fewer slippage planes.

5-2 Grains and Grain Boundaries

With few exceptions, commercial metals are composed of aggregates of many crystals or grains. Only in the case of some advanced materials, such as iron whiskers, for example, are the materials composed of a single crystal. Thus metals are polycrystalline aggregates of many

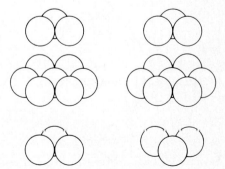

Fig. 5-1 Difference in positioning of adjacent atomic planes in face-centered crystal structures (left) and hexagonal structures (right).

78   Industrial and Engineering Materials

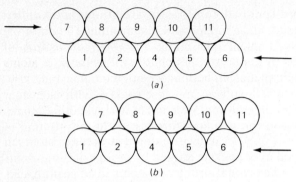

Fig. 5-2 Plastic deformation in metals caused by close-packed atoms moving from one "pocket" location to another. (a) Top layer moves to right as indicated in (b).

individual crystals, each with its regular array of atoms (Fig. 5-3), which fit together to form a structure much like a mass of soap bubbles.

The sizes of individual grains vary from one metal to another. In some brasses, for example, the grains are large enough to see with the naked eye, whereas in most metals they are not visible without the aid of a microscope. Figure 5-4 shows photomicrographs of fine- and coarse-grain alloys. Grains within a piece of metal are not the same

Fig. 5-3 Schematic drawing showing an aggregate of crystals (the irregular shapes), and grain boundaries between individual crystals or grains. The small squares represent unit cells.

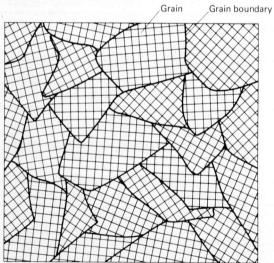

size, and size can vary from one location to another. Because of grain irregularities and size variability, estimates of size are made by comparison with a set of standards. Grain size importantly influences many mechanical and physical properties. For example, as grain size becomes smaller, yield strength increases, and surface smoothness of formed parts improves.

The regions between the grains are known as grain boundaries (see Fig. 5-3). The role grain boundaries play in a metal's performance is complex. They can be a source of either strength or weakness. For instance, impurities collecting in them can lead to accelerated deterioration by the phenomenon known as stress corrosion. On the other hand, second-phase particles may precipitate at the boundaries and serve to improve strength and hardness. Or, the grain boundary itself can be a barrier to dislocation movement or arrest crack propagation.

5-3 *Crystal Imperfections*

A commercial metal is seldom, if ever, composed of perfect crystals, and imperfections scattered throughout the crystal lattices account for many of a metal's performance characteristics under applied stresses. There are two principal types of crystal imperfections: point defects and line defects or dislocations. Grain boundaries (discussed above) can also be considered a crystal imperfection.

Point defects occur at a single lattice point and affect the immedi-

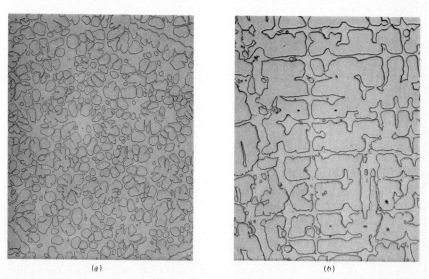

Fig. 5-4 Different grain sizes in brass (100×): (a) fine grain, (b) coarse grain. (Ledgemont Lab, Kennecott Copper Corp.)

ately adjacent atoms. The imperfection can be a vacancy (Fig. 5-5a) in the lattice—that is, atoms can be missing—or it can be an interstitialcy (Fig. 5-5c), in which case an extra atom is present between the atoms located at normal lattice points. The interstitial atom can be the same as the other atoms in the crystal or it can be a different atom. In either case, the presence of extra atoms distorts the lattice and creates an internal strain which in turn leads to increased hardness and strength and decreased ductility.

When vacancies are present in a metal, they facilitate diffusion mechanisms (see Sec. 5-5) by providing easy paths for atoms to move from one site to another within crystals. Also, vacancies themselves can move and cluster together and become the source of cracks.

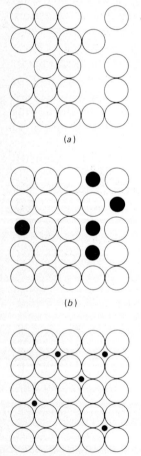

Fig. 5-5 Crystal-lattice imperfections and solid solutions: (a) vacancies, (b) substitutional atoms in vacancies, and (c) interstitial atoms between lattice points.

Dislocations are essentially an interruption in the regular linear or planar arrangement of atoms in the lattice. There are a number of different types of line imperfections, some quite complex, including screw dislocations, bent dislocations, and curved edge dislocations. The most common and simplest is the edge dislocation. As shown in Fig. 5-6, the dislocation disturbance is located at ABCDE and is caused by the presence of an extra plane of atoms indicated as BFG. Line imperfections such as this create slip planes (XY) in crystals, allowing portions of a crystal to shift with respect to other crystals.

Crystal dislocations are a major reason why the strength of metals is many times lower than the theoretical strength based on bonding strength within perfect crystals. Likewise, the presence of dislocations is considered the major agent of plastic flow or deformation because the slip planes dislocations create allow movement of atoms within crystals under the application of stress. On the other hand, dislocations can interact with each other and form networks that act as barriers to further movement, leading to hardening and embrittlement.

5-4 *Metal Alloys*

With the exception of the highest purity elemental metal, most commercial metallic materials are alloys composed of two or more different elements. Even in commercially pure grades of metals, small amounts of impurity elements are present, either from the ore or introduced during the refining process.

We can define a metal alloy, then, as a combination of two or more different elements (atoms), the major one of which is a metal. The metallic element present in the largest amount, by weight, is called the base or parent metal; other elements present are termed alloying elements or alloying agents. Most alloys are composed of two or more

Fig. 5-6   Schematic drawing of an edge dislocation and slip plane.

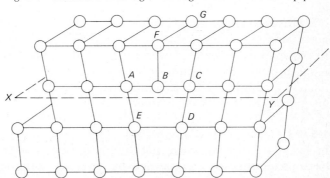

metallic elements. However, there are important exceptions, as in the case of steel, in which the presence of carbon atoms importantly influences steel's properties.

Solid Solutions. Structurally, there are two kinds of metal alloys—single phase and multiphase. Single-phase alloys are composed of crystals with the same type of crystal structure (Fig. 5-7a). They are formed by "dissolving" together different elements to produce a solid solution. The crystal structure of a solid solution is normally that of the base metal. The atoms of the alloying element (solute) join the base metal (solvent) either as substitution atoms or as interstitial atoms. In the former case, alloying atoms occupy some of the lattice sites normally occupied by the host atoms (Fig. 5-5b). In the latter case, alloying atoms move in between the host atoms (Fig. 5-5c). In substitutional solid solutions, the solute and solvent atoms are of approximately similar size. In the interstitial type, the solute atoms must be small enough to fit between the atoms.

The nature of solid solutions has important effects on many alloy properties. For example, strength and hardness increase with the amount of solute present, but ductility usually decreases. Electrical conductivity also is generally lowered by the presence of the solute element.

Multiphase Alloys. In contrast to single-phase alloys, multiphase alloys are mixtures rather than solid solutions. They are multicomponent systems of aggregates of two or more different phases (Fig. 5-7b). The individual phases making up the alloy are different from each other in their composition or structure. Solder, in which the metals lead and tin are present as a mechanical mixture of two separate phases, is an example of the simplest kind of multiphase alloy. In contrast, steel is a complex alloy composed of different phases, some of which are solid solutions.

Multiphase alloys far outnumber single-phase alloys in the industrial materials field, chiefly because they provide greater property flexibility. Properties of multiphase alloys are dependent on many factors including the composition or structure of the individual phases, the relative amounts of the different phases, and the positions of the various phases relative to each other.

5-5 *Structural Transformations*
In Sec. 5-4, it became clear that the presence of alloying elements in a base metal modifies properties by forming solid solutions and/or

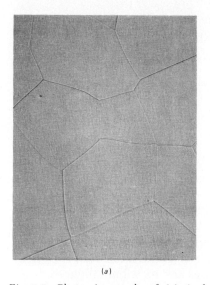 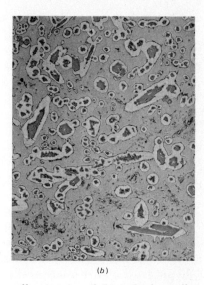

(a)   (b)

Fig. 5-7 Photomicrographs of: (a) single-phase alloy (50×), and (b) multiphase alloy (100×). (Ledgemont Lab, Kennecott Copper Corp.)

various types of constituent phases. It is also possible to modify properties, often drastically, by bringing about changes in the microstructure of a given material. At the microscopic level, these transformations involve changes in solid solutions and phases brought about by a variety of mechanisms, such as diffusion, solid-state reactions, and recrystallization. At the macrolevel, these transformations are produced by heat treatments and/or plastic deformation of the metal.

Allotropy. A useful transformational characteristic of some metals is that they can change from one crystalline structure to another. The transformation can be brought about by a change in temperature (or heat energy) or by a combination of a change in temperature and pressure. The properties of the various allotropic phases of a metal can differ considerably. Thus, by alloying or heat treatment, appearance of the phases can be controlled in order to develop desired properties. However, only a few metals are allotropic. The most important one is iron. Two of its three allotropic forms make possible the whole science and technology of steel. Others are tin and titanium.

Diffusion. As already discussed, many imperfections exist in crystal lattices, some of which cause vacant atom sites; also, "foreign" or interstitial atoms are often present. The result is that though the bulk form of a metal is essentially rather stable, it is relatively unstable at the microstructural level. Because of energy differences within the lattice, atoms tend to shift positions to fill vacancies, and interstitial atoms move about from one site to another. At a solid's surface, where the atoms are usually free to move, diffusion can take place more rapidly and extensively. Diffusion in solids is temperature dependent and generally occurs more rapidly at elevated temperatures.

The major significance of diffusion is that, at the microscopic level, metals can intermingle in a variety of ways. Therefore diffusion and solid-solution mechanisms are the basis for producing metal alloys and for modifying properties by thermal treatment.

Solid-Solution Solubility. Alloy solid solutions have limited solubility. Therefore, under proper conditions, supersaturated solutions can be produced and manipulated to precipitate out elements for the purpose of altering properties. Generally, heat treatment is used to precipitate out hard and/or intermetallic phases and distribute them throughout the metal matrix.

Recrystallization. Another transformation process involves the formation and growth of new crystals with the dissolution of the old ones. In order for recrystallization to take place, the crystals must be deformed—usually as a result of plastic deformation—and the metal must be raised to the recrystallization temperature, which is, very roughly, 40 percent of the melting point (see Sec. 5-11). The principal purpose of recrystallization is to restore ductility to metals that have been embrittled by cold working.

Producing Transformations. There are two principal ways of bringing about the transformations we have just discussed. One is by changes in thermal energy in a metal, such as the application or withdrawal of heat, which results in a temperature change in the metal. This process is commonly referred to as *heat treatment*. The other way is by application of mechanical energy to deform the metal. When applied to a metal at or near room temperature, the process is termed *cold working*; when applied at elevated temperatures, it is referred to as *hot working*. These two approaches—metal deformation and heat treatment—will be covered in the next two sections.

Metal Deformation

5-6 *Effects on Properties*

When metals in the solid state are subjected to mechanically applied forces required to form or shape them, some of their properties can be significantly altered. When plastic deformation occurs above the recrystallization temperature, as in hot working, there are no significant changes in mechanical properties. However, other changes do take place. Density, for example, increases because hot working eliminates cavities and pores. Also, grain structure is refined and greater homogeneity of structure is obtained since inclusions and secondary phases, such as carbides, tend to elongate in the direction of rolling (Fig. 5-8a). Likewise, hot working may develop preferred orientation of crystals in the direction of rolling.

5-7 Work Hardening

When metals are plastically deformed cold, the working distorts the grain, or crystal, structure (Figs. 5-8b and 5-9b), causing an increase

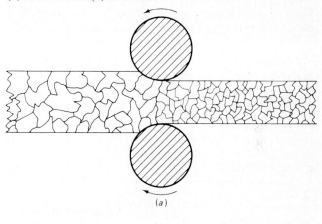

Fig. 5-8 Effects of mechanical work on grain size and shape: (a) hot work and (b) cold work.

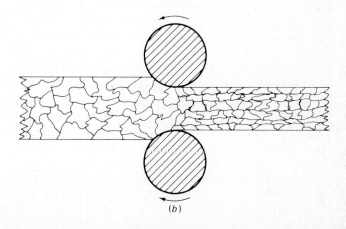

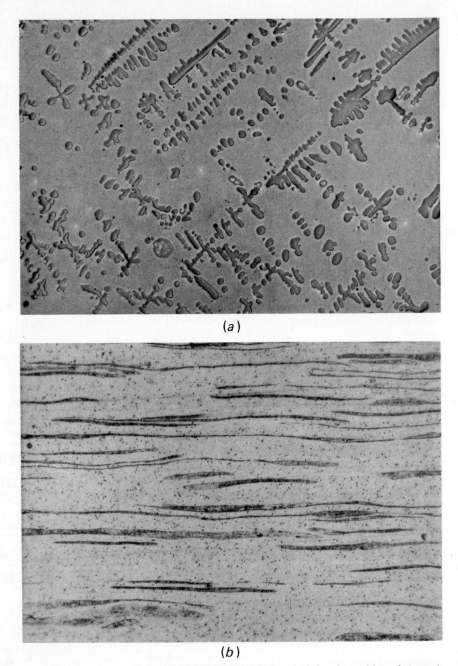

Fig. 5-9 Photomicrographs of a metal alloy (a) before and (b) after cold work (100×). (Ledgemont Lab, Kennecott Copper Corp.)

in strength and hardness and a reduction in ductility (Fig. 5-10). This phenomenon is referred to as work or strain hardening. The simplest demonstration of this effect is the hardening of a piece of wire when it is bent back and forth. After hardening, the wire becomes more difficult to bend, and, with more bending, it usually breaks.

Although strain hardening is often undesirable, it can be effectively used for strengthening metals. A number of metallic materials cannot be hardened by heat treatment, and therefore cold working is the only method available to improve mechanical properties. Also, when simple shapes, such as bar, sheet, and tubing, are available in the cold-worked condition, improved properties are usually gained more economically than if heat treatment were used.

Figure 5-10 shows schematically how cold working alters mechanical properties. It is evident that the increase in hardness and strength and the decrease in ductility is a function of the amount of cold work. The amount or degree of cold work performed on a piece of metal is expressed in terms of the percentage of reduction of thickness resulting from the operation. When the amount of reduction required to shape a piece of metal to final form is large, or when strain-hardening

Fig. 5-10 Effects of cold working on hardness, strength, and ductility.

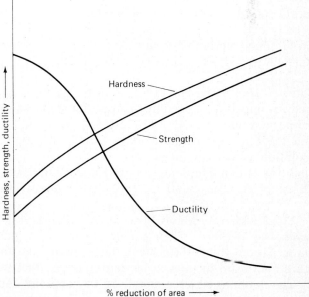

effects are undesirable, process annealing is used to relieve internal stresses and to produce recrystallization (Sec. 5-11).

Heat Treatments

Heat treatments have several functions. They can produce structural, or phase, changes in a metal and thus alter its properties—principally mechanical properties; they can alter the size of the grains (crystals); and they can relieve residual stresses.

Heat-treatment processes, or cycles, of the simplest kind involve four controlling factors: (1) the temperature to which the material is heated, (2) the length of time the material is held at that temperature, (3) the temperature to which the material is cooled, and (4) the rate of cooling. Frequently, heat-treatment processes require the metal to be heated and cooled in stages or the treatment cycle to be repeated. A brief description is given below of commonly used metal heat-treatment techniques.

5-8 *Hardening or Strengthening*

As the name implies, the purpose of these heat treatments is to increase strength and/or hardness. It is accomplished by raising the alloy to a temperature at which desired solid-state reactions or phase changes occur, and then rapidly cooling or quenching the material. There are a number of different quenching mediums used. Water provides the most rapid cooling speed. Oil cools the material more slowly, and air most slowly of all. For extremely rapid cooling, brine or water sprays are used. Many hardening-treatment cycles include a tempering step after quenching, which is covered in Sec. 5-10.

Quench Hardening and Hardenability. Simple quench hardening of ferrous metals, practiced by the traditional blacksmith, is still the most important and widely used hardening process. It consists of heating the metal to a temperature above 1333°F, at which an interstitial solid solution of carbon in gamma iron (austenite) is formed, and then holding at that temperature until the desired degree of transformation takes place. The metal is then cooled rapidly to form the desired structure (martensite) and hardness.

Thickness, or section size, is an important factor in hardening processes because, when a piece of metal is quenched, heat is removed from the surface layers faster than from the interior. This in turn can result in uneven hardening across that section of the piece.

The depth to which a metal can be hardened depends on its rate of hardening, which is referred to as hardenability. The faster the hardening rate, the thicker the section that can be hardened with a given quenching treatment. The term, it should be noted, is not related to the level of hardness a metal can achieve. Hardenability varies widely among various steels and depends chiefly on the presence of carbon and alloying elements (see Sec. 6-3).

Hardenability is measured by a specific test, the Jominy end quench test, in which a test bar undergoes standard hardening treatment, after which hardness readings are taken at specified distances from the directly quenched end of the bar (Fig. 5-11). The hardness readings, which are plotted as a function of distance from the quenched end, produce a so-called hardenability curve. The hardenability curves shown in Fig. (5-12) represent steels with high, medium, and low hardenability.

Martempering is a special method of hardening that minimizes the internal stresses of the hardening mechanism by quenching the heated steel in a molten salt bath and holding it at just above the temperature at which martensite (see Sec. 6-3) begins to form. After being uniformly heated throughout, the steel is air cooled to room temperature. Conventional tempering may then follow.

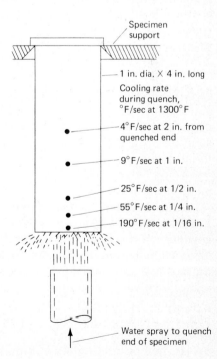

Fig. 5-11 Jominy end quench test setup for determining hardenability.

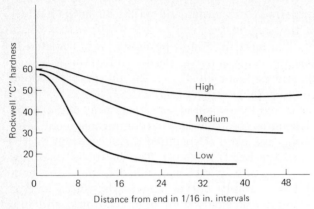

Fig. 5-12 Hardenability curves for steels with high, medium, and low hardenability.

Austempering is another special hardening process. It was developed to produce a combination of high strength and ductility in treated steels. The quenching medium is oil or molten salt at a temperature between 300 and 800°F. The steel is held in the quenching medium until a microstructure known as bainite (see Sec. 6-4) is developed, after which it can be cooled to room temperature at any rate. The transformation period can take several hours or more depending on the steel and desired properties.

Precipitation and Age Hardening. A number of alloys, both ferrous and nonferrous, can be hardened by this three-step process that depends on precipitating an alloy constituent from a supersaturated solid solution. The precipitate is usually an intermetallic compound that forms a finely divided phase on the grain boundaries. It is a hard constituent that hardens and strengthens without serious embrittlement of the alloy.

In the first step, called solution treatment or anneal, the alloy is heated to and held at a temperature where it will exist as a homogeneous solid-solution phase. In the second step, the metal is cooled, or quenched, rapidly enough so as to retain the solution and thus prevent the appearance of a second phase. The third step, called aging, involves holding the alloy at a temperature that allows the second phase to precipitate out of the solid solution. If the aging takes place at room temperature, it is referred to as natural aging; when it takes place above room temperature, it is artificial aging.

The time of holding at solution temperature varies widely depending on the alloy, section thickness, and desired properties. In the case of aluminum alloys, for example, holding time can vary from 10 min for a thin sheet to 10 or 12 hr for heavy parts. Likewise, the aging period varies considerably depending on the alloy and temperature. Some aluminum alloys require months to age at room temperature as compared to only 24 hr when aged a few hundred degrees above room temperature.

5-9 *Surface Hardening*
There are a number of heat-treating processes used for altering only the surface of metals, especially ferrous alloys. They can be divided into two groups—those in which there are no surface additives and those in which a substance is added.

In the former group are flame hardening and induction hardening. The metal surface is rapidly heated by an intense flame or by electrical inductance and then quenched to produce a hardened surface. The other group of techniques involves heating the metal in the presence of a substance that reacts and/or diffuses into the surface. After the diffusion step, the metal may or may not be quenched to achieve the desired surface hardness, depending on the additive substance. For example, in carburizing, a carbonaceous material supplies carbon to the surface, and quenching produces a hardened surface layer. In nitriding, the metal's surface, when heated in an ammonia atmosphere, reacts with nitrogen to form inherently hard metallic nitrides.

There are a number of other surface thermal treatments using the same principle. Referred to as diffusion coating or cementation coating, each uses a different substance, including aluminum (calorizing), chromium (chromizing), zinc (sheradizing), silicon (siliconizing) and nickel phosphorus.

5-10 *Tempering*
Quench hardening is accompanied by several problems. One is residual internal stresses caused by the uneven cooling over the cross section of the quenched piece of metal. Another is the inherent brittleness of the martensitic microstructure produced by quenching. The purpose of tempering is to relieve the stresses and reduce brittleness. This is accomplished by reheating the quenched steel to some temperature below the transformation range, holding it there for a relatively short time, and then cooling in air. The tempering of steel is usually done between 300 and 375°F, and holding time ranges from a few minutes to about 1 hr. In effect, tempering softens the hard martensite by precipitation of carbide particles, which allows the martensite's lattice to lose some of its distortion.

5-11 Annealing

Annealing is a general term that is often used to cover a number of widely differing heat treatments performed to achieve various end results. However, a common characteristic is that the metal is heated and/or cooled slowly. In this respect, tempering is related to annealing. In contrast to hardening treatments, annealing usually, but not always, tends to soften the metal.

When the term annealing is related to ferrous materials, it implies full annealing. When applied to nonferrous materials, it implies process annealing. Full annealing consists of heating steel to above the recrystallization temperature to produce a transformation to austenite. The steel is then slowly cooled, usually in the furnace. The heating removes the previous grain structure. The new grain structure depends on the type of transformation and the cooling rate. Annealing produces a refined grain, increases ductility, and improves machinability.

Normalizing is similar to full annealing, except that cooling is done in still air at room temperature. It refines grain, and it gives somewhat higher strength than annealing in addition to increasing ductility. It is used on castings to change undesirable microstructures.

Homogenizing anneal, also similar to full anneal, is used to evenly distribute alloying elements to produce a completely homogeneous structure. The treatment temperature is considerably above the transformation temperature, and holding time is usually several hours. A coarse-grain structure is produced unless special alloying elements are present.

Process annealing, which is applicable to both ferrous and nonferrous metals, is used to soften the stock for further cold work. It involves heating the metal close to, but below, the transformation range. Three functions are performed by the treatment (Fig. 5-13): (1) internal stresses are relieved; (2) as the temperature rises, the distorted grains of cold-worked metal are replaced with new grains by recrystallization; and (3) depending on temperature and time, the grain size will increase with additional loss in strength and a gain in ductility.

Stress relieving, also sometimes called stress-relief annealing, is a heating process that reduces or eliminates residual stresses in metals without changing the microstructure or degrading the material's properties. The treatment can be applied to both ferrous and nonferrous metals in which stresses have been induced by quenching, machining, cold working, casting, and welding.

Spherodizing, applied principally to steel, involves heating and cooling the metal so as to cause the carbides to assume a rounded or

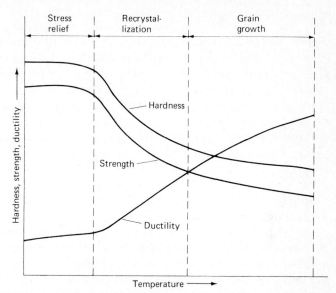

Fig. 5-13 Schematic diagram of the effects of process annealing on hardness, strength, and ductility.

globular form. The resulting structure, small, globular carbide in a ferrite matrix, is the softest and most ductile structure obtainable in steel by heat treatment. The temperature used is immediately below the lower transformation temperature of 1333°F, and holding time for complete spherodization, which requires a long time, depends on the specific steel involved.

Metal-Processing Methods

The cycle of processes that converts metals into parts and products starts with the raw materials after they are extracted from minerals. Metallic raw materials are usually produced in two processing steps. First, beneficiation processes, which involve such operations as crushing, roasting, separation, leaching, and other chemical reactions, are used to refine the ore. Second, various additional processes, such as smelting and the addition of alloying elements, produce the basic raw-metal forms, such as slabs and ingots, ready for processing into intermediate forms, such as wrought mill products, or into castings or fabricated forms, parts and products.

Two major functions are performed by the processes used to convert raw materials into finished objects. The first and major function is to form, shape, and fabricate the material into the intended part.

This may or may not include the joining together of the same or dissimilar materials. The second function is to alter or improve material properties. This function may be accomplished as an integral part of a form-producing process, or it may be achieved by means of a separate processing operation, such as in heat treatment.

Metal-forming and shaping processes can be conveniently classified into two broad types—those which are performed on the material in a liquid state (casting) and those in which the material is in a solid or plastic condition during the processing operations.

5-12 *Casting*

Casting is used to produce both finished parts as well as intermediate forms for further forming into finished parts. In casting, the desired shape is produced by pouring or injecting molten metal into a prepared mold cavity where it solidifies. In most casting processes, there are four major elements or steps: (1) making of an accurate pattern of the part, (2) making a mold from the pattern, (3) introducing the liquid metal into the prepared mold, and (4) removing the part from the mold, and finishing operations, if needed. There are about a half-dozen major metal-casting processes. The common distinguishing characteristic is the type of mold used.

Sand Casting. Molds made of sand with varying amounts of resinous binders are used in this process. It is most widely used for iron and steel castings, and is applicable for parts ranging from a few pounds to tons in weight. Sand castings of great complexity can be readily produced. The molds are destroyed on removal.

Permanent-mold Casting. This process makes repeated use of molds, which are made in two or more parts and are hinged and clamped for easy removal of the finished castings. The materials that can be cast are generally limited to aluminum and magnesium, although gray iron, zinc, and lead are sometimes cast by this process. The average size of permanent-mold castings varies from about ½ lb to about 12 lb.

Plaster-mold Casting. This process utilizes molds made from a slurry of gypsum. The castings have fine finishes, and surface detail can be faithfully produced. Sizes of plaster-mold castings usually fall within the range of ¼ lb up to 15 or 20 lb. The process is particularly suitable for thin wall parts—down to about $1/16$ in. thick. Most plaster-mold castings are made of copper alloys. Other metals often used include aluminum and magnesium.

Investment Casting. This process, frequently referred to as the lost-wax process, was first used in China centuries ago, and Cellini employed it to make art objects in the sixteenth century. Patterns, called investments, are made by forming wax, frozen mercury, or plastic in a master die. The investments are then covered with a slurry of ceramic material. After the ceramic mold hardens, the investment is melted out, leaving the cavity into which the molten metal is poured. Investment casting is relatively expensive, but it is suitable for almost any fine configuration and for wall thicknesses as thin as $\frac{1}{64}$ in. Castings weighing anywhere from a few ounces to over 25 lb can be produced in most castable metals.

Die Casting. This process is best suited of all the metal-casting processes for high production. It involves forcing molten metal under pressure into closed metal dies. Because of the use of solid dies, the range of complexity is less than with sand casting. Most die castings are made of zinc and aluminum. Other metals that are die cast include copper alloys, lead, and tin. Aluminum parts approaching 100 lb and zinc parts as heavy as 200 lb can be die cast.

Electroforming. This is essentially a plating process for producing parts. A mandrel or core pattern of the part is placed in a plating bath, and metal is deposited on the surface to the desired thickness. Then the part is removed from the bath and the core and deposited metal are separated. The finished part is a self-supporting structure with inside dimensions matching those of the mandrel or core.

5-13 *Shaping and Forming*
Metals in their solid state are formed into desired shapes by the application of force or pressure. Processing of metals in the solid state can be divided into two major stages. In the first, the raw material in the form of large ingots or billets is hot worked, usually by rolling, forging, or extrusion, into smaller shapes and sizes. In the second stage, these shapes are processed into final parts and products by one or more smaller-scale hot or cold forming processes.

Rolling. This is the most common primary metal-forming process. Hot metal ingots or billets are passed through a series of rolls that work the metal in successive stages into the desired cross-sectional shapes and sizes. Plates, sheets, strips, bars, and rods are rolled as well as special shapes, such as I beams, channels, and angles.

Extrusion. This is a process used for producing long lengths of rod, tubing, and other cross-sectional shapes. Hot or cold metal is forced through a die having the desired cross section. The process is widely used for producing intricate shapes chiefly in copper alloys, aluminum, and magnesium.

Forging. This is one of the most widely used metal-forming processes. Basically the process involves the working of either hot or cold metal into the desired shape by impact or pressing. Most metals can be forged. Forging processes are generally more expensive than casting, but produce parts with higher strength and toughness.

The simplest and oldest method is open-die forging, in which the hot metal piece is manipulated by an operator as it is shaped by a mechanical hammer. This method is limited to large, heavy shapes such as shafts and die and gear blanks. Drop forging is a refinement of the open-die process. Hot billets are forged into an open die by the repeated impact of a mechanical hammer. Often a series of progressive dies is needed to attain the final shape. Drop forgings vary in size from about 1 oz to several tons. Upset forging is used to make relatively small to medium-size parts from rod stock. The end of the heated rod stock, while gripped between portions of a die, is struck by another die. The upsetting action reduces the length and increases the diameter of the rod and thus produces the desired shape.

Cold Heading. This process is similar in principle to upset forging. Cold wire stock is fed automatically to the dies and is then shaped by upsetting. Extremely high production rates are reached in the manufacture of a wide variety of small parts. Swaging is the opposite of cold heading. It is a cold-working process in which a rotary hammering action reduces the diameter and increases the length of a piece of rod or tubing. Press forging employs pressure rather than impact to squeeze the hot metal, placed between dies, into the desired shape. As in drop forging, progressive dies are frequently used. This process is most commonly used for forgings weighing 25 to 30 lb. Impact extrusion can be considered either as a cold forging or extrusion process. A plug of metal, placed in a die, is subjected to the pressure of a punch, causing the metal to flow into the space between the punch and the die. The process is widely used for collapsible tubes made of soft metals.

Stamping and Pressing. These are another large family of metal-forming processes. Included are blanking, pressing, stamping, and drawing, all of which are used to cut and/or form metal plates, sheets, and strips. The steps common to all stamping and pressing operations are the preparation of a flat blank and shearing or stretching the metal into a die to attain the desired shape.

In drawing, the flat stock is either formed in a single operation or progressive drawing steps may be needed to reach the final form. In spinning, flat disks are dished by a tool as they revolve on a lathe.

Stamping involves placing the flat stock in a die and then striking it with a movable die or punch. Besides shaping the part, the dies can perform perforating, blanking, bending, and shearing operations. Almost all metals can be stamped. In general, stampings are limited to metal thicknesses of ⅜ in. or less. Pressing and drawing operations can be performed on cold metals up to ¾ in. thick and up to about 3½ in. on hot metals.

In recent years, many new press techniques have been developed. A number make use of rubber pads, bags, and diaphragms as part of the die or forming elements. Some involve stretch forming over dies. Others combine forming and heat-treating operations. And still other methods, known as high-energy rate forming, employ explosives or electrical or magnetic energy to produce shock waves that form the material into the desired shape.

Powder Metallurgy (P/M). This process involves the production of parts from metals in powder form. The metal powder is compacted in a die to the desired form and then heated (sintered) to fuse the powder particles together. A secondary press operation, known as coining, is sometimes performed on the sintered part. Also P/M parts can be impregnated with oil or plastic and infiltrated with low-melting metals. Most metals, and many combinations of materials, can be formed by the P/M process. Parts range in size from about ⅛ to 25 in.² in projected area, and between ¹⁄₃₂ and 6 in. in length. Metal strip can also be produced from metal powders. The metal powder, chiefly copper or nickel, is cold rolled into a continuous strip that passes through a sintering furnace and then through a hot-rolling mill to produce the final thickness and density.

5-14 *Machining*

Machining involves a large number of methods which shape metal by removing portions of the material to achieve the final shape. In almost all, a tool is forced against the material to be shaped. The tool, which is harder than the material to be cut, removes the unwanted material in the form of chips. Thus the elements of machining are a cutting device, a cutting force, a means of holding and positioning the workpiece, and usually a lubricant (or cutting oil). Materials used as cutting tools must have a combination of strength, toughness, hardness, and wear resistance at the relatively high temperatures generated by the cutting action. The common cutting-tool materials are high-speed steel, sintered tungsten carbides, diamonds, ceramics, and cobalt-base alloys.

The oldest and most common machining process is turning. The workpiece (material) is simply rotated while a cutting tool moves parallel to it, cutting away the material to reduce its diameter. Shaping and planing methods are used to cut and remove material from plane surfaces. The cutting edge takes successive cuts straight across the surface of the workpiece in a predetermined pattern to achieve the desired shape. The milling process, in which cutting is performed by rotating multiple tool cutters, is also used for facing plane surfaces as well as for cutting grooves and slots. Holes are cut into materials by various drilling, boring, and tapping operations.

Grinding, honing, and lapping are methods of removing material with abrasives. The abrasive action may be performed by abrasives bonded in belts or rotating wheels (grinding), or by fine abrasive stone (honing), or by a soft material with abrasive particles embedded in it (lapping). In ultrasonic machining, abrasive particles remove material by high-velocity bombardment of the workpiece. A mixture of air or inert gas under pressure can also be used to direct a high-velocity stream of fine abrasive particles against the workpiece.

There are four basic chipless-machining processes. In chemical milling, the metal is removed by the etching reaction of chemical solutions on the metal. Although usually applied to metals, it can also be used on plastics and glass. Electrochemical machining uses the metal-plating principle in reverse. The workpiece, as the anode in the electrolytic circuit, is caused to deplate to conform to a tool (the cathode) having the inverse shape of that desired. The process of electrodischarge machining and grinding erodes or cuts the metal by the action of high-energy electric sparks or electrical discharges. And, finally, the laser beam has recently found limited application as a cutting tool.

5-15 Joining

Joining is the process of permanently (or sometimes temporarily) bonding or attaching materials to each other. In most metal-joining processes, a bond between two pieces of material is produced by application of one or a combination of two kinds of energy—thermal and mechanical force or pressure. A bonding or filler material (the same as, or different from, the materials being joined) may be used. There are three broad groups of processes that use thermal energy: pressure welding, fusion welding, and brazing and soldering.

Pressure Welding.   Forge welding is the oldest-known welding process. It involves the heating of the joint areas to a plastic condition. The heated surfaces are placed in contact with each other and bonded together by hammering or pressing pressure. In oxyacetylene pressure welding, a gas flame heats the abutting ends of the pieces to be joined. In resistance welding, the joint surfaces are heated by a heavy, localized electric current before mechanical pressure is applied to produce the finished bond.

Fusion Welding.   Most metals can be joined by one or more fusion methods. In fusion welding, the edges of the materials to be joined are melted so they flow together to achieve localized coalescence. Upon cooling, a permanent bond is produced. A filler material (the same as, or different from, the materials being joined) may be melted in the joint area as part of the bonding process. The heat energy used to bring the metals to the fusion point is provided by any one of a number of sources—electric arc, electric current, atomic hydrogen stream, burning fuels or gases, electron beam, exothermal chemical reaction, and, most recently, laser beams.

There are over 45 different methods of arc welding. The most common are metal arc, submerged arc, gas-shielded arc, and carbon arc. Along with arc welding, gas-welding processes are the most widely used fusion methods. The most common gas-welding processes are oxyacetylene, oxyhydrogen, and air-acetylene welding.

Electroslag welding is a relatively new method in which molten slag creates a high heat that melts both the filler metal and the metals being joined. Electron-beam welding and laser-beam welding, two other recent developments, utilize electron bombardment and a laser beam, respectively, to generate the heat of fusion.

In thermit welding, which is widely used for continuously joining rails, a mixture of metallic oxide and finely divided aluminum are ignited and react exothermally to produce a superheated metal that is poured into a mold around the joint area, much like in sand casting.

Brazing and Soldering.   These joining methods are used primarily to join metals, although such nonmetallics as ceramics, glass, and graphite are brazable. Soldering and brazing differ from fusion welding in that the edges or surfaces of the materials being joined are kept below their melting temperature during the bonding process. Only the filler metal, which is always used and which is different from the base metals, is in a molten condition. The part or the joint area of the metals is heated to above the melting temperature of the filler metal. The molten filler metal flows into the joint by capillary attraction and coalesces with the base metals to produce the bond. By definition, brazing involves the use of filler metals that melt at 800°F or higher, and soldering employs filler metals that melt below 800°F. A third method, known as braze welding, is the same as brazing except that it does not utilize capillary attraction to distribute the filler metal.

Many nonferrous metals, especially copper, nickel, and precious metals, are easily brazed or soldered. Aluminum brazing is common, and some steels can be brazed. Brazing filler materials most commonly used include alloys of silver, copper, and gold. The commonly used solders include tin alloyed with antimony or silver or lead, lead silver, cadmium alloyed with silver or zinc, and bismuth alloys.

Any one of a number of heating methods can be used for brazing and soldering. These include torch or flame, furnace or oven, electrical induction, molten metal or salt bath, electrical resistance, hot block or iron, and radiant lamp.

Review Questions

1. What microstructural feature or characteristic in metals accounts for:
   (a) Their good electrical conductivity?
   (b) Their relatively high strength and stability?
2. Explain two ways in which grain boundaries affect the strength of a metal.
3. Explain how movements of atoms within a metal cause plastic flow.
4. What is the microstructural difference between a single-phase and a multiphase alloy?
5. Identify the parent, or base metal, and the alloying elements in the following alloys:
   (a) 20 percent zinc, 10 percent lead, 56 percent copper, 12 percent nickel, 2 percent tin.
   (b) 32 percent chromium, 23 percent nickel, 39 percent iron, 2 percent manganese, 2 percent silicon, 2 percent other elements.
6. For each of the following structural-transformation mechanisms, give the structural change that takes place in the metal:
   (a) Allotrophy.   (c) Solid-solution solubility.
   (b) Diffusion.   (d) Recrystallization.

7. Explain the microstructural changes that occur in some metals when they are hot worked.
8. What property changes occur when a bar of steel is cold rolled?
9. List the critical controlling factors in heat-treatment processes.
10. Rockwell C hardness readings taken on three metal specimens (A, B, and C) after undergoing the Jominy end quench test are given in the table below.
    (a) Which specimen has the highest hardenability?
    (b) Which has the lowest hardenability?
    (c) Which specimen would you estimate has the highest strength?

	Distance from Quenched End, in.				
	1/4	1/2	1	2	3
A	65	64	62	56	50
B	64	40	30	27	25
C	65	50	35	32	30

11. What hardening process would you use:
    (a) To minimize internal stresses?
    (b) To increase strength of heat-treatable aluminum alloy?
    (c) To develop a bainite structure in the hardened steel?
12. What is the function of a tempering treatment?
13. After hardening, which annealing treatment would you use on steel:
    (a) To relieve internal stresses and soften for additional cold-working operations?
    (b) To relieve internal stresses without changing the metal's properties?
    (c) To obtain an austenitic structure?
14. What fundamental microstructural characteristic of metals allows many of them to be formed in the solid state?
15. Name the mold material used in each of the five major casting processes.
16. Which metal-forming process or processes is most likely to be used to produce the following objects?
    (a) Ash trays.
    (b) Nails.
    (c) Car fenders.
    (d) Stock for aluminum storm windows.
    (e) Self-lubricating bearings.
    (f) Structural building shapes.

17. What is the difference between conventional machining techniques and chipless machining?
18. Which of the three broad groups of joining processes involves the highest temperatures?
19. What is the principal difference between soldering and brazing?
20. List four heat sources used in fusion welding.

Bibliography

Bennett, A., et al.: *Crystals, Perfect and Imperfect,* Walker and Co., New York, 1965.

Slade, E.: *Metals in the Modern World,* Doubleday & Co., Inc., Garden City, N.Y., 1968.

Smith, C. O.: *The Science of Engineering Materials,* Prentice-Hall, Inc., Englewood Cliffs, N.J., 1969.

Van Vlack, L.H.: *Elements of Materials Science,* Addison-Wesley Publishing Co., Inc., Reading, Mass., 1964.

FERROUS METALS

Despite the rapid rise in the use of light metals and polymeric materials, ferrous alloys, particularly steel, remain the predominant structural material of our civilization. The production of irons and steels still exceeds by a considerable margin that of the next most widely used metals, aluminum and copper, as well as that of polymeric materials. However, on the basis of volume instead of weight, production of polymers and aluminum is rapidly approaching steel production.

A major reason for the preeminence of ferrous materials has been the development of a technology that has produced thousands of different alloys and grades that provide a range of properties not found in any other family of materials. Strengths, for example, range from less than 50,000 psi in conventional constructional grades to 500,000 psi in the newest ultrahigh-strength alloys; many steels rust quickly in the atmosphere, while others resist attack from the strongest acids; and some irons and steels are brittle, while many are ductile and tough. Furthermore, most ferrous metals are readily formed, and their properties can be altered in many ways by either thermal treatments or mechanical working.

Composition and Structure

The basic ingredient of all ferrous materials is, of course, the element iron. By definition, metallic materials containing at least 50 percent iron are classified as ferrous metals or alloys. They range from cast irons and carbon steels, with over 90 percent iron, to specialty iron alloys, containing a variety of other elements that add up to nearly half the total composition. Pure iron has a specific gravity of 7.86 and a melting point of 2800°F. Commercially pure iron, often referred to by the trade name Armco Ingot Iron, contains very small amounts of carbon, manganese, phosphorous, sulfur, and silicon. It has limited use, compared to other ferrous materials. It is very ductile, has good resistance to atmospheric oxidation, high magnetic permeability, and relatively low electrical resistivity.

Except for this commercially pure iron, all other ferrous materials, both irons and steels, are considered to be primarily iron-carbon alloy systems. Although the carbon content is small (less than 1 percent in steel and not more than 4 percent in cast irons) and often less than other alloying elements, it nevertheless is the predominant factor in the development and control of most mechanical properties.

6-1 *Pure Iron and Allotropic Forms*

Because pure iron is allotropic, it can exist as a solid in two different crystal forms (Fig. 6-1). From subzero temperatures up to 1670°F, it has a body-centered cubic structure and is identified as alpha (α) iron. Between 1670 and 2552°F, the crystal structure is face-centered cubic. This form is known as gamma iron (γ). At 2552°F and up to its melting point of 2802°F, the structure again becomes body-centered cubic. Because this last form, called delta (δ) iron, has no practical use, we will only be concerned here with the alpha and gamma forms.

The transformation from one allotropic form to another is reversible. Thus, when iron is heated to above 1670°F, the alpha body-centered cubic crystal changes into face-centered cubic crystals of gamma iron. When cooled below this temperature, the metal again reverts back to a body-centered cubic structure. These allotropic phase changes inherent in iron make possible the wide variety of properties obtainable in ferrous alloys by various heat-treating processes.

6-2 *Carbon's Role*

There are a number of alloying elements used in irons and steels, but carbon is the key one (Fig. 6-2). The element carbon can be present with iron to form any one or more of three different phases (see Sec. 6-3). Because carbon atoms are small relative to iron atoms, they can enter the iron crystal lattice and produce an interstitial solid solution. However, carbon's solubility in iron is limited. In gamma iron, the maximum is about 2 percent. In alpha iron at room temperature, only about 0.005 percent carbon is soluble in the body-centered cubic structure. Because of its limited solubility in iron, when more than 0.005 percent is present, the carbon can occur in either of two other forms. It can form an intermetallic compound, iron carbide (Fe_3C) phase, often termed cementite, or it can exist as uncombined carbon in the form of graphite flakes or nodules.

The distinction between steel and cast iron is made on the basis of the phase forms in which carbon is present in the iron. In cast irons, at least some of the carbon is present in uncombined form as graphite. In steel, the carbon is combined with iron, and in most steels it is present in the solid solution ferrite, and as iron carbide.

6-3 *Basic Phases*

Graphite, iron-carbon solid solutions, and the iron-carbide intermetallic compound are the three basic phases of which ferrous alloys are composed (Fig. 6-3).

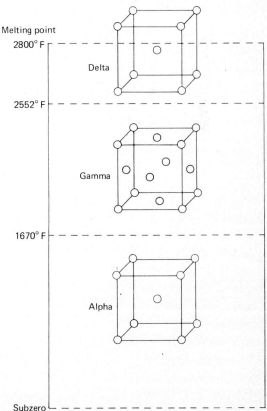

Fig. 6-1 Three allotropic forms of iron.

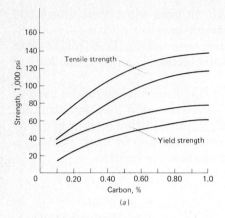

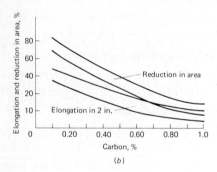

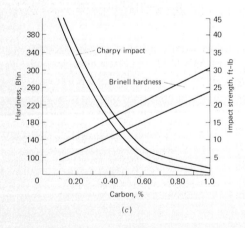

Fig. 6-2 Effect of carbon content on mechanical properties of rolled (hot) steel: (a) tensile and yield strength, (b) elongation and reduction in area, and (c) hardness and impact strength.

Graphite in ferrous alloys is often referred to as free carbon or uncombined carbon. It is seldom found in steels because their carbon contents do not exceed 2 percent. Graphite, however, is almost always present in cast irons, where it usually appears as flat flakes or as nodules of spheroidal shape. The flakes, having sharp edges, act as inclusions and cause stress-concentration areas which lead to decreases in strength and ductility.

Ferrite is a dilute solid solution of carbon in alpha iron. Because the maximum amount of carbon that can be present is not more than 0.025 percent, ferrite is close to pure iron and has a body-centered cubic crystal structure. It is magnetic and relatively soft (50 to 100 Bhn) and ductile (30 to 40 percent elongation). It has a tensile strength of between 40,000 to 50,000 psi.

Iron carbide, or cementite, as it is often called, is an intermetallic compound of about 7 percent carbon and 93 percent iron. It is extremely hard (about 650 Bhn), brittle, and has a tensile strength of over

Fig. 6-3 Photomicrographs showing basic phases in ferrous alloys: (a) ferrite, (b) austenite grains (outlined areas), (c) carbide (cementite) grain-boundary phase in pearlite, (d) martensite (needlelike areas) in retained austenite. (Bethlehem Steel Corp.)

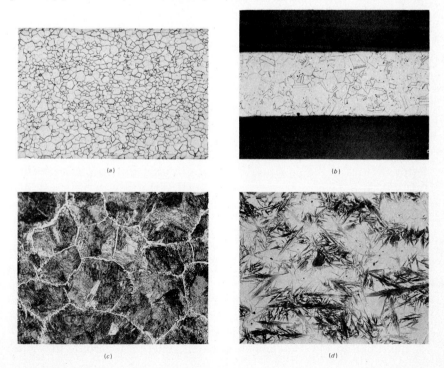

300,000 psi. Cementite commonly appears in ferrous metals either in pearlite (see below) and/or independently, usually at the grain boundaries.

Austenite is the name given to solid solutions of carbon in gamma iron. Carbon can be present in amounts up to 2 percent. In simple binary steels, such as carbon steels, austenite can only be present at temperatures between about 1330 and 2715°F. It is only present at room temperature in a few alloy steels, notably austenitic stainless steels. Its mechanical properties depend on the carbon content, heat treatment, and the alloying elements used to produce austenite at room temperature. In general, it is usually soft and ductile, unless cold worked. It is less ductile than ferrite, tough, and has strengths between 100,000 and 150,000 psi. It is essentially nonmagnetic.

The major significance of austenite is that other constituents are its transformation products. By controlled heating and cooling, several different microconstituents are possible. The specific transformation product obtained depends principally on the alloy content, including carbon and other elements, and the cooling rate from above the critical temperature.

Martensite is a supersaturated solid solution of carbon in iron. It is produced by hardening processes that involve quenching from above the critical temperature, where austenite exists. The supersaturation distorts the normal body-centered cubic structure, which accounts for martensite's extreme brittleness, high tensile strength, and hardness. It surpasses cementite in these properties.

6-4 *Microconstituents*

The basic phases just discussed appear independently, as is always the case with graphite, or they can combine with each other to form other microconstituents. However, as a two-component alloy system, ferrous alloys in stable equilibrium are limited to two phases. While this would appear to be restrictive, great versatility in structure and properties is possible for two reasons: (1) The relative amounts of the basic phases in steels vary widely with carbon content, and (2) several different microconstituents can be formed by various solid-state transformations (Fig. 6-4).

Pearlite, perhaps the most important single constituent in steel, is a mixture of about 88 percent ferrite and 12 percent iron carbide. The carbide is present as laminas or plates in a matrix of ferrite. The carbide provides strength and hardness, while the ferrite contributes ductility.

Tempered martensite, which is a product of tempering in the range of about 660 to 1112°F, consists of small spheroidal carbide in ferrite.

Ferrous Metals 109

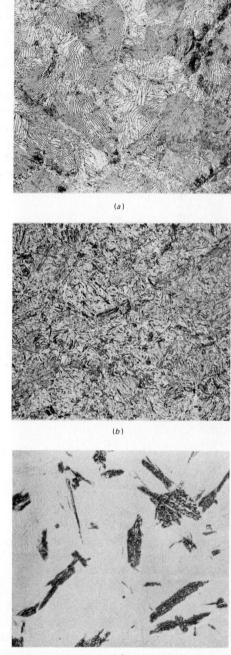

Fig. 6-4 Photomicrographs showing microconstituents in ferrous alloys: (a) pearlite, (b) tempered martensite, and (c) bainite (dark areas) in as-quenched martensite. (Bethlehem Steel Corp.)

The size of carbides increases from quite small at the lower end of the tempering range up to large globules at the upper end. Tempered martensite generally provides the maximum impact toughness in steels.

Bainite is a dispersion of iron carbide in ferrite. It occurs when certain steels are given a step quench (isothermal quench). Although not as strong as martensite, bainite combines high strength with reasonable ductility, which can be varied somewhat by choice of transformation temperature.

Ledeburite is a mixture of austenite and cementite that occurs primarily in cast irons with a carbon content of 2 percent or more. Because the cementite content is about 50 percent, this constituent is hard and brittle.

6-5 Alloying Elements

Besides carbon, a number of other elements are always present in steels and irons. Some of these, such as sulfur and phosphorus, manganese and silicon, are unavoidably present in small amounts and may or may not be advantageous, depending on the specific steel and/or desired properties. Others are intentionally added to alter structure, properties, and processing characteristics.

Like carbon, a number of alloying elements are soluble in alpha iron and therefore form solid solutions with iron to produce alloys with improved strength, ductility, and toughness. Also, carbon, besides forming an intermetallic compound with iron, combines with many alloying elements, including molybdenum, chromium, vanadium, boron, titanium, and tungsten. These alloy carbides as well as iron-alloy carbides usually are extremely hard and lack toughness.

Some alloying elements are added to prevent or restrict grain growth. One way they do this is by forming oxides or carbides that produce a network at the austenite grain boundaries that inhibits grain growth. Aluminum is considered the most effective alloying element in this respect. Others are zirconium, vanadium, chromium, and titanium.

Structurally, the addition of alloying elements almost always affects the austenite-ferrite transformation mechanism by changing the temperature at which the transformation from gamma to alpha iron takes place. Some lower and some raise the critical temperature. For example, the proper amount of nickel can expand austenite's stability down to room temperature. Other alloying elements raise the transformation temperature, and, in some cases, entirely eliminate the austenite zone.

Another important role alloying elements play in the austenite-ferrite transformation is to improve hardenability. In hardening treatments, because cooling of the metal's interior takes place by conduction (Sec. 5-6), time is needed for the transformation mechanism to proceed through the entire cross section. Many alloying elements, such as nickel, boron, and chromium, by influencing and altering the transformation rate or time period, allow ferrous alloys to be hardened to any desired depth. In this respect, there are two classes of alloying elements: (1) those that tend to stabilize austenite or lower the transformation temperature—nickel, manganese, carbon, boron, and copper; (2) those that tend to stabilize ferrite or raise the transformation temperature—chromium, molybdenum, silicon, tungsten, vanadium, columbium, tin, and titanium.

The compositional and structural changes produced by alloying elements change and improve the physical, mechanical, and processing properties of irons and steels. The effects on properties are discussed at some length in the sections covering specific ferrous materials in this chapter. Table 6-1 gives a short summary of how alloying elements alter the structure and properties of ferrous alloys.

Table 6-1 *Effects of Alloying Elements in Steels*

Element	Effects
Aluminum	Deoxidizes; restricts grain growth; aids surface hardness in nitriding
Boron	Increases hardenability, the increase being greater at lower carbon levels
Carbon	Increases strength and hardness; decreases ductility
Chromium	Increases hardenability, strength at room and elevated temperatures, resistance to corrosion, oxidation, and abrasion
Cobalt	Holds cutting edge at elevated temperature; increases hardness
Columbium	Inhibits intergranular corrosion in high-chromium and chromium-nickel stainless and heat-resisting steels
Copper	Increases corrosion resistance and strength
Lead	Increases machinability
Manganese	Renders sulfur innocuous; increases hardenability, strength, hardness, abrasion resistance; increases rate of carbon penetration in carburizing
Molybdenum	Increases hardenability, hardness, and strength at room and elevated temperatures, resistance to shock and corrosion; enhances creep strength; counteracts tendency toward temper brittleness

(continued)

Table 6-1 *Effects of Alloying Elements in Steels* (continued)

Element	Effects
Nickel	Increases toughness and shock resistance (especially at subzero temperatures); strengthens as rolled and annealed steels; renders high-chromium steels austenitic; improves resistance to heat and corrosion
Nitrogen	Increases strength and hardness; reduces ferritic grain size; as nitride, hardens surface; promotes austenite formation
Phosphorus	Increases strength, hardness, machinability, atmospheric-corrosion resistance, wear resistance, and electrical resistivity
Selenium	Increases machinability
Silicon	Deoxidizes; increases strength and oxidation resistance; decreases core (watt) loss in magnetizing silicon steel electrical sheets with alternating current
Sulfur	Increases machinability
Tantalum	Like columbium and titanium, stabilizes carbon to inhibit intergranular corrosion in high-chromium and chromium-nickel stainless and heat-resisting steels
Titanium	Deoxidizes; scavenges; has greatest known tendency to form carbide; see tantalum
Tungsten	Hardens and strengthens at room and high temperatures; increases hardenability; forms hard, abrasion-resistant particles in tool steels
Vanadium	Increases strength, ductility, resiliency, and endurance limit; promotes fine grain; forms carbide, nitride, and oxide; dissolves in ferrite; when dissolved, increases hardenability; improves strength and hardness at elevated temperatures; prevents age hardening in low-carbon rimmed steel
Zirconium	Deoxidizes; scavenges; combines with oxygen, sulfur, nitrogen; formation of zirconium nitride reduces age hardening in deep-drawing steels; addition of over 0.10% Zr usually results in fine grain

SOURCE: J. W. W. Sullivan, "How to Select Wrought Steels," *Materials Engineering*, January 1955.

6-6 Production Types

Steelmaking processes and methods used to produce mill products, such as plates, sheets, and bars, have an important effect on a steel's properties and characteristics.

Deoxidation Practice. Steels are often identified as to the degree of deoxidation resulting during steel production:

1. Killed steels, because they are strongly deoxidized, are characterized by high composition and property uniformity. They are used for forging, carburizing, and heat-treating applications.
2. Semikilled steels have variable degrees of uniformity, intermediate between those of killed and rimmed steels. They are used for plates, structural sections, and galvanized sheets and strips.
3. Rimmed steels are only slightly deoxidized during solidification. Carbon concentration is highest at the center of the ingot. Because the ingot's outer layer is relatively ductile, these steels are ideal for rolling. Sheets and strips made from rimmed steels have excellent surface-quality and cold-forming characteristics.
4. Capped steels have a thin low-carbon rim which gives them surface qualities similar to those of rimmed steels. Their cross-sectional uniformity approaches that of semikilled steels.

Melting Practice. Steels are also classified as air melted, vacuum melted, or vacuum degassed:

1. Air-melted steels are produced by conventional melting methods, such as open hearth, basic oxygen, and electric furnace.
2. Vacuum-melted steels are produced by induction vacuum melting and consumable electrode vacuum melting.
3. Vacuum-degassed steels are air-melted steels that are vacuum processed before solidification. This produces steels with lower gas content, fewer nonmetallic inclusions, and less center porosity and segregation. They are more costly, but have better mechanical properties, such as ductility and impact and fatigue strengths.

Rolling Practice. Steel mill products are produced from various primary forms such as heated blooms, billets, and slabs. These primary forms are first reduced to finished or semifinished shape by hot-working operations. If the final shape is produced by hot-working processes, the steel is known as hot rolled. If it is finally shaped cold, the steel is known as cold finished, or, more specifically, as cold rolled or cold drawn. Hot-rolled mill products are usually limited to low and medium, nonheat-treated, carbon-steel grades. They are the most economical steels, have good formability and weldability, and are used widely for large structural shapes. Cold-finished shapes, compared to hot-rolled products, have higher strength and hardness and better surface finish, but are lower in ductility.

Steel

6-7 Plain Carbon Steels

By definition, plain carbon steels are those that contain up to about 1 percent carbon, not more than 1.65 percent manganese, 0.60 percent silicon, and 0.60 percent copper, and only residual amounts of other elements, such as sulfur (0.05 percent maximum) and phosphorus (0.04 percent maximum). They are identified by means of a four-digit numerical system established by the American Iron and Steel Institute (AISI). The first digit is the number 1 for all carbon steels. A 0 after the 1 indicates nonresulfurized grades, a 1 for the second digit indicates resulfurized grades, and the number 2 for the second digit indicates resulphurized and rephosphorized grades. The last two digits give the nominal (middle of the range) carbon content in hundredths of a percent. For example, for grade 1040, the 40 represents a carbon range of 0.37 to 0.44 percent. If no prefix letter is included in the designation, the steel was made by the basic open hearth, basic oxygen, or electric furnace process. The prefix B stands for the acid Bessemer process. The letter L between the second and third digits identifies leaded steels, and the suffix H indicates that the steel was produced to hardenability limits.

Properties. For all plain carbon steels, carbon is the principal determinant of many performance properties. As Fig. 6-2 shows, carbon has a strengthening and hardening effect. At the same time, it lowers ductility, as evidenced by a decrease in elongation and reduction of area. In addition, a rise in carbon content lowers machinability and decreases weldability. The amount of carbon present also affects physical properties and corrosion resistance. With an increase in carbon content, thermal and electrical conductivity decline, magnetic permeability decreases drastically, and corrosion resistance is lowered.

Carbon-Steel Grades. Plain carbon steels are commonly divided into three groups, according to carbon content:

Low carbon—up to 0.30 percent.
Medium carbon—0.31 to 0.55 percent.
High carbon—0.56 to 1 percent.

Ferrous Metals   115

Fig. 6-5 Parts cold formed from carbon-steel bar. (American Iron and Steel Institute.)

Low-carbon steels are the AISI grades 1005 to 1030. Sometimes referred to as mild steels, they are characterized by low strength and high ductility, and are nonhardenable by heat treatment except by surface-hardening processes. Because of their good ductility, low-carbon steels are readily formed into intricate shapes. Cold work increases strength and decreases ductility. Where necessary, annealing is used to improve ductility after cold working. These steels are also readily welded without danger of hardening and embrittlement in the weld zone. Although low-carbon steels can not be thoroughly hardened, they are frequently surface hardened by various methods (carburizing, carbonitriding, and cyaniding, for example) which diffuse carbon into the surface. Upon quenching, a hard, abrasion-resistant surface is obtained.

Low-temperature carbon steels have been developed chiefly for use in low-temperature equipment and especially for welded pressure vessels. They are low-carbon (0.20 to 0.30 percent), high-manganese (0.70 to 1.60 percent), silicon (0.15 to 0.60 percent) steels, which have a fine-grain structure with uniform carbide dispersion. They feature moderate strength with toughness down to −50°F.

Medium-carbon steels are the grades 1030 to 1055. They usually are produced as killed, semikilled, or capped steels, and are hardenable by heat treatment. However, hardenability is limited to thin sections or to the thin outer layer on thick parts. Medium-carbon steels in the quenched and tempered condition provide a good balance of strength and ductility. Strength can be further increased by cold work. The highest hardness practical for medium-carbon steels is about 550 Bhn (Rockwell C55). Because of the good combination of properties, they are the most widely used steels for structural applications, where moderate mechanical properties are required.

High-carbon steels are the grades 1060 to 1095. They are, of course, hardenable with a maximum surface hardness of about 710 Bhn (Rock-

Table 6-2 *Mechanical Properties of Typical Plain Carbon Steels*

AISI TYPE & Condition	Tensile Strength, 1,000 psi	Elongation, % in 2 in.	Hardness, Bhn	Impact Strength, Izod, ft-lb
1010, as-rolled	40–65	25–50	110–140	—
1020, as-rolled	60–70	35–40	125–150	60–80
1030, quenched & tempered	75–120	17–35	180–490	—
1050, quenched & tempered	95–150	10–30	190–320	16–50
1080, quenched & tempered	116–190	10–25	220–390	10–12
1095, quenched & tempered	120–190	10–26	230–400	5–6

NOTE: The range of values represents differences in heat treatments.

Fig. 6-6 This energy-absorbing bumper system includes a plated steel bumper and boxlike, high-strength reinforcing member-and-spring shock absorbers. (American Iron and Steel Institute.)

well C64) achieved in the 1095 grade. These steels are thus suitable for wear-resistant parts. So-called spring steels are high-carbon steels available in annealed and pretempered strips and wires. Besides their spring applications, these steels are used for such items as piano wire and saw blades.

Free machining steels are low- and medium-carbon grades with additions of sulfur (0.08 to 0.13 percent), sulfur-phosphorus combinations, and/or lead to improve machinability. They are grades 1108 to 1151 for sulfur grades, and 1211 to 1215 for phosphorus-and-sulfur grades. The presence of relatively large amounts of sulfur and phosphorus causes some reduction in cold formability, weldability, and forgeability, as well as a lowering of ductility, toughness, and fatigue strength.

6-8 *Low-Alloy Carbon Steels*
Low-alloy steels are roughly defined as steels that do not have more than 5 percent total combined alloying elements. They are designated by the same AISI system used for plain carbon steels. The last two digits show the nominal carbon content. The first two digits identify the major alloy element(s) or group. For example, 2317 is a nickel-alloy steel with a nominal carbon content of 0.17 percent.

Alloy Effects. Table 6-3 lists the standard AISI low-alloy steels along with the principal alloying elements. As the table shows, one or more of the following elements are present: manganese, nickel, chromium, molybdenum, vanadium, and silicon. Of these nickel, chromium, and molybdenum are the most frequent. (In Sec. 6-5 and Table 6-1, the effects of alloying elements in ferrous alloys were discussed.)

Whereas surface hardness attainable by quenching is largely a function of carbon content, the depth of hardness depends in addition on alloy content. Therefore a principal feature of low-alloy steels is their enhanced hardenability compared to plain carbon steels.

Like plain carbon steels, however, low-alloy steels' mechanical properties are closely related to carbon content. In heat-treated, low-alloy steels, the alloying elements contribute to the mechanical properties through a secondary hardening process that involves the formation of finely divided alloy carbides. Therefore, for a given carbon content, tensile strengths of low-alloy steels can often be double those of comparable plain carbon steels.

Standard Grades. A majority of low-alloy steels are produced in surface hardening (carburizing) and through hardening grades. The former are comparable in carbon content to low-carbon steels. Grades such as 4023, 4118, and 5015 are used for parts requiring better core properties than obtainable with the surface-hardening grades of plain carbon steel. The higher-alloy grades, such as 3120, 4320, 4620, 5120, and 8620, are used for still better strength and toughness in the core.

Most through, or direct, hardening grades are of medium-carbon content and are quenched and tempered to specific strength and hardness levels. They can be divided into three classes:

Tensile Strength, psi	Hardness, Bhn
275,000–300,000	550–600
175,000–225,000	350–450
125,000–170,000	260–350

Alloy steels also can be produced to meet specific hardenability limits as determined by end quench tests. Identified by the suffix H, they

Table 6-3 *Low-Alloy Steel Grades*

AISI No.	Alloying Elements, %
13XX	Manganese, 1.75
23XX	Nickel, 3.5
25XX	Nickel, 5
31XX	Nickel, 1.25; chromium, 0.65
33XX	Nickel, 3.50; chromium, 1.55
40XX	Molybdenum, 0.25
41XX	Chromium, 0.95; molybdenum, 0.20
43XX	Nickel, 1.80; chromium, 0.80; molybdenum, 0.25
46XX	Nickel, 1.80; molybdenum, 0.26
47XX	Nickel, 1.05; chromium, 0.45; molybdenum, 0.20
48XX	Nickel, 3.50; molybdenum, 0.25
50XX	Chromium, 0.30 or 0.60
51XX	Chromium, 0.70 to 1.05
52XX	Chromium, 0.40 to 1.60; carbon, 0.95 to 1.10
61XX	Chromium, 0.70 to 1.10; vanadium, 0.10 to 0.15
81XX	Chromium, 0.40; nickel, 0.30; molybdenum, 0.12
86XX	Nickel, 0.55; chromium, 0.50; molybdenum, 0.20
87XX	Nickel, 0.55; chromium, 0.50; molybdenum, 0.25
92XX	Silicon, 2.00; chromium, 0.10 to 0.40
93XX	Nickel, 3.25; chromium, 1.20; molybdenum, 0.12
98XX	Nickel, 1.00; chromium, 0.80; molybdenum, 0.25

afford steel producers more latitude in chemical composition limits. The boron steels, which contain very small amounts of boron, are also H steels. They are identified by B's after the first two digits of their designation.

A few low-alloy steels are available with high carbon content. These are mainly spring-steel grades 9260, 6150, 5160, 4160, and 8655, and bearing steels 52100 and 51100. The principal advantages of low-alloy spring steels are their high degree of hardenability and toughness. The bearing steels, because of their combination of high hardness, wear resistance, and strength, are used for a number of other parts, in addition to bearings.

Special Grades. Special low-temperature low-alloy steels are also available. The three most common grades have a carbon content of 0.12 to 0.20 percent and nominal nickel contents of 2.25, 3.50, and 9 percent. They have relatively high strength and very good toughness at temperatures from −75 to as low as −320°F, and therefore are widely used for pressure vessels and gas-storage tanks.

Steels with exceptionally good magnetic properties can be classified as low-alloy steels. Known as electrical steels, they contain from 0.5 to 4.5 percent silicon, have high magnetic permeability, high electrical resistance, and low hysteresis loss. Grain-oriented and nonoriented grades are available, with the latter grades subdivided into low-,

Table 6-4 *Mechanical Properties of Typical Low-Alloy Carbon Steels*

Type & Condition	Tensile Strength, 1,000 psi	Elongation, % in 2 in.	Hardness	Impact Strength, Izod, ft-lb
Carburizing Grades:				
Quenched & Tempered				
4320	215	12	60 (case) R (C)	28
4620	118	20	60 (case) R (C)	56
8620	175	13	62 (case) R (C)	28
Through-hardening Grades:				
Quenched & Tempered				
1340	100–140	20–26	212–285 Bhn	50–90
4130	100–165	16–25	195–330 Bhn	40–90
8740	115–180	13–23	230–250 Bhn	35–90

NOTE: The range of values represents differences in heat treatments.

Table 6-5 *Distinguishing Characteristics and Typical Uses of Selected Low-Alloy Steels*

Identification	Distinguishing Characteristics	Typical Uses
Medium manganese (Mn 1.75%)	Strength and workability	Machinery: logging, road, agricultural
Straight chromium (Cr 0.95%)	Strength and workability	Springs, shear blades, wood-cutting tools
3½% nickel (C 0.30, Ni 3.5%)	Toughness	Rock drill and air-hammer parts, crankshafts
Carbon vanadium (C 0.50, V 0.18%)	Resists impact	Locomotive parts
Carbon molybdenum (C 0.20, Mo. 0.68%)	Resists heat	Boiler shells, high-pressure steam equipment
High silicon (Si 4.0%)	Electrical efficiency	Transformers, motors, generators
Silicon manganese (Si 2.00, Mn 0.75%)	Springiness	Automobile and railroad car springs
Chromium nickel (Cr 0.60, Ni 1.25%)	Surface readily hardened	Automobile ring gears, pinions, piston pins, transmissions
Chromium vanadium (Cr 0.95, V 0.18%)	Strength and hardness	Automobile gears, propeller shafts, connecting rods
Chromium molybdenum (Cr 0.95, Mo 0.20%)	Resists fatigue, impact, heat	Aircraft forgings and fuselages
Nickel molybdenum (Ni 1.75, Mo. 0.35%)	Resists fatigue	Railroad roller bearings, automobile transmission gears
Manganese molybdenum (Mn 1.30, Mo. 0.30%)	Resists impact and fatigue	Dredge buckets, rock crushers, turbine parts
Nickel chromium molybdenum (Ni 1.75, Cr 0.65, Mo. 0.35%)	Resists twisting	Diesel engine crankshafts

SOURCE: J. W. W. Sullivan, "How to Select Wrought Steels," *Materials Engineering*, January 1955.

medium-, and high-silicon grades. The standard AISI designation is the letter M followed by a number that originally stood for the specified core loss for the grade.

Nitriding steels are low- and medium-carbon steels with combinations of chromium and aluminum or nickel, chromium and aluminum. After nitriding, these steels have extremely high surface hardnesses of about 92 to 95 Rockwell N. The nitride layer also has considerable resistance to corrosion from alkali, the atmosphere, crude oil, natural gas, combustion products, tap water, and still salt water. Nitrided parts usually grow about 0.001 to 0.002 in. during nitriding. The growth can be removed by grinding or lapping, which also removes the brittle surface layer. Most uses of nitrided steels are based on resistance to wear. The steels can also be used in temperatures as high as 1000°F for long periods without softening. The slick, hard, and tough nitrided surface also resists seizing, galling, and spalling. Typical applications are cylinder liners for aircraft engines, bushings, shafts, spindles and thread guides, cams, and rolls.

6-9 Cast Carbon and Low-Alloy Steels

The general nature and characteristics of cast steels are, in most respects, closely comparable to wrought steels. Cast and wrought steels of equivalent composition respond similarly to heat treatment and have fairly similar properties. A major difference is that cast steel is more isotropic in structure. That is, its properties tend to be more uniform in all directions than wrought steel's properties, which generally vary, depending on the direction of hot or cold working.

Cast plain carbon steels can be divided into three groups similar to wrought steels: low-, medium-, and high-carbon steels. However, cast steel is usually specified by mechanical properties, primarily tensile strength, rather than composition. Standard classes are 60,000, 70,000, 85,000, and 100,000. Low-carbon grades, used mainly annealed or normalized, have tensile strengths ranging from 55,000 to 65,000 psi. Medium-carbon grades, annealed and normalized, range from 70,000 to 100,000 psi. When quenched and tempered, strength exceeds 100,000 psi.

Ductility and impact properties of cast steels are comparable, on average, to those of wrought carbon steel. However, the longitudinal properties of rolled and forged steels are higher than those of cast steel. Endurance-limit strength ranges between 40 and 50 percent of ultimate tensile strength.

Low-alloy steel castings are considered to be in the low-alloy category if their total alloy content is less than about 8 percent. Although many alloying elements are used, the most common are manganese, chromium, nickel, molybdenum, and vanadium. Small quantities of titanium and aluminum are also used for grain refinement. Carbon

Table 6-6 *Mechanical Properties of Typical Cast Steels*

Class & Condition	Tensile Strength, 1,000 psi	Elongation, % in 2 in.	Hardness, Bhn	Impact Strength, Charpy, ft-lb
Carbon Grades:				
60,000, annealed	63	30	131	12
80,000, normalized & tempered	82	23	163	35
100,000, quenched & tempered	105	19	212	40
Alloy Grades:				
65,000, normalized & tempered	68	32	137	60
80,000, normalized & tempered	86	24	170	48
105,000, normalized & tempered	110	21	217	58

content is generally under 0.40 percent. The standard categories of low-alloy cast steels for specification purposes, in terms of tensile strength, are 65,000, 80,000, 105,000, 150,000, and 175,000. For service at elevated temperatures, however, chemical compositions as well as minimum mechanical properties are often specified.

6-10 *High-Strength Structural Steels*

This arbitrary classification encompasses three groups of steels noted primarily for their structural strength characteristics: high-strength low-alloy steels, high-yield-strength quenched-and-tempered steels, and ultrahigh-strength steels. Tensile strengths available in these three groups range from roughly 70,000 psi in the high-strength low-alloy grades to over 400,000 psi in the ultrahigh-strength steels.

High-Strength Low-Alloy Steels (HSLA). In general, these steels contain from 0.05 to 0.33 percent carbon, 0.2 to 1.65 percent manganese, and small additions of other elements (such as chromium, columbium, copper, molybdenum, and nickel) which dissolve in a ferritic-matrix structure to provide high strength and corrosion resistance. Their cost and strength falls between that of structural carbon steels and quenched-and-tempered steels. They are available in most commercial forms and are usually supplied and used in the as-rolled condition, although they are also available in other conditions.

Fig. 6-7 A 112-in. hemisphere of ultra-high-strength steel for use in a deep-submergence vehicle. (American Iron and Steel Institute.)

Most HSLA steels after a short period of exposure to air develop a pebbly, rusty looking surface film which markedly reduces the corrosion rate. Because this protective film turns to a deep purple brown after several years, HSLA steels are being used as exposed structural members in large buildings. Other advantages are their high ratios of yield to tensile strength, and good toughness, abrasion resistance, and weldability.

HSLA steels are covered by several different specifications. The American Society for Testing and Materials (ASTM), for example, classifies them into six groups based on chemical composition and mechanical properties. The Society of Automotive Engineers (SAE) covers 12 grades with emphasis on toughness and fabricability.

HSLA steels have minimum yield points between 40,000 and 70,000 psi, and ultimate tensile strengths from 60,000 to 85,000 psi. In general, the fatigue endurance limit for 10^6 cycles is about 50 to 60 percent of their ultimate tensile strength. While room-temperature toughness is

Table 6-7 *Mechanical Properties of Typical HSLA Steels*

ASTM Type & Forms	Tensile Strength, 1,000 psi	Yield Strength, 1,000 psi	Elongation, % in 2 in.
A94: Plate, bar, & shapes (1⅛ to 2 in.)	72	47	21
A242, A440, A441: Plate & bar (¾ to 1½ in.)	67	46	—
A375: Sheet & strip	70	50	22

roughly comparable to other carbon steels, their lower transition temperature gives them better notch toughness at low temperatures. HSLA steels are more resistant to abrasion than structural carbon steels. Of all HSLA steels, the intermediate manganese grades are the best in this respect.

Most grades of HSLA are two to eight times more resistant to atmospheric corrosion than plain carbon steels, due to the presence of such elements as nickel, copper, phosphorus, and chromium. In fresh- and saltwater environments, however, there is little improvement in corrosion resistance over plain carbon steels.

In general, the cold formability of HSLA steels is not as good as that of plain carbon steels, but conventional techniques can be used. Most HSLA steels are readily welded by conventional arc, gas, and resistance welding processes.

Quenched-and-Tempered Low-Alloy Steels. As contrasted to the HSLA steels, quenched-and-tempered steels are usually treated at the steel mill to develop optimum properties. Generally low in carbon, with an upper limit of 0.2 percent, they have minimum yield strengths from 80,000 to 125,000 psi. Some two dozen types of proprietary (trademarked) steels of this type are produced. Many are available in three or four different strength or hardness levels. In addition, there are several special abrasion-resistant grades.

Mechanical properties are significantly influenced by section size. Hardenability is chiefly controlled by the alloying elements. Roughly, an increase in alloy content counteracts the decline of strength and toughness as section size increases. Thus specifications for these steels take section size into account.

In general, the higher-strength grades have endurance limits of about 60 percent of their tensile strength. Although their toughness is acceptable, they do not have the ductility of HSLA steels. Their atmospheric-corrosion resistance in general is comparable, and, in some grades, it is better. Most quenched-and-tempered steels are readily welded by conventional methods.

Ultrahigh-Strength Steels. These are the highest strength steels available. Arbitrarily, steels with tensile strengths of around 200,000 psi or higher are included in this category, and, surprisingly, more than 100 alloy steels can now be thus classified. They differ rather widely among themselves in composition and/or the way in which the ultrahigh strengths are achieved. A number of them are also discussed in other sections as part of other steel classes or groups.

Medium-carbon low-alloy steels were the initial ultrahigh-strength steels, and, within this group, a chromium-molybdenum (4130) grade and a chromium-nickel-molybdenum (4340) grade were the first developed. Later, others of this type were developed. These steels have yield strengths as high as 240,000 psi, and tensile strengths approaching 300,000 psi. They are particularly useful for thick sections because they are moderately priced and have high hardenability.

Several types of stainless steels are capable of strengths above 200,000 psi, including a number of martensitic, cold-rolled austenitic, and semiaustenitic grades. The typical martensitic grades are types 410, 420, and 431, as well as certain age-hardenable alloys. The cold-rolled austenitic stainless steels work-harden rapidly and can achieve 180,000-psi yield and 200,000-psi ultimate strength. The strength can also be increased by cold working at cryogenic temperatures. Semiaustenitic stainless steels can be heat-treated for use at yield strengths as high as 220,000 psi and ultimate strengths of 235,000 psi.

Maraging steels, a relatively new family, contain 18 to 25 percent nickel plus substantial amounts of cobalt and molybdenum. Some newer grades contain somewhat less than 10 percent nickel and between 10 and 14 percent chromium. Because of the low carbon (0.03 percent maximum) and nickel content, maraging steels are martensitic in the annealed condition, but are still readily formed, machined, and welded. By a simple aging treatment at about 900°F, yield strengths of as high as 300,000 and 350,000 psi are attainable, depending on specific composition. In this condition, although ductility is fairly low, the material is still far from being brittle.

Two types of ultrahigh-strength, low-carbon, hardenable steels have been developed in recent years. One, a chromium-nickel-molybdenum steel, named Astralloy, with 0.24 percent carbon is air-hardened to a yield strength of 180,000 psi in heavy sections when it is normalized and tempered at 500°F. The other type is based on the iron-chromium-molybdenum-cobalt system and is strengthened by a precipitation hardening and aging process to levels of up to 245,000 psi in yield strength.

Finally, high-alloy quenched-and-tempered steels are another group that have extrahigh strengths. They contain 9 percent nickel, 4 percent cobalt, and from 0.20 to 0.30 percent carbon, and develop yield strengths close to 300,000 psi and ultimate strengths of 350,000 psi. Another group in this high-alloy category resembles high-speed tool steels, but are modified to eliminate excess carbide, thus considerably improving ductility. These so-called matrix steels contain tungsten, molybdenum, chromium, vanadium, cobalt, and about 0.5 percent carbon. They can be heat-treated to ultimate strengths of over 400,000 psi—the highest strength presently attainable in steels, except for heavily cold-worked plain high-carbon steel strips used for razor blades and drawn wire for musical instruments, both of which have tensile strengths as high as 600,000 psi.

6-11 *Stainless Steels*

Stainless steels constitute a large and widely used family of iron-chromium alloys known for their corrosion resistance—notably their "nonrusting" quality. This ability to resist corrosion is attributable to a surface chromium-oxide film that forms in the presence of oxygen. The film is essentially insoluble, self-healing, and nonporous. A minimum chromium content of 12 percent is required for the film's formation, and 18 percent is sufficient to resist the most severe atmospheric-corrosive conditions. However, for other reasons, the chromium content of stainless steels goes as high as 30 percent. Other elements, such as nickel, aluminum, silicon, and molybdenum, may also be present.

There are nearly 50 compositions that are recognized by the American Iron and Steel Institute (AISI) as standard stainless grades. They are divided into three categories, according to the major microstructure present: austenitic, ferritic, and martensitic. In addition to the classic grades, a number of specialty stainless steels have been developed over the years. Principal among these are precipitation-hardening compositions and grades that are hardenable by cold work.

Corrosion Resistance. Because stainless steels owe their corrosion resistance to the presence of an oxide film, it is important to know what factors promote its formation. As already mentioned, chromium is the major contributing factor. Aluminum and silicon also contribute; and nickel, in amounts of around 6 to 8 percent, broadens the film-forming

Fig. 6-8 Modern chair with stainless steel frame. (TCI Co.)

range, especially in nonoxidizing environments. Molybdenum is beneficial in martensitic and particularly in austenitic grades where it improves resistance to pitting and seawater corrosion.

Corrosion resistance of stainless steels to chemical environments depends on the presence of oxidizing conditions. Thus they have good resistance to nitric and chromic acids, but are attacked by hydrochloric and hydrofluoric acids, which are reducing in nature. Sulfuric acid is a borderline chemical. They are also attacked by all the halogen salts—chlorine, bromine, fluorine, and iodine.

To maintain the corrosion-resistance qualities in service, the integrity of the oxide film must be preserved. Therefore surfaces must be clean, smooth, and free from contamination by foreign substances such as dirt, grease, and metal particles generated by fabricating operations.

Properties. Besides nonrusting qualities, stainless steels offer a wide range of mechanical properties in the annealed and hardened conditions. Even annealed, they rank with many structural steels. Ductility, which is also good, varies significantly among the various types. In the annealed condition, the austenitic grades have elongations roughly 2½ to 3 times those of the martensitic and ferritic types.

Austenitic grades show excellent performance at elevated temperatures, and are widely used for parts in which load-carrying ability is required above 1200°F. Of all the commonly used metals, only specialized superalloys and refractory metals surpass these grades in high-temperature strength. However, because of high coefficients of expansion and low thermal conductivity, their high-temperature ductility and thermal-shock resistance are low.

The strength of all stainless steels increases with decreasing temperature. However, ductility varies with structure or type. Austenitic grades retain most of their ductility and impact strength down to at least −320°F. In contrast, ferritic and martensitic grades show marked reduction in ductility and impact strengths below about −40°F.

All stainless steels have low thermal conductivity and are relatively poor conductors of electricity. Their electrical resistivity is high enough so that certain grades are used as low-cost resistance alloys. For example, type 310 has resistance of 30 microhms/in.[3] compared to only 4.3 for low-carbon steel.

In general, wrought stainless steels can be formed and fabricated by all the common steel-processing methods. The differences in workability among the various types stem from differences in microstructure. The austenitic grades, which are more ductile than ferritic and martensitic steels, are particularly well suited to cold forming, but cannot be readily hot formed. The ferritic grades are sufficiently ductile to be formed by practically all standard processes. Although the martensitic grades can be formed readily, because of their lower ductilities they are not as workable as the other two types.

All grades of stainless steel are relatively easy to machine and have the machining characteristics of medium-carbon low-alloy steels. The ferritic and martensitic grades, with similar machining properties, are somewhat more easily machined than austenitic grades. In the ferritic, martensitic, and austenitic groups, there are two modifications of free-machining steels.

Table 6-8 *Mechanical Properties of Typical Stainless Steels*

Type & Condition	Tensile Strength, 1,000 psi.	Elongation, % in 2 in.	Hardness, Rockwell B
302			
Annealed	85–90	50–60	80
Cold worked	110	10–20	240
301			
Annealed	100–118	50–60	85
Cold worked	185	7–10	46 (C)
430			
Annealed	75	25–30	85
Cold worked	75–90	15–25	95
410			
Annealed	70–75	25–35	80
Hardened & tempered	90–190	15–30	180–390
420			
Annealed	95	25	195 Bhn
Hardened & tempered	230	8	500 Bhn
17-7PH			
Solution heat-treated, & hardened	200–235	6–9	—
AM 350			
Solution heat-treated, & hardened	206	13	—

NOTE: The range of values represents differences in forms and heat treatments.

Austenitic Grades. Austenitic grades have the widest use of all stainless steels. There are two principal types: chromium-nickel and chromium-nickel-manganese. The majority of them make up the AISI 300 series. The principal grade in this series, type 302, is best known as 18-8, where the numbers stand for 18 percent chromium and 8 percent nickel.

As the grade name implies, the structure of these steels is predominantly austenitic at all temperatures from cryogenic to the melting point. Thus they can be hardened only by cold working, and in the nonhardened condition they are nonmagnetic. In the cold-worked condition, tensile strengths normally run between 100,000 and 180,000 psi. Typical annealed strengths range up to about 80,000 psi. They are tough and ductile even at low temperatures, are readily cold worked, and weldability is good to excellent. Austenitic grades have excellent high-temperature structural properties.

Because of their good formability and mechanical properties, austenitic stainless steels are widely used for transportation equipment, including aircraft and railways, household items, marine hardware, chemical-processing equipment, and architectural applications.

The chromium-nickel grades are subject to intergranular corrosion if held for a period of time in the temperature range from about 800 to 1600°F. This is caused by precipitation of chromium carbide into the grain boundaries, which thus robs adjacent areas of chromium and makes these areas less corrosion resistant than the rest of the metal. When this occurs, the metal is said to be sensitized. This condition often develops adjacent to welds.

When carbide precipitation occurs (in welding, for example), it can be eliminated by heating the steel to 1800 to 2050°F and quenching, which redissolves the carbides. Carbide precipitation can be minimized by using extra-low-carbon grades (304L and 316L), or it can be avoided with stabilized grades (347 and 321), which contain columbium or titanium.

Ferritic Grades. Ferritic stainless steels, the AISI 400 series, contain approximately 12 to 28 percent chromium, but no nickel. The basic composition is type 430 with 14 percent chromium, widely used for automotive trim. Since ferritic grades are low in carbon, not easily hardened by heat treatment, and only moderately hardened by cold working, they are lower in strength than the other two major stainless grades. They are always magnetic and retain their basic microstructure up to the melting point.

Because of high chromium content, the ferritic grades have good corrosion resistance and excellent resistance to oxidation at high temperatures. However, they are susceptible to embrittlement, particularly the higher chromium compositions, at temperatures in the 750 to 1000°F range.

Martensitic Grades. These stainless-steel grades are essentially straight chromium and, with two exceptions, contain no nickel. Like ferritic grades, they are in the AISI 400 series, except for a couple of grades in the 500 series. The basic and general-purpose composition is type 410 with nominally 12 percent chromium. Martensitic stainless steels differ from the other two major types in that they can be hardened by heat treatment and are always magnetic. When heated to 1650°F and then quenched, the austenite transforms to martensite. In the heat-treated condition, they are the strongest of the standard grades, with strengths approaching 300,000 psi. However, they are less resistant to corrosion than the austenitic and ferritic types. They are easily hot worked and can also be cold worked with ease since they have a low rate of work hardening.

Precipitation-hardening Grades. These relatively new stainless steels were introduced some 20 years ago and are commonly referred to as PH steels. And as the name implies, they owe their mechanical properties to precipitation-hardening reactions. The combination of properties is the key feature of these steels. They possess strength, corrosion resistance, and fabricability to an extent not available in the standard stainless steels. Strengths range from 125,000 to nearly 300,000 psi, depending on heat treatment; they are about equivalent to 18-8 stainless in corrosion resistance; and they are readily weldable without requiring preheating.

6-12 Corrosion- and Heat-resistant Cast Steels

Corrosion- and heat-resistant cast steels are arbitrarily defined as steels containing a minimum of 8 percent nickel and/or chromium. They are almost exclusively used in applications involving either a corrosive environment or a continuous or intermittent service at above 1200°F. The corrosion-resistant grades are often thought of as cast stainless steels, and many of them have corresponding grades with similar compositions and properties. These cast ferrous high alloys are divided into two major series and are specified by a designation code established by the Alloy Casting Division of the Steel Founders' Society of America. The H series are primarily heat-resistant grades; the C series are corrosion-resistant grades.

Table 6-9 *Mechanical Properties (As-cast) of Typical Heat-resistant Cast Alloys*

| Type | Room Temperature | | 1800°F | |
	Tensile Strength 1,000 psi	Elongation, % in 2 in.	Creep Stress, 0.001%/hr; 1,000 psi	100 hr Rupture Stress, 1,000 psi
HC	70	2	—	—
HD	85	16	0.9	2.5
HH	80–85	15–25	1.1–2.1	3.1–4.0
HK	75	17	2.7	4.5
HT	70	10	2.0	4.5
HW	68	4	1.4	3.6

Heat-resistant grades. Heat-resistant high-alloy grades have both high strength and good chemical stability at temperatures between 1200 and 2200°F. Although not immune to corrosive media, their rate of corrosion is low compared to low-alloy steels and cast irons. There are three groups of heat-resistant grades, according to composition and microstructure:

1. Iron-chromium grades (types HB, HC, HD). These contain 8 to 30 percent chromium and under 7 percent nickel. They are predominantly ferritic and therefore have relatively low hot strength. They are seldom used in critical load-bearing parts at temperatures above 1200 or 1400°F. They have excellent resistance to oxidation and to sulfur-containing atmospheres.
2. Iron-chromium-nickel grades (types HE, HF, HH, HI, HK, HL). These contain 18 to 32 percent chromium and 8 to 22 percent nickel, with the chromium content always higher than the nickel. Being partially or completely austenitic, they have greater high-temperature strength and ductility than the ferritic grades. They are characterized by good high-temperature strength, hot and cold ductility, and resistance to oxidizing and reducing conditions. They are particularly suited for parts that operate in atmospheres high in sulfur. They also have good weldability and generally good machinability.
3. Nickel-chromium-iron grades (types HN, HT, HU, HW, HX). These contain 10 to 20 percent chromium and 30 to 70 percent nickel. They are austenitic, and the nickel predominates. If it were not for their relatively high cost and the problem of corrosion in high-sulfur atmospheres, these alloys could be used for practically all applications up to 2100°F. They have good hot strength, carburization resistance, and thermal-fatigue resistance. Therefore they are widely used for load-bearing parts subject to large temperature differentials and cyclic heating. Although their resistance to high-sulfur atmospheres is low, they will withstand reducing and oxidizing atmospheres.

Corrosion-resistance grades. Corrosion-resistant grades are intended for continuous or intermittent service in corrosive environments at temperatures less than 1200°F. They have a minimum of 8 percent alloy content and are commonly referred to as cast stainless steels. In general, corrosion resistance of corresponding cast and wrought alloys is comparable. At room temperature, all corrosion-resistant grades tolerate food products, oxidizing salts and acids, and ordinary water. As temperatures and concentrations of corrodants increases, the choice of grade narrows. There are two groups of corrosion-resistant grades:

1. Iron-chromium grades. These are comparable to the 400 series of wrought stainless steels. Some of the grades (type CA) are martensitic and hardenable by heat treatment, and others (types CB and CC) are ferritic and virtually nonhardenable. The martensitic grades have their best corrosion resistance when fully hardened. The ferritic grades are normally supplied in the annealed condition. The iron-chromium grades are generally highly resistant to oxidizing solutions and are used for parts and equipment in chemical plants processing nitric acid and nitrates. Deaerated or reducing conditions are unfavorable to them. In general, the ferritic grades have greater resistance to most corrosive environments than the martensitic grades.
2. Iron-chromium-nickel grades (types HA, HC, HD). These are austenitic grades, generally comparable to the 300 series of wrought stainless steels. Because austenitic steels undergo no change in phase, heat treatment has only a minor effect on mechanical properties. However, these alloys must be properly heat-treated to ensure complete solution of carbides for maximum corrosion resistance. Within this group, grades CH-20 and CK-20 are suitable for parts exposed to strong, hot, weakly oxidizing solutions such as sulfurous, sulfuric, acetic, and phosphoric acids. Three grades (CD-4MCu, CF-8M, and CN-7M) are the only standard grades suitable for handling hot chlorides and hydrochloric and hydrofluoric acids (after careful evaluation of exact service conditions).

6-13 *Iron-base Superalloys*

The term *superalloy* broadly applies to iron-base, nickel-base, and cobalt-base alloys that combine exceptional high-temperature mechanical properties with excellent oxidation resistance. The operational temperature range for superalloys is roughly between 1000 and 2000°F. In general, iron-base alloys are the least costly and the most easily worked of the three major groups. They are used at the lower end of the range. While room-temperature strength may exceed that of other superalloys, it falls off rapidly above 1200°F. Thus most uses are in the 1000 to 1500°F range.

While many of the iron-base superalloys now have standard designations in the AISI 600 steel series, they are still better known by trade names such as A-286, Discaloy, N-155, and 16-25-6. These and the other alloys are chromium-nickel-iron steels. They are of two major types. One group consists of nonheat-treated chromium-nickel-iron steels commonly used in the as-worked or in the hot- or cold-worked condition. They are primarily low in carbon with one or more carbide-forming elements such as molybdenum. They have higher creep stress than regular austenitic stainless steels. They also have excellent surface stability provided proper precautions are taken to prevent carbide precipitation.

Alloys in the other group are generally heat-treated. Optimum high-temperature properties are obtained by precipitation-hardening treatments. Their high-temperature strengths are somewhat higher than those of the nonheat-treated group. Major use of the iron-base superalloys is in jet engines for such parts as ducts, bolts, exhaust covers, turbine blades, and tail cones.

6-14 *Tool and Die Steels*

Tool steels can be broadly defined as special steels used to form or cut another material. Hundreds of tool steels have been developed over the years to meet the special requirements of the manufacturing industries. With some exceptions they are medium- or high-carbon alloy steels with high hardness, abrasion resistance, and resistance to softening at elevated temperatures. Following is a brief discussion of each of the major classes of tool steels and the standard identification symbols.

1. Water-hardening (W) steels. These are the oldest of the tool steels. They are high-carbon steels with carbon ranging from about 0.90 to 1.10 percent. Hardened by water quenching, they are the hardest of tool steels, and have the best machinability ratings, good toughness, and from fair to good wear resistance. Inherent disadvantages are susceptibility to cracking during hardening, poor resistance to heat above 350°F, and large dimensional change in hardening.
2. Shock-resisting (S) steels. These extremely tough steels are relatively low in carbon (0.45 to 0.65 percent) and alloy content, and are oil-hardening. They were developed for tools in which toughness and strength are of prime importance. The silicon-manganese type is most widely used. Typical uses are pneumatic tools, hand chisels, cold cutters, punches, and heavy-duty shear blades.
3. Cold-worked steels. These are basically carbon tool steels to which small quantities of alloying elements have been added. They are among the most important and widely used of all tool steels. They are used primarily in manufacturing operations in which wear resistance and deformation in heat treatment are the most important criteria. Type 0 is the least expensive. Type A is suited for intricate

shapes and dies with close tolerances. Type D is noted for exceptionally good wear resistance. Typical uses of cold-worked steels are blanking, forming, coining, and trimming dies; thread-rolling dies; and cold-forming rolls.
4. Hot-worked (H) steels. Because they are used in operations involving the forming, shearing, or punching of metals at high temperatures, they have the best heat resistance of the tool steels. While low in carbon, they are relatively high in total alloy content, including chromium, tungsten, vanadium, and molybdenum. They are noted for their ability to withstand a combination of heat, pressure, abrasion, and shock.
5. Mold steels (P). These specialty steels are used almost exclusively for zinc die casting and plastic molds. Of low-carbon content, they are easily machined and can be polished to a high luster.
6. Special-purpose (L) steels. These steels are used in a wide variety of applications that do not necessarily fall within the other classifications. They are used for cold-worked tool pins, jigs, shims, and hand stamps.
7. High-speed steels. High-speed tool steels are highly alloyed; some have as much as 25 percent total alloy content as well as high percentages of carbon. The major alloying element is tungsten, which can run as high as 20 percent. Their chief use is as machining (cutting) tools. Even though used as cutting tools, they are not noted for keen edges. Most of them have fair to good toughness, good to exceptional heat resistance, and good wear resistance. They are also used for extrusion dies, burnishing tools, and blanking punches and dies.

6-15 *Austenitic Manganese Steel*
This is a tough, nonmagnetic steel alloy that has exceptionally high resistance to abrasion and wear. The nominal composition is 12 to 13 percent manganese and 1.2 percent carbon. Nickel can also be present up to about 4 percent. Austenitic manganese steel is produced as rolled shapes or in the form of castings. Improved mechanical properties normally produced by heat treatment in alloy steels are developed in manganese steel by work hardening. A maximum hardness of about 550 Bhn can be obtained by repeated impacts. A major use of this steel is for railway switches and crossovers, where it provides maximum wear resistance by work hardening under train wheel impacts.

6-16 *Ferrous-Base Powders*
Iron, primarily sponge iron, is the most common base metal for powder metallurgy (P/M) parts. Pure iron powders are seldom used alone for

structural parts. Small additions of carbon in the form of graphite and/or copper are used to improve performance properties.

Iron-copper powders contain from 2 to 11 percent copper. Small amounts of graphite are sometimes added. Copper increases strength properties, improves corrosion resistance, and tends to increase hardness, but it lowers ductility somewhat. Densities of around 6 g/cm^3 are most common for iron-copper parts, although densities approaching 7.0 can be achieved. Strengths range between 30,000 and 100,000 psi depending on density and heat treatment.

Fig. 6-9 Parts made of metal powders. (Metal Powder Industries Federation.)

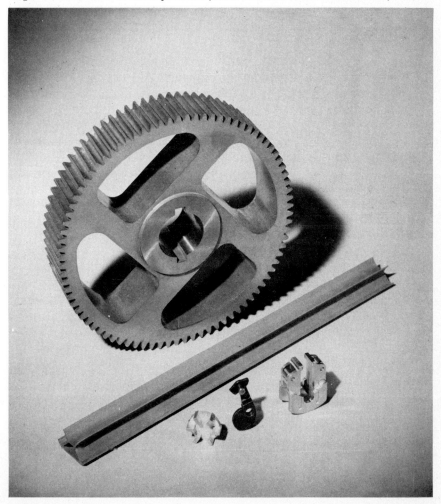

Table 6-10 *Mechanical Properties (As-sintered) of Typical Iron-base P/M Parts*

Type	Density, g/cm³	Tensile Strength, 1,000 psi	Elongation, %	Hardness, Rockwell B	Impact Strength, ft-lb
99 Iron (sponge)	5.7–6.1	19	5	20 (H)	4
	7.3	40	12	20	
99 Iron, 1 carbon	6.1–6.5	35	1.0	50	1
	7.0	60	3.0	—	2
90 Iron, 10 copper	5.8–6.2	30	0.5	—	—
92 Iron, 7 copper, 1 Carbon	5.8–6.2	50	0.5	70	3
	6.8	83	1.0	73	4

Iron-carbon powders contain up to 1 percent graphite. When pressed and sintered, internal carburization results and produces a carbon-steel structure, although some free carbon remains. In general, iron-carbon steel P/M parts will have densities of around 6.5 g/cm³. However, densities of over 7.0 are used to produce higher mechanical properties. These carbon-steel P/M parts have higher strength and hardness than those of iron, but they are usually more brittle. As-sintered strengths range from about 35,000 to 70,000 psi depending on density. By heat treatment, strengths up to 125,000 psi are achieved.

The mechanical properties of ferrous powder parts can be considerably improved by impregnating or infiltrating them with any one of a number of metallic and nonmetallic materials, such as oil, wax, resins, copper, lead, and babbitt.

In addition to the above powders, which are used for the bulk of P/M applications, a number of specialty ferrous alloys are available. These include 3 to 9 percent silicon irons, iron-nickel alloys, alloy steels (2, 4, and 7 percent nickel steels and 4600-series steels). A range of stainless-steel powders are finding increasing use. These include types 302, 303, 304, 316, 330, 410, 430, and 17-4 PH.

Iron-base P/M parts can range in size from about 0.10-in. thick and ⅛-in. diameter to 2-in. thick and over 2 ft in diameter. Because they can be mass produced at relatively low cost, iron-base P/M parts are used in such high-volume products as appliances, business machines, power tools, and automobiles. Typical parts are gears, bearings, rotors, valves, valve plates, cams, levers, ratchets, and sprockets. Also, recent developments have made possible the production of forging preforms or blanks out of low-alloy steel powders and some of the superalloys.

6-17 *Wrought Irons*
Wrought iron is one of the oldest of the ferrous metals. It is, in effect, a composite material, being composed of glasslike iron-silicate (slag) fibers or platelets in an iron (ferrite) matrix. Carbon content seldom exceeds 0.035 percent, and manganese is held to a maximum of 0.06 percent. Besides plain wrought iron, alloy grades are produced with additions of either nickel, molybdenum, copper, or phosphorus to increase strength.

Strengths of wrought irons range from about 40,000 to 50,000 psi for plain wrought iron to about 60,000 psi for the alloy grades. All grades have good ductility and toughness. The good resistance to corrosion of wrought iron has been demonstrated by long years of service life in many outdoor applications such as architectural hardware, bridge railings, and railroad and marine parts. It is also used for brine coils, condenser tubes, and sprinkler systems.

Wrought iron is an easy material to forge and bend. It is readily welded and the slag content provides a self-fluxing action during the welding operation.

6-18 *Prefinished Steels*
As the name implies, these steels are given a special surface finish at the mill or by custom producers. The surface can be (1) a mechanical one, such as a brushed or textured finish; (2) a treated surface, such as a phosphate coating; or (3) an applied coating, either metallic or organic. Table 6-11 lists the general characteristics of common types of prefinished steels.

Cast Irons

Cast iron is the generic name for a group of metals that are basically ternary alloys of carbon and silicon with iron. Included are gray, ductile, white, malleable, and high-alloy irons.

6-19 *Composition and Structure*
The borderline between steel and cast iron is 2 percent carbon, which is the carbon content of saturated austenite. However, most cast irons have at least 3 percent total carbon, and normally the upper limit is 3.8 to 4 percent. Carbon is present in cast irons in two forms—as graphite, often referred to as free carbon, and as iron carbide (cementite).

The large amount of carbon and the presence of some of it as graphite are major distinguishing characteristics of cast irons' distinctive

Table 6-11 Prefinished Steels

Type	Key Reasons for Using	Major Uses	General Characteristics
Aluminized	Heat and corrosion resistance at lower cost than more expensive base metals	Auto mufflers, combustion chambers, major appliances, heater tubes, furnace parts, heat-treating equipment	Al coating about 2 mil thick with intermetallic layer on low-carbon steel sheet and strip; provides oxidation and corrosion resistance to $1000°+F$; strength and formability are reduced
Galvanized	Corrosion resistance at lower cost than more expensive base metals	Auto body parts, roofing and siding, ducts, road guard rails, gutters and downspouts, awnings, fences, cables, pails, storage tanks	Zinc coating with intermed layer of Fe-Zn compounds on low-carbon steel sheet and strip; best corrosion resistance is in rural atmosphere, but also used in industrial and marine atmospheres
Tinplate	Nontoxicity and corrosion protection at low cost; enhances solderability; promotes bonding to other metals	Food packaging, milk cans, food grinders, kitchen utensils, electronic parts, underground cable insulation	Tin coating on low-carbon steel sheet and strip; foil resists stain and tarnish indoors, in rural atmosphere, and in contact with foods; excellent formability and solderability
Terneplate	Corrosion resistance, especially to certain acids and solvents at low cost; improves solderability	Roofing, auto gas tanks, gaskets, heating and ventilating equipment, caskets, electronic chassis, condenser cans, printed circuits	Coating of about 80–85 Pb/15–20 Sn alloy on low-carbon steel sheet and strip; high resistance to H_2SO_4 and good atmospheric corrosion resistance; excellent formability and solderability
Nickel- and chrome-plated	Decorative appeal; corrosion and heat resistance	Electrical appliances, housewares, reflectors, display stands, ornaments, kitchen utensils	Up to about 0.06-in.-thick coating of Ni or Cr on steel sheet and strip; smooth, lustrous, nontarnishing surface; high reflectivity; heat resistance 500–1000°F

Type	Properties	Applications	Specifications
Copper- and brass-plated	Decoration; copper plated also used as base for further plating and to improve sheet formability	Light fixtures, reflectors, jewelry, curtain rods, wall tiles, wall decorations, trim, door hardware	Up to about 0.050-in.-thick coating of Cu or brass on steel sheet and strip; Cu-coated steel usually lacquered to inhibit tarnish
Chromized	Heat and corrosion resistance	Liners, heads, baffles, auto muffler tubes, furnace parts, flame spreaders, laundry chutes, fuel tanks	Diffused Cr-enriched surface 0.0015 to 0.003 in. thick on low-carbon steel sheet and strip; high oxidation resistance to about 1400°F; superior corrosion resistance
Clad or metal-bonded	Combines special property of cladding metal (e.g., corrosion resistance, thermal and electrical properties, decoration) with high strength and low cost of steel	Electronic parts, gaskets, radiator tanks, electrical contacts and switches, heater plates, heat-exchanger fins and tube plates, cookware, chemical equipment	Pure metals or alloys metallurgically bonded to carbon, alloy, and stainless-steel sheet, strip, plate, and wire. Some common cladding metals over steels include Al, brass, Cu, Pb, Ni, precious metals, Sn, Ti
Prepainted	Decoration (variety of colors, patterns, textures); color and thickness uniformity; eliminates postfabrication finishing and painting costs; corrosion, wear, and heat resistance	Containers, indoor and outdoor furniture, awnings, appliances and fixtures, auto dashboard panels and trim, shelving, counter tops, storm doors, window frames, lockseam tubing, TV cabinets, arch uses, truck trailers, toys	Prepainted low-carbon steel sheet and strip most common. Stainless steels also available; paint coatings permanently bonded to one or both sides. Wide variety of paints available
Plastic-bonded steel	Decoration (variety of colors, patterns, textures); color and thickness uniformity; corrosion and wear resistance; reduces postfabrication finishing and painting costs; dielect properties	Radio, TV, stereo cabinets, appliances, building panels, billboards, office-machine housings, auto instrument panels, furniture, machinery housings, vending machines	Primarily PVC (some PVF) laminated to low-carbon, galvanized, and stainless steels, tinplate and other plated steels; vinyl thickness from 0.004 to 0.025 in. (0.008–0.012 in. most common); available in most colors

SOURCE: J. A. Vaccari, "Wrought Steel Primer," in *Materials Engineering*, mid-Sept. 1972.

properties. Also, as we shall see, each of the five major cast-iron types differs from the others in the form in which carbon is present (Fig. 6-10).

The high carbon content makes molten iron very fluid, thus providing excellent castability. The precipitation of carbon as graphite during casting solidification counteracts the normal contraction of cooling metal, thus producing sound castings. The graphite also provides excellent machinability and damping qualities and adds lubrication to wearing surfaces. And, in some cast irons (white), where most of the carbon is present as iron carbide, it provides good wear resistance.

Besides carbon, silicon, from 0.5 to 3.5 percent, is a major alloying element in cast irons. Its major function is to promote formation of graphite and to provide the desired as-cast microstructures.

The matrix structures of cast irons, where any graphite present is embedded, vary widely depending not only on casting practice and cooling rate but also on the shape and size of casting. Furthermore, it is possible to have more than one kind of matrix in the same casting. Also, the matrix structure can be controlled by heat treatment, but once graphite is formed, it is not changed by subsequent treatments. The matrix can be entirely ferritic. It differs from the ferrite found in wrought carbon steels because the relatively large amount of silicon produces a structure that makes the iron free machining. Addition of alloys can produce an acicular (needlelike) matrix. Hardening treatments yield a martensitic matrix. Other possible matrix structures are pearlite and ledeburite.

Because the same composition in a cast iron can produce several different types of structure, cast irons are seldom specified by composition. Within each major type, standard grades are classified by minimum tensile strength.

6-20 *Gray Irons*

Gray irons are the most common, least costly, and most widely used of the cast irons. They also hold the honor of being the cheapest of all engineering materials.

Types. The distinguishing microstructural characteristic is the presence of graphite flakes in a matrix of, usually, ferrite, pearlite, and austenite. The graphite flakes occupy about 10 percent of the total volume of the metal so that gray irons have a lower density than steels.

Alloying produces a broader range of properties than is possible in unalloyed types. Common elements added are chromium, copper, nickel, molybdenum, and vanadium.

Standard gray irons are classified according to tensile strengths

Ferrous Metals 143

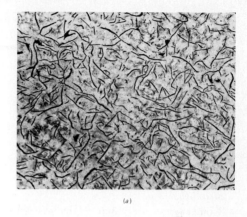

(a)

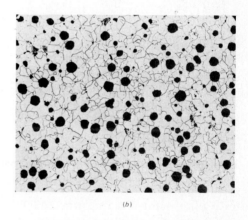

(b)

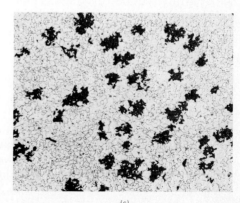

(c)

Fig. 6-10 Microstructure of three major cast-iron types (the black areas are graphite): (a) gray iron, (b) ductile iron, and (c) malleable iron. (Gray and Ductile Iron Founders Society.)

Table 6-12 *Selected Properties of Gray, Malleable, and Ductile Cast Irons*

Property	Type		
	Gray	Malleable	Ductile
Tensile strength, 1,000 psi	20–80	48–120	60–160
Yield strength, 1,000 psi	15–60	30–95	41–135
Compressive strength, 1,000 psi	3–5 × tensile strength	48–95	40–135
Elongation, (in 2 in.), %	0–3	1–26	1–26
Modulus of elasticity, 10^6 psi	12–22	25	24–26
Density (68°F), lb/in.^{3}	0.25–0.266	0.258–0.274	0.25–0.28
Coefficient of thermal expansion, 10^{-6} in./in.	5.8 (32–212°F)	6.6 (70–750°F)	7.5 (70–1100°F)
Electrical resistivity (68°F), microhm-cm	75–120	28.8–34.4	55–70

SOURCE: American Foundrymen's Society, *Cast Metals Handbook*, Cleveland, Ohio, 1957.

into nine main classes from 20,000 to 65,000 psi, in increments of 5,000 psi. A complicating factor is that the tensile strength varies with casting section thickness. Thus strengths of class 20 (20,000 psi), for example, can range from 13,000 to 40,000 psi depending on the thickness of the test bar. This strength increase occurs as section thickness decreases because of the faster cooling rates in thinner sections.

Properties and Characteristics. Gray cast irons possess the highest fluidity of the ferrous casting metals and are thus well suited to the production of relatively intricate and thin-walled castings. In addition, solidification shrinkage is low, ranging from 0 to about 2 percent compared to about 5 percent for steel and malleable iron. Also, the machinability is superior to virtually all grades of steel.

Gray irons are notably low in tensile strength but high in compressive strength—about three to four times the equivalent values of tensile strength (Table 6-13). Their compressive strengths are higher than those of most nonferrous materials and are about equal with those of nonheat-treated low-alloy steels. They retain their strength properties down to below −300°F and up to about 800°F.

Unlike that of most ferrous metals, the modulus of elasticity of gray irons is not constant (12×10^6 to 20×10^6 psi); it decreases with increasing strain and varies with the specific grade of iron. Because their impact resistance is lower than that of most other cast irons, gray irons are not suitable for applications where extreme overloading might be encountered. However, they have good damping capacity, which increases as the amount of flake graphite increases.

Gray irons have excellent wear resistance because of the presence of graphite. Hardened gray iron is used to obtain maximum wear properties. Although not generally considered a corrosion-resistant material, gray irons offer better corrosion resistance than most carbon steels. The rusting action forms a relatively adherent protective coating that offers fairly good resistance to the atmosphere, soil, acids, and alkalies. For example, some cast-iron water and gas pipes have been in service for more than 100 years.

Fig. 6-11 The body of this plug valve is a ductile iron casting with a lining of a fluorocarbon plastic (Teflon). (Gray and Ductile Iron Founders Society.)

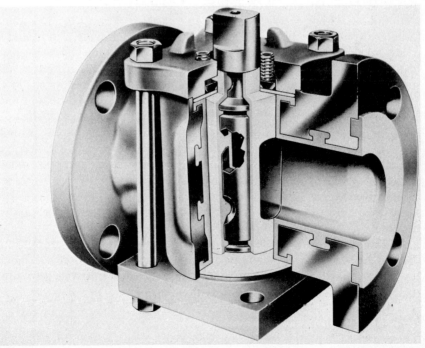

Table 6-13 *Mechanical Properties (As-cast) of Typical Gray Cast Irons*

Type	Tensile Strength, 1,000 psi (min.)	Compressive Strength, 1,000 psi	Modulus of Elasticity, 10^6 psi	Hardness, Bhn (min.)	Impact Strength, Izod, ft-lb
20	20	83	12	140	—
40	40	140	17	200	31
60	60	185	20	230	75

Because of gray irons' low cost and excellent castability, they are produced in a wide size range for a great variety of parts and components. The largest single use may be for automotive cylinder blocks. Huge machine and equipment bases, large gears, heavy compressors, diesel engine castings, heavy rolls, high-pressure cylinders, press and crusher frames, flywheels, and brake drums are other common uses.

High-Alloy Gray Irons. Gray irons containing over about 3 percent in alloying elements are classified as high alloys. The alloying elements are silicon, nickel, chromium, copper, and aluminum, either singly or in combination. One type of high-silicon iron contains 14 to 17 percent silicon. Other grades contain 4 to 6 percent. The 14 to 17 percent grades are used for handling corrosive liquid chemicals. Their shock resistance is poor and they are virtually unmachinable. The 4 to 6 percent silicon irons have excellent resistance to scaling and growth in use in temperatures up to 1650°F, but their thermal-shock resistance is low.

There are a number of austenitic gray-iron alloy grades that contain 18 percent or more of nickel, up to 4 percent chromium, and up to 7 percent copper. Because they have good scaling and growth resistance up to about 1500°F, and are much tougher and more resistant to thermal shock than silicon- and chromium-alloy white irons, they are used where heat, corrosion, and wear resistance are important. For example, traditionally these alloy irons have served as stove tops as well as for heavy-duty industrial products such as flood gates, seawater valves, turbocharger housings, and caustic pumps and valves.

High-aluminum irons, which contain 18 to 25 percent aluminum, have excellent scaling resistance along with good corrosion resistance in the presence of hot acids. However, brittleness and the difficulty in producing sound castings limits their use.

6-21 *Ductile Irons*

Ductile irons, which have been in commercial use only about 25 years, have the basic compositions of gray irons, but differ in the shape of the graphite particles. In contrast to the flat flakes present in gray irons, the graphite in ductile irons is nodular or spheroidal. Because of this distinctive feature, ductile irons have also been called nodular irons and SG (spheroidal) irons.

In both structure and some properties, ductile irons resemble steel. They are sometimes considered a high-silicon steel with the addition of graphite. The spherical graphite nodules have a minimum effect on the matrix, which can be varied considerably by foundry practice from completely pearlitic to totally ferritic. Also, by hardening heat treatments, martensitic or banitic structures can be produced to give high strength and hardness. Alloying elements also influence structure as they do in steels. For example, nickel strengthens the matrix without forming carbides, and molybdenum increases strength, hardness, and hardenability.

Standard Grades. The ASTM classifies and codes ductile irons according to mechanical properties. A three-number designation is used to specify minimum tensile strength, yield strength, and elongation. For example, the designation 60-40-18 indicates the following minimums: 60,000-psi tensile strength, 40,000-psi yield strength, and 18 percent elongation.

There are five such ASTM standard grades: Types 60-40-18 and 65-45-12 have maximum toughness and machinability and are used for valve and pump bodies, gear housings, and farm-machine parts. Type 80-55-06 has maximum strength in the as-cast condition, and, in general, good creep strength. It can be surface hardened for use in applications requiring resistance to wear and good strength, as in mining machinery and automotive and diesel engine parts. Also, its heat

Table 6-14 *Mechanical Properties of Typical Ductile Cast Irons*

Type	Tensile Strength 1,000 psi	Elongation, % in 2 in.	Hardness, Bhn	Impact Strength, Charpy, (Notched), ft-lb
Standard Grades				
60-40-18	60–80	18–30	150–187	10–15
80-55-06	80–100	6–10	180–248	2–5
120-90-02 (Heat-treated)	120–175	2–7	240–300	—
Austenitic Grades				
D-2	58–62	8–20	140–200	12
D-5	55–60	20–40	130–180	17

resistance is an advantage in moderate high-temperature service. Type 100-70-03 is usually normalized and tempered to develop a combination of strength, ductility, and wear resistance. It also responds well to surface hardening and is used for high-strength gears and automotive and machine components. Type 120-90-02 provides maximum strength and wear resistance and is used for pinions, gears, rollers, and slides.

There is also a heat-resistant grade that contains more silicon than the other grades. It has the highest resistance to scaling, but has the lowest room-temperature ductility. It is used chiefly for parts in metal and glass-producing machinery and furnaces.

High-Alloy Ductile Grades. Also referred to as austenitic ductile iron, these grades are highly alloyed, with nickel ranging from 18 to as high as 36 percent, and chromium running between 1.5 and 5 percent. Their relationship to the standard ductile irons is similar to that of wrought stainless to low-alloy steels. They were developed primarily for corrosion- and heat-resistant applications. Some of the grades have elevated-temperature strengths comparable with that of 18-8 stainless steel. These austenitic grades are nonmagnetic and readily welded. Typically, they are used in the chemical industry for impellers, pumps, and valves; in paper mills; in saltwater service; and for parts requiring dimensional stability.

6-22 *Malleable Irons*
Malleable irons are a family of cast ferrous alloys consisting mainly of iron, carbon, and silicon. They are produced chiefly as sand- or shell-mold castings. Production involves two major steps. A white-iron casting, which is a hard, brittle combination of cementite and pearlite, is produced. The white-iron casting is then converted into malleable iron by a two-stage annealing treatment. In the first stage, the casting is heated to between about 1550 and 1750°F and held at that temperature for a given length of time, depending on composition, maximum section thickness, and temperature. In the second stage, the casting is slow cooled through a temperature range of from about 1400 down to 1300°F. Total time for the complete annealing cycle varies from 100 hr to 12 hr.

Characteristics and Properties. Among the most useful characteristics of malleable irons are high toughness, good resistance to atmospheric corrosion, excellent machinability, and good casting properties. Charpy impact strengths run from about 14 to 17 ft-lb (unnotched).

Table 6-15 *Mechanical Properties of Typical Malleable Cast Irons*

Type	Tensile Strength, 1,000 psi	Elongation, % in 2 in.	Hardness, Bhn	Impact Strength Charpy (notched), ft-lb
Ferritic	50–53	7–10	110–145	16
Pearlitic	75–100	2–7	170–269	14
Alloyed	58–65	15–20	135–155	—

High toughness also is evident by high compressive strength, which ranges around 200,000 psi, and in the excellent damping capacity, which runs about three times that of cast steel.

The presence of a thin surface skin of ferrite, which is resistant to most ordinary atmospheres and salt water, makes these castings suitable for outdoor products such as pole line hardware, farm-implement parts, and railway rolling-stock components (Table 6-16).

Malleable irons, specifically the ferritic grades, are generally considered to be the most machinable and free cutting of any irons and steels of comparable or higher strength. And, because of high ductility, they can be cold formed. But, like most cast ferrous metals, they are difficult to weld.

Compared to other ferrous cast metals, malleable irons have better fluidity than steel, but not as good as gray irons. Because annealing

Table 6-16 *Atmospheric-Corrosion Resistance of Several Ferrous Materials*

Material (Unmachined)	Average Weight Loss after 3 Yr, g/cm²		
	Rural[1]	Marine[2]	Industrial[3]
Malleable iron	3.44	6.04– 7.68	5.17–6.63
Nodular iron	3.44	6.04– 7.68	5.17–6.63
Mild steel (1020)	5.70	24.75–22.01	8.85–8.18
Copper-chromium-nickel steel	3.83	6.11– 6.68	4.77–5.53
Copper-alloyed steel	5.05	10.12–15.45	8.76–7.43

[1] State College.
[2] Point Reyes and Kure Beach.
[3] Neward and East Chicago.

is an integral part of the production cycle, malleable castings are homogeneous and stress free.

Ferritic Malleable Irons. Ferritic malleable irons are composed of a ferrite matrix interspersed with nodules of carbon. As a class, they have high toughness and ductility and excellent machinability. There are two standard grades—32510 and 35018.

Of the two, grade 32510 has the higher carbon content, ranging from 2.30 to 2.65 percent. The high carbon makes this grade extremely fluid and therefore suitable for thin and/or intricate castings. Sections as thin as $\frac{1}{32}$ in., in small areas, can be successfully cast. This grade is mainly used for automotive parts, agricultural implements, and electrical products.

Grade 35018 has lower carbon content (2 to 2.45 percent), slightly higher strength, and somewhat higher ductility than grade 32510. However, in casting properties and annealability, the two grades are about equal. Typical applications of grade 35018 are railroad, high-pressure, and oil-field castings.

Pearlitic Malleable Irons. Pearlitic malleable irons are produced either from the same white-iron compositions as the ferritic grades or from compositions containing small alloying additions. They differ from ferritic malleable irons in that some of the carbon is retained in combined form by a special heat treatment and/or by the alloy addition. Although commonly referred to as pearlitic malleable irons, this group includes irons in which the combined carbon (ranging from 0.3 to 0.8 percent) is present either as pearlite or martensite.

There are seven standard grades, based on required minimum tensile, yield, and elongation values. Grade 80002 is oil quenched and is always martensitic. Grade 60003 may be either martensitic or pearlitic, depending on the quenching medium. All the other grades are pearlitic.

In general, pearlitic malleables have higher strength, hardness, rigidity, and abrasion resistance than ferritic grades. However, they have lower ductility and shock resistance and are not as easily machined. An important characteristic is that they are readily surface, local and through hardened.

Pearlitic malleable castings are particularly suitable where wear resistance or relatively high strength is required. They are widely used in the automotive, agricultural equipment, and ordnance industries for such parts as gears, sprockets, and rocker arms.

Alloyed Malleable Irons. Compared to the ferritic and pearlitic malleable grades, special alloyed malleable irons are used in relatively small amounts. They are produced in the same way as the ferritic malleables. They derive their special properties from additions of copper and molybdenum, which are not normally present in significant quantities in ferritic malleable iron.

The two major kinds of special malleable irons are copper-alloyed and copper-molybdenum-alloyed. Copper-alloyed malleable irons contain about 0.25 to 1.75 percent copper. A copper content of about 0.25 to 0.5 percent improves resistance to atmospheric corrosion, and especially to sulfurous atmospheres. A higher copper content of 0.5 to about 1.75 percent improves strength properties up to around 10 percent, while elongation is slightly decreased. The strengths of these irons can be further increased by precipitation hardening.

Copper-molybdenum-alloyed malleables contain up to 0.5 percent molybdenum and from 0.5 to 1.25 percent copper. They provide higher strengths than other ferritic grades without an appreciable sacrifice in ductility.

6-23 *White Irons*
In white cast iron there is virtually no graphite. Rather, the carbon is present in a matrix of fine pearlite as large particles of iron carbide, thus providing a material that is high in compressive strength, very hard, and abrasion resistant, but low in tensile strength and impact resistance. Hardness of unalloyed white iron ranges between 300 and 575 Bhn, but tensile strength is only 30,000 psi. For somewhat higher hardness, higher tensile and impact strength, and other special service properties, white irons are alloyed with nickel, chromium, and molybdenum. For example, nickel and chromium in a white cast iron commonly known as Ni-Hard, provide a martensitic matrix. Hardness is

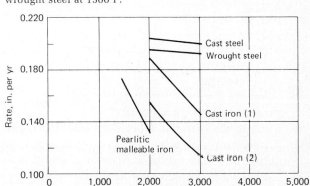

Fig. 6-12 Oxidation rates of various cast ferrous metals and wrought steel at 1300°F.

increased to around 600 Bhn and tensile and impact strength are more than doubled. The high-chromium and chromium-molybdenum-alloyed white irons combine excellent wear resistance with oxidation and corrosion resistance at elevated as well as normal temperatures.

By controlling composition and cooling rates, castings can be made with a white-iron surface layer and a core of gray iron, thus providing a duplex structure that combines excellent abrasion resistance with relatively good toughness. This metal is known as chilled iron.

Castings of white and chilled iron are mainly used in applications that involve resistance to wear and abrasion. Typical uses include parts for crushers, grinders, slurry pumps, railroad car wheels and brake shoes, rolling mill rolls, and machinery handling abrasive materials.

Review Questions

1. What base metal and what alloying element are present in all ferrous materials?
2. Name the allotropic form of iron and the type of crystal structure that exists in the following temperature ranges:
   (a) 2552 to 2802°F.
   (b) Below zero and up to 1670°F.
   (c) 1670 to 2552°F.
3. Name the three forms in which carbon is present in ferrous materials.
4. What is the normal range of carbon content for steels and cast irons?
5. Which iron-carbon solid solution in ferrous materials contains the least carbon?
6. Which standard group of stainless steels is essentially nonmagnetic? why?
7. What major influence or influences does each of the following phases and constituents have on the properties of ferrous materials?
   (a) Martensite.   (d) Austenite.
   (b) Pearlite.   (e) Graphite.
   (c) Cementite.
8. What are the principal functions alloying elements perform when present in ferrous materials?
9. Explain the difference between killed and rimmed steels.
10. Three wrought carbon steels are the same except that the carbon contents are as follows: A: 0.20 percent; B: 0.35 percent; and C: 0.60 percent. Which steel is

(a) Highest in yield strength?
(b) Lowest in hardness?
(c) Highest in ductility?
(d) Most brittle?
(e) Highest in hardenability?
(f) Best in weldability?
(g) Most easily hot and cold worked?
Which steel would you choose
(h) For good balance between strength and ductility?
(i) To obtain maximum strength by heat treatment?
(j) For low-temperature service?
11. State a principal effect that each of the following alloying elements has on a steel's properties.
(a) Silicon.   (d) Manganese.
(b) Nickel.   (e) Molybdenum.
(c) Chromium.
12. How are standard cast steels classified and usually specified?
13. Why can high-strength, low-alloy steels be used in structures exposed to outdoor conditions?
14. What steels have the highest attainable tensile strength?
15. What accounts for the unusually good corrosion resistance of stainless steels as compared to that of most other steels?
16. Which of the four major groups of stainless steels are
(a) Always magnetic?
(b) Can only be hardened by cold work?
(c) Lowest in strength?
(d) Generally the least corrosion resistant of the stainlesses?
(e) Highest in strength (two)?
17. Name the three major groups of heat-resistant cast steels and give their identifying microstructural characteristic.
18. What are three differences between iron-base superalloys and other major types of superalloys?
19. What are three qualities, or properties, that tool steels must have?
20. How can the porosity in iron-powder parts be used to advantage?
21. What is the general relationship between density and strength in iron-powder parts?
22. State the form or shape in which graphite is present in gray, ductile, and malleable irons.
23. Which cast iron (or cast irons) is identified with the following?
(a) Lowest in cost.
(b) Hard and abrasion resistant, but low in strength and toughness.

(c) Good resistance to outdoor atmospheres.
(d) Most similar to steel of the cast irons.

Bibliography

American Society for Metals: *Metals Handbook*, 8th ed., Novelty, Ohio.
Bethlehem Steel Company: *Modern Steels and Their Properties*, 1959.
Everhart, J. L.: "Cast Ferrous Metals," *Materials Engineering*, January 1970.
"Materials Selector Issue," *Materials Engineering*, Mid-September 1972.
"Metals Reference Issue," *Machine Design*, Feb. 17, 1972.
Sullivan, J. W. W.: "How to Select Wrought Steels," *Materials Engineering*, July 1955.
Vaccari, J. A.: "Wrought Steel Primer," *Materials Engineering*, September 1967.

NONFERROUS METALS

By definition, metallic materials that do not have iron as their major ingredient are considered to be in the nonferrous family. While tonnage consumption of ferrous metals far surpasses that of nonferrous materials, when considered individually, the use of such nonferrous materials as aluminum and copper is greater than that of some single groups of ferrous alloys.

There are roughly a dozen nonferrous metals in relatively wide industrial use. At the top of the list is aluminum, which next to steel is the most widely used structural metal today. It and magnesium, titanium, and beryllium are often characterized as light metals because their density is considerably below that of steel.

Copper is the second nonferrous material in terms of consumption. There are two major groups of copper alloys: brass, which is basically a binary-alloy system of copper and zinc, and bronze, which was originally a copper-tin-alloy system. Today, the bronzes include other copper-alloy systems.

Zinc, tin, and lead, with melting points below 800°F, are often classified as low-melting alloys. Zinc, whose major structural use is in die castings, ranks third to aluminum and copper in total consumption. Lead and tin are rather limited to applications where their low melting points and other special properties are required.

Another broad group of nonferrous alloys is referred to as refractory metals. Such metals as tungsten, molybdenum, and chromium, with melting points above 3000°F, are used in products that must resist unusually high temperatures. Although nickel and cobalt have melting points below 3000°F, they are often considered as refractory metals also, because of their excellent heat resistance and their use in heat-resistant alloys.

Finally, the precious, or noble, metals family has the common characteristic of high cost. In addition, they have in common unusually high corrosion resistance and generally high density.

Aluminum

Aluminum is the most abundant metallic element in the earth's crust. It is present as aluminum oxide (alumina) in the ore (bauxite), usually in amounts of from 40 to 60 percent. Aluminum is extracted from the ore by means of an electrolytic process in which the alumina is first dissolved in a molten electrolyte, and then an electric current is passed through it, causing the metallic aluminum to be deposited on the cathode.

7-1 *Major Characteristics and Properties*

Aluminum and its alloys, like most ductile metals, have a face-centered cubic crystal structure. They offer an attractive combination of properties—light weight, excellent atmospheric-corrosion resistance, high electrical and thermal conductivity, and ease of fabrication. Because their density is about one-third that of steel, some aluminum alloys have strength-to-weight ratios exceeding those of many steels and most nonferrous materials. Of the common conductor metals, aluminum is second only to copper on an equal-volume basis, but twice that of copper on an equal-weight basis. Due to the natural formation of a tough oxide film on the surface, aluminum materials resist corrosion in many atmospherical and chemical environments. The relatively good ductility and low hardness make aluminum easy to form and finish. In addition, aluminum is noted for its nontoxicity, nonmagnetic, and nonsparking properties.

As temperatures decrease below room temperature and down to −320°F, aluminum alloys increase in strength and toughness, thus making them attractive for cryogenic applications. However, all aluminum alloys lose strength rapidly as temperatures increase. Except for a few alloys, they are not used in products that are continuously exposed to temperatures above 600°F.

Another attractive feature of aluminum is its workability. Its face-centered cubic crystal structure has many slip systems, which make the metal very ductile and easily shaped. Consequently, aluminum is available in a great variety of cold-rolled sheets and strips and in extruded forms in many intricate shapes. Foil can be obtained as thin as 0.00025 in. Aluminum can also be hot worked. However, in the case of heat-treated alloys, the elevated working temperatures required can sometimes significantly lower strength properties.

Because aluminum metal has low strength (about 13,000 psi ultimate), alloying and/or cold working is required in order to develop useful properties for structural applications. The addition of alloying elements results in strength increase through two hardening mechanisms: solid-solution strengthening and precipitation (age) hardening.

7-2 *Types and Conditions*
Aluminum materials are classified in three major ways:

1. By composition, they are divided into commercially pure aluminum and aluminum alloys.
2. By treatment, they are divided into nonheat-treatable and heat-treatable alloys.
3. By production methods, they are divided into wrought and cast shapes or forms.

All wrought-aluminum materials are identified by means of a four-number system developed by the Aluminum Association. The number 1 identifies commercially pure aluminum grades. The last two digits give the aluminum content over 99 percent in hundredths of percent. The second digit designates the degree of control over impurity limits.

The first-digit designations for the various aluminum-alloy groups are copper, 2; manganese, 3; silicon, 4; magnesium, 5; magnesium and silicon, 6; zinc, 7; other elements, 8; and unused series, 9. The last two of the four digits identify the alloying elements in a particular alloy group. The second digit indicates modifications of the original alloy: a zero denotes the original alloy; integers 1 to 9 are assigned consecutively to alloy modifications. Experimental alloys carry the standard four-digit number preceded by X until the alloy becomes standard.

Alloys in the 1, 3, and 5 series, which owe their strength to the hardening effects of manganese and magnesium, respectively, are the nonheat-treatable classes of aluminum alloys. However, they can be strengthened by cold work. Alloys in the 2, 6, and 7 series are the heat-treatable classes. Although some of the alloys in the 4 series are heat-treatable, most of them are used only for brazing sheet and welding wire.

Fig. 7-1 Aluminum electrical cable for conducting electricity at cryogenic temperatures. Cable can transmit over 3,500 million voltamperes. (The Aluminum Association.)

A standard system of letters and numbers is used to indicate the processed condition of aluminum alloys. The designation, called *temper*, follows the alloy-identification number. The word temper is used here in a broader sense than in its connection with the heat treatment of steel. In relation to aluminum alloys, temper covers the heat-treated and/or production or fabricated condition of the alloy. The letter H, for example, indicates cold work (strain hardening), and the letter T indicates the heat-treated condition. In addition, a number following the letter specifies the degree and/or combination of cold work and heat treatment. Thus the temper designation also gives a rough indication of an alloy's mechanical properties. Table 7-1 provides a summary of the aluminum-alloy temper-designation system.

Table 7-1 *Temper Designations for Aluminum Alloys*

Temper Designation	Condition	Characteristics
F	As-fabricated	Normal wrought or cast production operations; no guaranteed properties
O	Annealed, recrystallized	Softest temper
H	Cold worked	Cold worked (strain hardened)
H1	Cold worked	Cold worked only
H12	Cold worked	Quarter-hard
H14	Cold worked	Half-hard
H18	Cold worked	Full-hard
H2	Cold worked and partially annealed	Work hardened and annealed to desired strength
H24	Cold worked and partially annealed	Half-hard
H28	Cold worked and partially annealed	Full-hard
W	Heat-treated, unstable	Natural aging alloys (at room temperature after heat treatment)
T	Heat-treated	Heat-treated with or without supplementary work hardening.
T2	Annealed (castings only)	
T3	Annealed, cold worked, naturally aged	
T4	Solution treated, naturally aged	
T5	Artificially aged (no solution treatment)	
T6	Solution treated, artificially aged	
T8	Solution treated, cold worked, artificially aged	
T9	Solution treated, artificially aged, cold worked	

The finish of as-supplied wrought-aluminum materials is also designated by a standard system of letters and numerals. Finishes are classified into three major groups: mechanical finishes, chemical finishes, and coatings. Each of these groups is designated by a letter, and specific finishes in each group are identified by a two-digit number. The sequence of operations leading to the final finish can be indicated by using more than one designation. Table 7-2 gives a summary of the major finish designations.

7-3 Nonheat-treatable Wrought Alloys

The initial mechanical properties of nonheat-treatable aluminum materials depend on the hardening and strengthening effect of elements such as manganese, silicon, iron, and magnesium, which are present alone or in various combinations. These alloys can be further hardened by cold work. There are three major nonheat-treatable groups (Table 7-3):

1000 Series. Commercially pure aluminum has a minimum aluminum content of 99 percent. The major difference between grades in the series is the level of two impurities—iron and silicon. The electrical conductivity grade, designated EC, with a 99.45 percent aluminum content, is widely used as an electrical conductor on the form of wire and bus bars. The 1000-series alloys are especially noted for high electrical and thermal conductivity and excellent corrosion resistance. In the annealed condition, they are relatively soft and weak. However, an annealed strength of 13,000 psi can be doubled by cold work, with a

Table 7-2 *Finish Designations for Aluminum Alloys*

Mechanical Finishes	Coatings
M1 As-fabricated	A1 Anodic
M2 Buffed	A2 Protective and decorative (less than 0.4 mil)
M3 Directional textured	A3 Architectural class II (0.4–0.7 mil)
M4 Nondirectional textured	A4 Architectural class I (0.7 mil and thicker)
Chemical Finishes	R1 Resinous and other organic coatings
	V1 Vitreous coatings, porcelain and ceramic types
C1 Nonetched cleaned	
C2 Etched	E1 Electroplated and other metal coatings
C3 Brightened	L1 Laminated coatings
C4 Chemical conversion coating	

Table 7-3 *Properties of Typical Nonheat-treatable Wrought-Aluminum Alloys*

Type & Condition	Tensile Strength, 1,000 psi	Elongation, % in 2 in.	Modulus of Elasticity, 10^6 psi	Hardness, Bhn	Electrical Resistivity, Microhm-cm
EC					
Annealed (0)	12	23	10	—	2.8
Hard (H19)	27	1.5	10	—	
1100					
Annealed (0)	13	40	10	23	2.92
Hard (H18)	24	10	10	44	3.02
3003					
Annealed (0)	16	35	10	28	3.45
Hard (H18)	30	7	10	55	4.31
5056					
Annealed (0)	42	35	10.3	65	5.90
Hard (H38)	60	15	10.3	100	6.40

sacrifice in ductility. Typical uses of commercially pure aluminum grades are sheet metal, foil, spun ware, chemical equipment, and railroad tank cars.

3000 Series. Manganese, up to about 1.2 percent, is the major alloying element in this series. It provides a moderate improvement over the 1000 series in mechanical properties without significant loss of corrosion resistance and workability. One of the alloys, 3004, also contains magnesium, giving additional strength improvement. Typical applications of this series are housings, cooking utensils, sheet metals, and storage tanks.

5000 Series. Magnesium is the main alloying element in this series, ranging from less than 1 up to about 5 percent. They are the strongest of the nonheat-treatable alloys, with strength increasing with magnesium content. However, high-magnesium grades are more difficult to hot work and are susceptible to stress corrosion above 150°F. Because of the presence of lower-density magnesium, the density of these alloys is less than that of pure aluminum. They are particularly resistant to marine atmospheres and various types of alkaline solutions. Typical applications are marine hardware, building hardware, appliances, welded structures and vessels, and cryogenic equipment.

162   Industrial and Engineering Materials

7-4 *Heat-treatable Wrought Alloys*
Heat-treatable aluminum alloys develop their final mechanical properties through the solid-solution hardening effects of alloying elements as well as by second-phase precipitation. Cold working is also sometimes employed to obtain optimum properties.

Two treatments are generally involved in the hardening process—solution heat treatment and age or precipitation hardening. The solution-heat-treatment cycle consists of: (1) heating the alloy up to between 800 and 1000°F and holding it at temperature to allow the alloy-

Fig. 7-2 Aluminum extrusions are produced in a range of cross-sectional shapes. This one is for a prototype auto bumper. (The Aluminum Association.)

ing elements to go into solid solution; and (2) quenching the alloy rapidly to hold the alloying element in solution. This treatment disperses the hardening element uniformly throughout the material. Age hardening, performed after solution heat treatment, is done either at room temperatures over an extended period (natural aging) or at a temperature somewhere between 240 and 450°F over a shorter period (artificial aging). This treatment further increases strength and hardness by the precipitation of hard, second-phase particles. The second phase, an intermetallic compound, appears as a fine network on the grain boundaries. Although loss of ductility is not great, the second phase penetrates the surface aluminum-oxide layer and consequently lowers corrosion resistance.

2000 Series. This is the oldest and probably the most widely used aluminum-alloy series. The principal alloying element, copper, combines with the base metal, aluminum, during heat treatment to form the hardening intermetallic compound $CuAl_2$. Copper content ranges from about 2 to 6 percent.

Until the introduction of the 7000 series, copper-aluminum alloys

Table 7-4 *Properties of Typical Heat-treatable Wrought-Aluminum Alloys*

Type & Condition	Tensile Strength, 1,000 psi	Elongation, % in 2 in.	Modulus of Elasticity, 10^6 psi	Hardness, Bhn	Electrical Resistivity, Microhm-cm
2014					
Annealed (0)	27	18	10.6	45	3.45
Heat-treated (T6)	70	13	10.6	135	4.31
2024					
Annealed (0)	27	20	10.6	47	3.45
Heat-treated (T3)	70	18	10.6	120	5.75
6061					
Annealed (0)	18	27	10	30	3.8
Heat-treated (T6)	45	12	10	95	4.31
6262					
Annealed (0)	—	—	10	—	3.9
Heat-treated (T9)	58	10	10	120	—
7075					
Annealed (0)	33	17	10.4	60	—
Heat-treated (T6)	83	11	10.4	150	5.7

were the highest strength aluminum alloys. They possess relatively good ductility but are more susceptible to corrosion than other aluminum alloys, particularly in the aged condition. Also, except for 2014 and 2219, they have limited weldability. Because of the corrosion problem, the sheet forms are often clad with commercially pure aluminum or special aluminum alloys.

Because they can be heat-treated to strengths up to 75,000 psi, these aluminum-copper alloys are used in structural applications. Alloys 2014 and 2024, the best known of the series, are widely used in the aircraft industry. Alloy 2014 is primarily a forging grade. Alloy 2024 was the first high-strength light-weight alloy.

6000 Series. Alloys in this series contain silicon and magnesium in approximately equal amounts up to about 1.3 percent. Small amounts of other metals, such as copper, chromium, or lead, are also present in some of the alloys to provide improved corrosion resistance in the aged condition, or to increase strength or electrical conductivity. As a group, alloys in this series have the lowest strengths of the heat-treatables, but possess good resistance to industrial and marine atmospheres. Typical uses include screw machine parts, moderate-strength structural parts, furniture, bridge railing, and high-strength bus bars.

7000 Series. This series, the newest standard group, has the highest strength of any aluminum-alloy group. High strength is obtained by the addition of 1 to 7.5 percent zinc, and 2.5 to 3.3 percent magnesium. Chromium and copper also contribute added strength, but tend to lower weldability and corrosion resistance. Alloy 7075, the most widely used grade, with a heat-treated (T6) tensile strength of about 80,000 psi, has many aircraft structural applications. Other lesser-used grades, such as alloy 7178 (T6), have tensile strengths from about 88,000 to over 90,000 psi.

7-5 *Cast Alloys*

Although there are a great many cast alloys available, less than 50 are in common use. Some are used for all three major types of casting—sand, permanent-mold, and die casting—while others were specifically developed for one of the casting processes. Both heat-treatable and nonheat-treatable alloys are available. Alloy compositions that do not respond to heat treatment are identified by an F following the alloy-designation number or by omission of a suffix. The heat-treatable alloys, which can be solution heat-treated and aged similar to wrought heat-treatable grades, carry the temper designations T2, T4, T5, T6, or

T7. Die castings are seldom solution heat-treated because of the danger of blistering.

As is true of most cast metals, the mechanical properties of aluminum castings are considerably lower than those of the wrought forms. With one or two exceptions, tensile strengths in the heat-treated condition do not exceed about 50,000 psi. Ductility and hardness are also lower.

Aluminum-Silicon Alloys. These are the most widely used aluminum-casting alloys, primarily because of their excellent castability. They find considerable application in marine equipment and hardware because of high resistance to saltwater and saline atmospheres. They are also used for decorative parts because of their resistance to natural environments and ability to reproduce detail.

Aluminum-Copper-Silicon Alloys. Aluminum-copper compositions, the earliest aluminum-casting alloys, have been largely replaced by those containing additions of silicon. Copper increases strength, hardness, and machinability, and the silicon provides excellent casting properties. These alloys are especially suited to the production of castings of intricate design with large differences in section thickness or requiring pressure tightness.

Table 7-5 *Properties of Typical Aluminum Casting Alloys*

Type & Conditon	Tensile Strength, 1,000 psi	Elongation, % in 2 in.	Hardness, Bhn	Electrical Resistivity, Microhm-cm
443 (Si-Al)				
As-cast	20	9	42	4.66
A413 (Si-Al)				
As-cast	35	3.5	80	5.56
A380 (Si-Cu-Al)				
As-cast	47	4	80	7.50
319 (Cu-Si-Al)				
As-cast	30	2.2	75	6.40
Solution treated & aged	38	2.5	88	6.40
355 (Cu-Si-Mg-Al)				
Solution treated & aged	39	3.5	85	4.80

Aluminum-Magnesium-Silicon Alloys. Aluminum-magnesium alloys are the most corrosion resistant of the casting alloys. Unfortunately, they are difficult to cast, and controlled melting and pouring practices are required. Castability is improved by the addition of silicon. Heat treatment produces mechanical properties that make them attractive for such applications as automotive and aircraft parts.

Other Alloys. Aluminum-zinc-magnesium alloys age at room temperature to provide relatively high tensile strength. They have good machinability and corrosion resistance, but are not recommended for elevated temperatures.

Aluminum-tin alloys, developed primarily as bearing alloys, have high load-carrying capacity and fatigue strength. Cast in sand or permanent molds, they are used for connecting rods and crankcase bearings.

7-6 *Aluminum Powder*

Roughly a half-dozen different aluminum-alloy powders are being used in the production of powder metallurgy (P/M) parts. The major alloying elements include copper, up to 4 percent; magnesium, up to 1 percent; and zinc and silicon, from 0.10 to 1 percent. One alloy contains 5.6 percent zinc.

Aluminum P/M parts offer natural corrosion resistance, light weight, and good electrical and thermal conductivity. Strengths range from about 15,000 to 50,000 psi depending on composition, density, and heat treatment. In general, average fatigue limits are about half those of the wrought alloys. This is directly related to the lower density of P/M parts. Corrosion resistance, however, is not markedly affected by the porosity. Still relatively new in production use, the largest use of aluminum powder is for oil-impregnated sleeve and spherical bearings.

Magnesium

Magnesium, a silvery white metal with a specific gravity of 1.74, is the lightest of the commercial metals. It was relatively unknown as an engineering material until World War II, when its development received a tremendous push because of military applications.

7-7 *Major Characteristics and Properties*

Magnesium is not used in its pure state for structural applications, but is alloyed with such metals as aluminum, zinc, rare earths, thorium, zirconium, and manganese. Although magnesium alloys are relatively low in strength, their low density gives a strength-to-weight ratio competitive with high-strength aluminum alloys. Magnesium is the easiest of all structural materials to machine. Other notable characteristics are relatively high electrical and thermal conductivity, and very high damping capacity. It is nonferromagnetic. Corrosion resistance is considered fairly good in a wide range of environments and chemical media. However, in marine atmospheres and salt water, magnesium quickly corrodes, and, since it is highly anodic, precautions must be taken against galvanic corrosion. Impurities, such as nickel and copper, cause rapid intergranular corrosion.

Under certain conditions, flammability can be a problem. In powder or finely divided form, magnesium reacts violently with oxygen and ignites at 900°F. Some alloys may ignite at temperatures as low as 800°F. Therefore, during machining and processing, where small chips or powder are generated, precautionary measures and special coolants and lubricants must be used. Casting and hot-processing operations also require special care.

Magnesium has a close-packed hexagonal crystal structure. Consequently, it has relatively poor room-temperature ductility as well as limited cold workability. However, above 400°F, formability improves significantly so that greater deformations are possible at elevated tem-

Fig. 7-3 Magnesium is used in the printing trade for mounting printing plates on rotary presses. (Brooks & Perkins, Inc.)

peratures than in a number of other metals at room temperature. Thus magnesium alloys are hot formed by virtually all methods used on other metals, except wire drawing. Magnesium alloys are available as plates, sheets, bars, extrusions, forgings, and drawn parts. Magnesium alloys also are readily cast by sand, permanent-mold, and die-casting methods.

Magnesium alloys have melting ranges from around 900 to about 1200°F, and are not normally used at temperatures over 250°F, except for several alloys that are useful somewhat above 500°F. Room-temperature mechanical properties are relatively low compared to aluminum, but on a property-to-weight ratio basis are quite comparable. They have good fatigue properties and perform well in parts involving a large number of cycles at relatively low stress levels.

An advantage of magnesium alloys is their excellent dimensional stability. Unlike many materials, including some metals, internal stresses and growth or shrinkage in magnesium parts—even after years of service—is negligible. However, magnesium's high coefficient of expansion [16×10^{-6} in./in./°F], which is among the highest of the metals and over twice that of many steels, counterbalances this advantage.

Although not as highly used as aluminum, magnesium alloys are used in a wide variety of applications. In the aerospace and defense fields, they are employed in parts in aircraft, missiles, ordnance vehicles, electronic and instrument cases, and satellite components. In the materials handling field, they are used for such items as hand trucks, conveyor parts, and grain shovels. Other industrial uses include storage tanks, ladders, portable tools, and moving parts in textile and printing equipment. And, in the consumer field, magnesium is used in appliances, furniture, luggage, and lawn mowers.

7-8 *Wrought Alloys*

The ASTM has developed a standard designation system for magnesium alloys. The first two letters indicate the major alloying elements followed by numbers representing the percentage of the two elements present.

There are three principal wrought-magnesium alloy groups. Many of the alloys are in the aluminum-zinc group (AZ). Aluminum content runs up to 10 percent, and zinc not more than 1.5 percent. These alloys can be precipitation hardened and are generally the strongest of the alloys. The most commonly used alloy in this group is AZ31B. The other two groups are the zinc-zirconium alloys (ZK) and the thorium alloys, which can also contain zirconium (HK) or manganese (HM). Be-

Table 7-6 *Properties of Typical Wrought-Magnesium Alloys*

Type & Condition	Tensile Strength, 1,000 psi	Elongation, % in 2 in.	Modulus of Elasticity, 10^6 psi	Hardness, Bhn	Electrical Resistivity, Microhm-cm
AZ31B Annealed, extruded	37	14	6.5	47	9.2
ZK60A T-5, Extruded	52	12	6.5	82	6.0
HK31A H-24	37	8	6.5	—	6.1
HM31A T-5	42	10	6.5	—	6.6
LA141A T-7	19	10	6.0	—	15.2

sides these three, a relatively new group of magnesium alloys, containing from 7 to over 12 percent lithium (LA), has been developed. Since lithium is the lightest of all metals, these alloys are even lighter than magnesium metal and, because they have a body-centered cubic structure, they are more ductile. Some of the newest of these alloys show ductility at temperatures as low as −450°F.

7-9 *Cast Alloys*

Magnesium alloys have excellent castability and are extensively used in the form of sand, permanent-mold, and die castings. Types and compositions of the casting alloys roughly parallel those of the wrought alloys already discussed. In addition, there are several alloys in which rare-earth metals are present. Their useful temperature range falls roughly between 350 and 500°F. Thorium-zirconium alloys with rare-earth additions can be used in thermal environments up to about 650°F.

With sand castings, sections as thin as 0.1 in. are feasible, while in die castings, wall thicknesses of 0.050 in. and less are possible. Because of rapid solidification in die castings, their mechanical properties are generally superior to those obtained in other types of castings.

Titanium

Titanium, one of the newer metals, was commercially available for the first time in 1952. That year only 1,000 tons were produced. In

1970, total production was around 20,000 tons, and it is expected to double by 1980. Titanium is the fourth most abundant metallic element in the earth's crust. Unfortunately, winning the metal from its ore is complicated and costly. Because of its affinity for oxygen, nitrogen, and hydrogen, titanium must be produced in a vacuum or in an inert-gas atmosphere. Originally, titanium cost around $15/lb. By the early 1970s, the price dropped to between $4 and $5/lb., depending on the alloy.

7-10 Major Characteristics and Properties

Titanium is one of the few allotropic metals (steel is another); that is, it can exist in two different crystallographic forms. At room temperature, it has a close-packed hexagonal structure, designated as the alpha phase. At around 1625°F, the alpha phase transforms to a body-centered cubic structure, known as the beta phase, which is stable up to titanium's melting point of about 3050°F.

Alloying elements promote formation of one or the other of the two phases. Aluminum, for example, stabilizes the alpha phase—that is, it raises the alpha to the beta transformation temperature. Other alpha stabilizers are carbon, oxygen, and nitrogen. Beta stabilizers such as copper, chromium, iron, molybdenum, and vanadium lower the transformation temperature, therefore allowing the beta phase to remain stable at lower temperatures, and even at room temperature.

Titanium's mechanical properties are closely related to these allotropic phases. For example, the beta phase is much stronger, but more brittle than the alpha phase. Titanium alloys therefore can be usefully classified into three groups on the basis of allotropic phases: the alpha, beta, and alpha-beta alloys.

Titanium and its alloys have attractive engineering properties. They are about 40 percent lighter than steel and 60 percent heavier than aluminum. The combination of moderate weight and high strengths, up to 200,000 psi, gives titanium alloys the highest strength-to-weight ratio of any structural metal—roughly 30 percent greater than aluminum and steel. Furthermore, this exceptional strength-to-weight ratio is maintained from −420 up to 1000°F.

A second outstanding property of titanium materials is corrosion resistance. The presence of a thin, tough oxide surface film provides excellent resistance to atmospheric and sea environments as well as a wide range of chemicals, including chlorine and organics containing chlorides. Being near the cathodic end of the galvanic series, titanium performs the function of a noble metal.

Other notable properties are a higher melting point than iron, low thermal conductivity, low coefficient of expansion, and high electrical resistivity.

Fabrication is relatively difficult because of titanium's susceptibility to hydrogen, oxygen, and nitrogen impurities, which cause embrittlement. Therefore elevated-temperature processing, including welding, must be performed under special conditions that avoid diffusion of gases into the metal. Thanks to extensive research and development supported by the federal government for a number of years after World War II, effective processing methods have been developed that minimize the fabricating problems.

Commercially pure titanium and many of the titanium alloys are now available in most common wrought mill forms, such as plate, sheet, tubing, wire, extrusion, and forging. Castings can also be produced in titanium and some of the alloys for surgical implants, marine hardware, and chemical equipment such as compressors and valve bodies. Two casting processes—investment casting and graphite-mold (rammed graphite) casting—are used. Because of titanium's highly reactive nature in the presence of such gases as oxygen, the casting must be done in a vacuum furnace. Although this limits the size of parts, titanium castings up to at least 60 in. in diameter have been produced.

7-11 Commercially Pure Titanium

There are about a half-dozen grades of commercially pure titanium, which have titanium contents of from 98.9 to 99.5 percent. Because the small amounts of impurities significantly affect mechanical properties, they can be considered as alloys of the alpha type. Although not nearly as strong as the more highly alloyed types, commercially pure titaniums have a broad range of strengths—from about 40,000 to nearly 100,000 psi. A special corrosion-resistant grade, with 0.15 to 0.20 percent palladium, has improved resistance to mildly reducing media such as dilute hydrochloric and sulfuric acids.

7-12 Titanium Alloys

Alpha Alloys. These alloys contain such alloying elements as aluminum, tin, columbium, zirconium, vanadium, and molybdenum in amounts varying from about 1 to 10 percent. They are nonheat-treatable, having good stability up to 1000°F and down as low as −420°F. They have a good combination of weldability, strength, and toughness.

The 5 percent aluminum and 2.5 percent tin alloy, perhaps the most widely used alpha alloy, has been employed in numerous space and aircraft applications. It has a strength at room temperature of 120,000

Table 7-7 *Properties of Typical Wrought-Titanium Alloys*

Type & Condition	Tensile Strength, 1,000 psi	Elongation, % in 2 in.	Modulus of Elasticity, 10^6 psi	Hardness, Rockwell	Impact Strength, Charpy V, ft-lb
Unalloyed					
98.9%, annealed	100	17	15	100R(B)	11
99.5%, annealed	40	30	15	70R(B)	40
Ti-5Al-2.5Sn					
Annealed	125	18	16	36R(C)	19
Ti-6Al-4V					
Annealed	138	12	16.5	36R(C)	20
Aged	170	8	—	—	10
Ti-1Al-8V-5Fe					
Aged	220	8	16.5	—	—

psi, acceptable ductility, and is useful at temperatures up to 800 and 1000°F. In addition, it has good oxidation resistance, and good weldability and formability.

Alpha-Beta Alloys. This is the largest and most widely used group of titanium alloys. Because these alloys are a two-phase combination of alpha and beta alloys, their behavior falls in a range between the two single-phase alloys. They are heat-treatable, useful up to 800°F, more formable than alpha alloys, but less tough and more difficult to weld.

The most popular alloy in this group is the 6 percent aluminum and 4 percent vanadium grade. Its volume of use equals that of all other titanium materials combined. It can be heat-treated up to 170,000 psi, has good impact and fatigue strength, and unlike other alpha-beta alloys, is weldable. The 6 percent aluminum, 6 percent vanadium, and 2 percent tin alloy is heat-treatable to higher strength than any other alpha-beta alloy (190,000 psi).

Beta Alloys. Although the beta alloys have exceptionally high strengths—over 200,000 psi—their lack of toughness and low fatigue strength limits their use. They retain an unusually high percentage of strength up to 600°F, but cannot be used at much higher temperatures, and they become brittle at temperatures below −100°F.

Beryllium

7-13 *Major Characteristics and Properties*
Beryllium is a recently emerged structural metal that has exceptional properties and serious deficiencies. It is one-third lighter than aluminum, its stiffness-to-weight ratio is roughly six times greater than that of the ultrahigh-strength steels, and its melting point approaches that of steel. It also has excellent thermal conductivity, is nonmagnetic, and a good conductor of electricity.

However, beryllium's hexagonal crystal structure combined with a high sensitivity to impurities results in an almost total lack of room-temperature ductility. This inherent brittleness seriously limits its otherwise outstanding structural service performance, as well as its fabrication. Ductility improves considerably between 390 and 750°F, but beryllium becomes brittle again above 930°F. Another limitation, its toxicity if inhaled or ingested, means that special precautionary measures are required in processing and handling, particularly when the metal is in powder or vapor form. And, finally, the cost of beryllium metal is high—$60 to $70/lb (1973).

7-14 *Shapes and Applications*
Most beryllium shapes and parts are produced from hot-pressed-powder block forms. Sheet, extrusion, rod, and bar are available. Beryllium foil, down to 0.0005 in. thick, is used in vacuum tubes. Wire is produced in diameters from 0.005 in. and larger down to as fine as 0.001 in. Beryllium mill products are used in specialized applications, such as nuclear systems, reentry vehicles, aircraft brakes, and satellite parts.

Because of the embrittlement problem, beryllium is most useful as an alloying element and as a composite constituent. In composites, it is used in the form of wire or particles in matrices of titanium or aluminum. For example, beryllium-wire-reinforced aluminum sheet used for pressure-bottle applications has a tensile strength of 85,000 psi, a modulus of elasticity of 25 million psi, and 5 percent elongation. Beryllium-reinforced titanium-alloy composites have strengths of 140,000 psi and a modulus of elasticity of 27 million psi.

Perhaps the best-known beryllium-aluminum composite is Lockalloy, in which beryllium particles are embedded in a ductile aluminum matrix. Developed specifically for aerospace applications, one such composite with 33 percent aluminum has a tensile strength of 61,000 psi and a modulus of elasticity of 29 million psi in the extruded and annealed condition.

Copper

Copper was probably the first metal discovered. It has been used over the centuries as both an art and an engineering material. Its outstanding engineering property is its electrical conductivity, which is second only to silver. In addition, copper and its alloys are noted for excellent corrosion resistance, ease of forming and finishing, and attractive color. Other distinguishing characteristics are their nonmagnetic properties and high thermal conductivity.

7-15 Major Characteristics and Properties

Copper has a face-centered cubic crystal structure, which accounts for its combination of moderate strength and good ductility. Alloying is required to increase strength, hardness, and most other mechanical properties. Hardness and strength are increased, but ductility is decreased by cold-working operations.

Copper-alloy mill products range widely in strength, from a low of 34,000 to a high of 200,000 psi for hard-temper beryllium-copper. Ductility and hardness vary just as widely, depending on the alloy and the temper. For example, commercial bronzes show elongations of as high as 45 percent compared to only about 3 percent in some yellow brasses. The rigidity or stiffness of copper materials (from 14×10^6 to 22×10^6 psi) falls roughly halfway between aluminum and steel. Fatigue strength also varies widely from excellent to poor. Notable for fatigue strength are some of the alloys used for springs, bellows, and diaphragms.

With melting ranges from about 1650 to 2250°F, copper alloys are not noted for their high-temperature mechanical properties. However, an exceptionally high thermal conductivity in some alloys, and particularly in the commercial coppers, suits them for heat-sink applications. At low temperatures, all copper alloys exhibit improved mechanical properties.

Copper and its alloys form adherent surface films that resist many corrosive environments. All alloys have good resistance to fresh water, steam, alkalies, and organic acids. Some alloys, however, are highly susceptible to dezincification, which results in a spongy, weak material. Copper and many of the alloys change color rapidly in air. Usually, the alloy's surface film darkens. In the case of copper, the color changes to a blue green, commonly referred to as patina.

7-16 Types and Forms

Besides some two dozen standard grades of commercial coppers, there are hundreds of wrought- and cast-copper-alloy compositions listed by the Copper Development Association (CDA). About 50 are frequently used. All copper alloys can be classified into five major groups by chemical composition. These are:

1. Brasses, which contain zinc as the principal alloying element.
2. Bronzes, which originally were confined to copper-tin alloys, but today include a variety of other alloys in which tin is only a minor element or is not present.
3. Cupronickels, in which nickel is the principal alloying element.
4. Nickel silvers, which are copper-nickel-zinc alloys. They actually contain no silver.
5. High-copper alloys, which contain less than 99.3 but more than 96 percent copper.

Copper alloys are produced in cast, wrought, and powder forms. Wrought shapes begin as billets, slabs, and wire bars, which are fabricated into usable forms such as sheet, strip, plate, rod, bar, tubing, extruded shapes, and forging by hot or cold forming.

The final or finished condition resulting from these processing operations is referred to as temper. Annealed (soft) tempers are defined by nominal grain size in millimeters. The larger the grain size, the softer the material. Rolled or drawn tempers of flat shapes—sheet and strip—are based on the percentage of reduction by rolling. The standard tempers are designated eighth-hard, quarter-hard, half-hard, three-quarter-hard, hard, extra-hard, spring, and extra-spring.

For rod (rounds and hexagons) and bar shapes (rectangles and squares), the tempers are quarter-hard, half-hard, hard, and extra-hard. For tubes, three tempers are recognized: drawn general purpose, hard, and light drawn.

Types of castings made with copper materials include sand, die, permanent-mold, plaster-mold, centrifugal, and investment castings. Where regularity of shape design permits, a number of foundry alloys are available in continuous cast form, either as solid or hollow shapes.

Cast alloys, approximately comparable in composition to the wrought grades, are produced in all five major copper-alloy groups. Lead is a constituent in many of the cast alloys for improved machinability; additives are used to improve strength and castability. Properties of cast-copper alloys are generally somewhat lower than those of annealed wrought alloys, and the range in grain size and in cold-worked properties obtainable with wrought products is not available in castings.

7-17 *Coppers*
The differences between the several types of copper metal available are related to the production method used or the presence of minor

alloy additions of silver, phosphorus, lead, or cadmium. Over 85 percent of all copper metal produced is electrolytically refined. The remainder, known as fire-refined copper, is refined by furnace processing.

There are three major types of commercial copper: tough-pitch, oxygen-free, and phosphorus-deoxidized.

Tough-pitch Copper. This is the standard copper of industry for electrical and many other uses. It contains about 0.04 percent oxygen. Its electrical conductivity is equivalent to 100 percent IACS conductivity, and therefore it is the standard with which other metals are compared.

Tough-pitch copper, while ductile enough for most forming operations, lacks ductility for severe drawing and edgewise bending. At temperatures above about 750°F in some atmospheres, in particular hydrogen, it is subject to hydrogen embrittlement.

Besides electrical applications, other common uses are building products, such as roofing and flashing; automobile radiators; cooking utensils; chemical-processing equipment; and printing rolls.

Phosphorus-Deoxidized Copper. This type is second to tough-pitch copper in volume of use largely due to its use for pipes and tubes for domestic water service and refrigeration equipment. Because of its phosphorus content, conductivity is only 85 percent IACS. However, it has excellent hot and cold workability and is not susceptible to hydrogen embrittlement. Typical applications are plumbing hardware; piping and tubing for water, gas, oil, and steam; heat exchangers; brewery and dairy tubs; and kettles and tubes.

Oxygen-free High-Conductivity Copper. This type is highest in purity (99.98 percent) and provides optimum electrical conductivity (101 percent IACS). It is the most expensive of the three major coppers and is limited largely to use in the electrical and electronic industries. A number of other copper grades are available. Among them are:

Silver-bearing Copper. The presence of very small amounts of silver in copper raises the softening temperature, thus improving elevated-temperature and creep strength, and permitting soldering operations without reduction of strength and hardness. Lake copper is a general term for silver-bearing copper having varying but controlled amounts of silver up to about 30 oz/ton. It's named for the native silver-bearing copper deposits in the Lake Michigan area.

Table 7-8 *Properties of Some Wrought Coppers*

Type & Condition	Tensile Strength, 1,000 psi	Elongation, % in 2 in.	Modulus of Elasticity, 10^6 psi	Hardness, Rockwell F	Electrical Resistivity, Microhm-cm
Oxygen-free					
Annealed	32–35	45–55	17	40–45	1.71
Hard	45–55	6–20	17	85–95	—
Electrolytic Tough-Pitch					
Annealed	32–35	45–55	17	40–45	1.71
Hard	50–55	6–20	17	85–95	—
Phosphorous-Deoxidized					
Annealed	32–35	45	17	40–45	2.03
Hard	55	8	17	95	—
Silver-bearing					
Annealed	40	34–40	17	70	1.92
Hard	62	—	17	97	—

Arsenical Tough-pitch Copper. This is a modified copper that contains arsenic, which raises the softening temperature and improves corrosion resistance in some environments. However, electrical conductivity is reduced to about 45 percent IACS.

Sulfur-bearing and Tellurium-bearing Coppers. These are coppers to which sulfur and tellurium have been added to increase machinability without appreciably lowering conductivity.

7-18 Brasses
Brasses are basically copper-zinc alloys whose zinc content ranges up to 40 percent. If the copper crystal structure is face-centered cubic, there will be up to 36 percent of zinc present. This solid solution, known as the alpha phase, has good mechanical properties, combining strength with ductility. Corrosion resistance is very good, but electrical conductivity is considerably lower than in copper. When above 30 to 36 percent of the alloy is zinc, a body-centered-cubic crystal structure is formed, known as the beta phase. This phase is relatively brittle and high in hardness compared to the alpha phase. However, ductility increases at elevated temperatures, thus providing good hot-working properties.

The mechanical properties of brasses vary widely. Strength and hardness depend on alloying and/or cold work. Tensile strengths of annealed grades are as low as 30,000 psi, although some hard tempers approach 90,000 psi. Although brasses are generally high in corrosion resistance, two special problems must be noted. With alloys containing a high percentage of zinc, dezincification can occur. The corrosion product is porous and weak. To prevent dezincification, special inhibitors—antimony, phosphorus, or arsenic—in amounts of 0.02 to 0.05 percent can be added to the alloy. The other problem is stress corrosion, or season cracking, which occurs when moisture condenses on the metal and accelerates corrosion.

Because brass alloys have relatively low melting points, they are suitable for die casting. Since the melting point decreases with zinc content, the high-zinc alloys are best for die casting.

Brasses are classified in various ways. Some are misleadingly named bronzes, such as commercial bronze and jewelry bronze. Here the brasses are divided into four broad groups:

Straight Brasses. These binary copper-zinc alloys are the most widely used of all copper-base alloys. Zinc content ranges from as low as 5 up to about 40 percent. Some of the common names of these alloys are: gilding metal (5 percent zinc), commercial bronze (10 percent zinc), jewelry bronze (12.5 percent zinc), red brass (15 percent zinc), yellow brass (35 percent zinc), and muntz metal (40 percent zinc). As the zinc content increases in these alloys, the melting point, density, electrical and thermal conductivity, and modulus of elasticity decrease while the coefficients of expansion, strength, and hardness increase. Work hardening also increases with zinc content. These brasses have a pleasing color, ranging from the red of copper in the low-zinc alloys through bronze and gold colors to the yellow of high-zinc brasses. The color of jewelry bronze closely matches that of 14-karat gold, and this alloy and other low brasses are used in inexpensive jewelry.

The low-zinc brasses have good corrosion resistance along with moderate strength and good forming properties. Red brass, with its exceptionally high corrosion resistance, is widely used for condenser tubing. The high brasses (cartridge and yellow brass) have excellent ductility and high strength and are widely used for engineering and decorative parts fabricated by drawing, stamping, cold heading, spinning, and etching. Muntz metal, primarily a hot-working alloy, is used where cold working is not required.

Leaded Brasses. These alloys have essentially the same range of zinc content as the straight brasses. Lead is present, ranging from less than

Table 7-9 *Properties of Typical Brasses*

Type & Condition	Tensile Strength, 1,000 psi	Elongation, % in 2 in.	Modulus of Elasticity, 10^6 psi	Hardness, Rockwell	Electrical Resistivity, Microhm-cm
Commercial bronze (220)					
Annealed	37	45	17	53(F)	3.92
Hard	61	5	17	70(B)	—
Red brass (230)					
Annealed	39	48	17	56(F)	4.66
Hard	70	5	17	77(B)	—
Yellow brass (268, 270)					
Annealed	46	65	15	58(F)	6.39
Hard	74	8	15	80(B)	—
Muntz metal (280)					
Annealed	54	45	15	80(F)	6.16
Half-hard	70	13	15	—	—
Medium-leaded brass (340)					
Annealed	51	53	15	72(F)	6.63
Hard	74	7	15	80(B)	—
Tin brass (405)					
Annealed	42	47	18	47(B)	4.3
Hard	62	10	18	59(B)	—
Spring	72	4.5	18	68(B)	—
Gilding brass (210)					
Annealed	34	45	17	46(F)	3.08
Hard	56	5	17	64(B)	—
Low brass (240)					
Annealed	43	52	16	57(F)	5.39
Hard	75	6	16	82(B)	—
Naval brass (464)					
Annealed	57	47	15	55(B)	6.63
Hard	75	20	15	82(B)	—

1 to 3.25 percent, to improve machinability and related operations. Lead also improves antifriction and bearing properties. Common leaded brasses include leaded commercial bronze (0.5 percent lead), medium-leaded brass (1 percent lead), high-leaded brass (2 percent lead), free-cutting brass (3.25 percent lead), and hardware bronze (1.75 percent lead). Free-cutting brass provides optimum machinability and is ideally suited for screw machine parts.

Tin Brasses. Adding tin to copper-zinc alloys improves corrosion resistance. Pleasing colors are also obtained when tin is added to the low brasses. Tin brasses in sheet and strip form, with 80 percent or more copper, are used widely as low-cost spring materials. Admiralty brass is a standard alloy for heat-exchanger and condenser tubing. Naval brass and manganese bronze are widely used for products requiring good corrosion resistance and high strength, particularly in marine equipment.

Silicon, Aluminum, and Manganese Brasses. Silicon red brass has the corrosion resistance of the low brasses, with higher electrical resistance than that of most brasses. It is especially suited for parts that must be resistance welded. Manganese brass serves the same purposes as silicon brass. Aluminum brass, a moderate cost alloy, is made chiefly in tube form for condensers and heat-exchanges subject to the corrosive and erosive action of high-velocity seawater.

7-19 Bronzes
The term bronze is generally applied to any copper alloy that has as the principal alloying element a metal other than zinc or nickel. Originally the term was used to identify those alloys that had tin as the only, or principal, alloying element.

The copper-tin bronzes are a rather complicated alloy system. The alloys with up to about 10 percent tin have a single-phase structure. Above this percentage, a second phase, which is extremely brittle, can occur, making plastic deformation impossible. Thus high-tin bronzes are used only in cast form. Tin oxide also forms in the grain boundaries causing decreased ductility, hot workability, and castability. Additions of small amounts of phosphorus, in production of phosphor bronzes, eliminate the oxide and add strength.

Because tin additions increase strength to a greater extent than zinc, the bronzes as a group have higher strength—from around 60,000 to 105,000 psi in the cold-worked high-tin alloys (phosphor bronze). In addition, fatigue strength is high.

Aluminum Bronzes. This group is made up of alpha-aluminum bronzes (less than about 8 percent aluminum) and alpha-beta alloys (8 to 12 percent aluminum plus other elements such as iron, silicon, nickel, and manganese). Because of the considerable strengthening effect of aluminum, in the hard condition these bronzes are among the highest strength copper alloys. Tensile strength approaches 100,000 psi. Such strengths plus outstanding corrosion resistance

Table 7-10 Properties of Typical Bronzes

Type & Condition	Tensile Strength, 1,000 psi	Elongation, % in 2 in.	Modulus of Elasticity, 10^6 psi	Hardness, Rockwell B	Electrical Resistivity, Microhm-cm
Phosphor bronze E (505)					
Annealed	40	48	17	—	—
Hard	75	8	17	75	3.59
Phosphor bronze D (524)					
Annealed	66	68	16	55	15.7
Hard	100	13	16	97	—
Aluminum bronze D (614)					
Annealed	72	35	17	—	12.0
Hard	82	30	17	90	—
Aluminum-silicon bronze (638)					
Annealed	82	36	16.7	86	17.4
Hard	120	7	16.7	98	—
Low-silicon bronze (651)					
Annealed	40	50	17	55	14.5
Hard	70	15	17	80	—
High-silicon bronze (655)					
Annealed	57	58	15	55	25
Hard	92	22	15	90	—
Manganese bronze (675)					
Annealed	65	33	15	65	7.2
Hard	82	25	15	—	—

make them excellent structural materials. They are also used in wear-resistance applications and for nonsparking tools.

Phosphor Bronzes. As a group, these alloys have a tin content of 1.25 to 10 percent. They have excellent mechanical and cold-working properties and a low coefficient of friction, making them suitable for springs, diaphragms, bearing plates, and fasteners. Their corrosion resistance is also excellent. In some environments, such as salt water, they are superior to copper. Leaded phosphor bronzes are available with improved machinability.

Silicon Bronzes. As a group, these alloys are similar to aluminum bronzes. Silicon content is usually between 1 and 4 percent. In some, zinc or manganese is also present. Besides raising strength, the presence of silicon sharply increases electrical resistivity. Aluminum-silicon bronze, an important modification, has exceptional strength and corrosion resistance and is particularly suited to hot working.

7-20 Other Alloys and Powders

Nickel-Silver Alloys. These alloys, which contain no silver as the name implies, actually are nickel brasses. They contain varying amounts of zinc (up to 40 percent) and nickel (up to 18 percent). Color ranges from ivory white in the lower-nickel alloys to silver white in the higher-nickel alloys. Their silver color plus excellent corrosion resistance and cold formability make them widely used for silver-plated hollow and flatware and for parts in optical goods, cameras, costume jewelry, and instruments. A low 18 percent nickel alloy is used for springs and other applications where high fatigue strength is needed.

Cupronickel Alloys. These are binary alloys of copper and nickel, with nickel ranging from 2.5 to 45 percent. The 10 and 30 percent nickel alloys are most commonly used. Corrosion resistance increases with nickel content. The cupronickel alloys are markedly superior in resistance to the corrosive and erosive effects of seawater. They are moderately hard, but quite tough and ductile.

The cupronickels have the unusual characteristic of electrical resistivity increasing with increasing nickel content while the temperature coefficient of resistance decreases. In the alloy with 45 percent nickel, named constantan, resistivity reaches a maximum value while the temperature coefficient drops to a minimum of nearly zero. Because of this, constantan is widely used as a thermocouple element and in other precision resistors.

High-Copper Alloys. This is a group of copper alloys with small amounts of either beryllium, chromium, or zirconium, plus other elements, added to produce alloys that are strengthened by precipitation-hardening treatment.

Beryllium-copper alloys (2 percent beryllium), the best known of the group, achieve tensile strengths as high as 200,000 psi. This high

Table 7-11 *Properties of Nickel-Silver, Cupronickel, and Beryllium-Copper Alloys*

Type & Condition	Tensile Strength, 1,000 psi	Elongation, % in 2 in.	Modulus of Elasticity, 10^6 psi	Hardness, Rockwell B	Electrical Resistivity, Microhm-cm
Nickel silver 65-18 (752)					
Annealed	56	35	18	40	29
Hard	85	3	18	87	—
Cupronickel 30 (715)					
Annealed	55	45	22	45	37.5
Hard	75	15	22	85	—
Beryllium copper (824)					
Annealed	70	40	18.5	—	—
Hard	115	2	18.5	—	—
Heat-treated	120	2	18.5	38 R (C)	9.6

strength combined with the highest modulus of elasticity of any copper alloy makes them an excellent spring material. Beryllium coppers also have fairly high electrical conductivity, their corrosion resistance is comparable to that of copper, and they have outstanding fatigue strength and wear resistance.

Copper-Alloy Powders. There is a rather large range of compositions of copper-alloy powders available, including brass, bronze, and copper-nickel powders.

Brass powders are the most widely used for P/M structural parts. Conventional grades are available with zinc content from around 10 to 30 percent. Sintered brass parts have tensile strengths up to around 35,000 and 40,000 psi, and elongations of from 15 to around 40 percent depending on composition, design, and processing. In machinability, they are comparable to cast- and wrought-brass stock of the same composition. Brass P/M parts are well suited for applications requiring good corrosion resistance, and where free-machining properties are desirable.

Copper-nickel, or nickel-silver, powders contain 10 or 18 percent nickel. Their mechanical properties are rather similar to the brasses, with slightly higher hardness and corrosion resistance. Because they are easily polished, they are often used in decorative applications.

Copper and bronze powders are used for filters, bearings, and electrical and friction products. However, bronze powders are relatively hard to press to densities that give satisfactory strength for structural parts. The most commonly used bronze powder contains 10 percent tin. The strength properties are considerably lower than iron-base and brass powders, being usually below 20,000 psi.

Zinc

Zinc is a relatively inexpensive metal in large-volume use. Among nonferrous metals, its total consumption is topped only by that of aluminum and copper. Its largest single use is as a zinc coating (galvanizing) on iron and steel. Its major structural use is in cast form. The low cost and low melting point (73°F) of zinc alloys make them especially suitable for die-cast parts widely used in automotive, appliance, and other consumer products. These alloys provide moderate strength and toughness plus excellent corrosion resistance in a number of environments.

Zinc has long been used as an anode for cathodic protection of steel piers, ships, pipelines, and many other steel structures. Since it is more active in a galvanic couple than iron or steel, it functions as the sacrificial anode and provides electrochemical protection against rust.

Zinc has a close-packed hexagonal crystal structure. Its mechanical properties are highly dependent on the degree of purity. Small amounts of impurities increase strength but decrease ductility. Also, impurities cause severe intergranular corrosion and dimensional instability. Therefore the impurities lead, cadmium, tin, and iron in high-quality zinc die-casting alloys should not exceed 0.01 percent.

7-21 *Casting Alloys*
The two principal die-casting alloys are ASTM AG40A (XXIII) and AC41A (XXV). Both contain a fraction of a percent of magnesium and 3.5 to 4.3 percent aluminum. AC41A also contains 0.75 to 1.25 percent copper.

The properties of both alloys are about the same. Tensile strength ranges between 40,000 and 47,000 psi. Impact strength exceeds that of most cast irons and aluminum castings. The alloys are close to the weight of steel and are relatively soft. Their excellent castability allows production of smooth-surface castings in sections as thin as 0.015 in. Machinability is good, but zinc alloys are difficult to weld

Fig. 7-4 Zinc has been used for many years for die-cast auto carburetors. (Zinc Institute, Inc.)

and solder. Applications of zinc die castings range in size from watch-hairspring wedges to automobile panels 6 ft long and 40 lb in weight.

Zinc alloys for permanent-mold and sand castings are similar to those used for die castings. Zinc-alloy sand castings are widely used for forming, drawing, stretching, blanking, and trimming large shapes made from light-gage sheet aluminum, magnesium, steels, and other metals. Some of these cast dies weigh several tons.

Table 7-12 *Properties of Die-cast and Forged Zinc*

Type & Condition	Tensile Strength, 1,000 psi	Elongation, % in 2 in.	Hardness, Bhn	Impact Strength, Charpy, ft-lb
AG40A (XXIII)	41	10	82	43
AC41A (XXV)	47.6	7	91	48
Forged zinc	50–72	13–18	100–120	—

Because of an aging process that takes place in cast zinc alloys, a shrinkage of about 0.0008 in./in. occurs. About two-thirds of the shrinkage takes place within few weeks after casting. The remaining shrinkage occurs over a period of several years. In copper-containing alloys, shrinkage is followed by expansion, which, however, is not as large as the initial shrinkage. The initial shrinkage period can be shortened by a low temperature (a few hours at 160 to 210°F) anneal. Long-term aging also affects mechanical properties. There can be a 10 to 25 percent decrease in strength with an equivalent increase in ductility. Also, a certain amount of creep occurs in zinc alloys even at room temperature, which must be considered where strength is a major design parameter.

7-22 Wrought Zinc

Wrought zinc is produced as sheet, strip, plate, rod, and wire in a number of different grades. The softer types contain a small amount of lead. Those containing small amounts of titanium and copper have better properties, in particular, a lower coefficient of thermal expansion and high creep resistance.

Several zinc alloys with about 12 percent aluminum have tensile strengths up to 75,000 psi. The most recently developed alloys are "superplastic" alloys that have elongations of 98 percent in the annealed condition. These alloys contain about 22 percent aluminum. One grade can be annealed and air-cooled to a strength of 71,000 psi. Parts of these alloys have been produced by vacuum forming and by a compression molding technique, similar to forging but requiring lower pressures.

Some major applications of wrought-zinc alloys are dry-cell battery cans, engraving plates, flashing and weatherstripping, marine hardware, novelties, and electronic condenser cans.

Low-Melting Metals

Low-melting metals and alloys are arbitrarily defined as those having melting points or melting ranges below 700°F. The metals included in this group are tin, lead, bismuth, antimony, cadmium, and indium.

The characteristics these metals and their alloys share are relatively high density, low electrical conductivity, low hardness, low strength, and usually high ductility. Also, because these metals exhibit almost no elasticity, they have a long plastic-deformation range. This range, coupled with low melting points, causes them to have considerable creep at room temperature. Thus the nature and properties of low-melting alloys make them unsuitable for structural uses. However, their softness and ductility give most of these metals wide use as bearing materials. Their relatively high corrosion resistance makes them suitable for use in a variety of environments. Their low melting points are an advantage in soldering and fusible alloys.

7-23 *Tin*
Tin is a white or silvery metal with a density close to that of steel. Commercial grades of tin contain small amounts of bismuth, antimony, lead, and silver, amounting to a total of 0.2 percent. Commercially pure tin is rarely used alone except as a coating material. Tin with the addition of 0.4 percent copper is used as foil and for collapsible tubes. A small amount of cold work somewhat increases hardness, but most tin alloys soften spontaneously at room temperature. A reduction of more than about 20 percent by cold work softens rather than hardens the alloys.

Tin alloys include white metal, containing 8 percent antimony, and pewter, containing 7 percent antimony and 2 percent copper. Various casting alloys containing antimony, copper, and lead are used as die castings and bearings. Typical tin-base babbitt bearing materials contain around 10 percent antimony and 6 percent copper. They have excellent bearing properties, relatively low fatigue strength, and high resistance to most oils.

The major use of tin and its alloys is in corrosion-resistant products. More than 40 percent of the world's tin is used as a coating for steels and copper. Tinfoil is used for liners for bottle caps and electrical condensers. Heavy-walled tin pipe and tin-lined copper pipe is used in the food and beverage industries. And, tin wire is used for electrical fuses and for packing glands in pumps of food machinery.

7-24 *Lead*
Besides having a low melting point, lead is also one of the heavier metals, with a density about 50 percent greater than that of steel. It has low strength and high creep. It is inherently corrosion resistant—especially to most acids and most natural environments.

Two principal grades of commercial lead are referred to as common lead and chemical lead. Typical uses include chemical-apparatus batteries and cable sheathing. Lead is the principal material used for X-ray and gamma-radiation shielding. It is also a leading material for vibration and sound control. Combined with asbestos, it has been used for years in building foundations to reduce transmission of ground vibrations. Sheet lead and sheets of leaded plastic are now being used to control noise and sound in airplanes, office equipment, and similar products.

Lead alloys cover a wide range of compositions for solders, fusible alloys, and bearing metals. Because of lead's relatively low cost compared to tin, many commercial soldering alloys contain at least 50 percent lead. Lead-base babbitt bearing alloys contain 75 to 90 percent lead, plus antimony and tin. The structure of these babbitts consists of hard, brittle antimony or tin-antimony crystals in a tin-antimony-lead matrix. The metal used for printing type is a lead-tin-antimony (and sometimes a little copper) alloy. Newer developments include alloys made by powder metallurgy. Some of these show tensile and creep strengths several times higher than normally obtained with lead.

Because of its malleability, lead and its alloys are readily rolled to any thickness down to 0.0005 in. Tin-coated lead can be produced by simply rolling lead and tin together. Lead is also one of the easiest metals to cast to reproduce fine detail, the major reason for its use as type. Other forms of lead and its alloys are ⅛- to 12-in.-thick cladding, powder, shot, and fibers.

7-25 Bismuth

Bismuth is a brittle metal with a high metallic luster. It is easy to cast, but not readily formed by working. It is one of the few metals that increases in volume on solidification. It also shows the greatest Hall effect (i.e., increase in resistance under influence of a magnetic field).

The major engineering application is as the principal ingredient in fusible alloys. These alloys contain from 40 to 50 percent bismuth plus combinations of lead, tin, cadmium, antimony, and indium. With a few exceptions, they have lower melting temperatures (115 to 340°F) than any of the individual constituents. Like bismuth, most of these alloys expand in volume as they solidify. This property enables reproduction of fine details, which makes the alloys useful for molds, patterns, and mandrels. Because of low melting points, they are extensively used for triggering devices in fire alarm and sprinkler systems, for safety plugs in compressed-gas tanks, and for automatic shutoffs in heating systems.

7-26 Antimony, Cadmium, and Indium

These three low-melting metals are seldom used alone. Their chief use is as an ingredient in low-melting alloys, in coatings, and in other metal alloys.

Antimony, like bismuth, expands on solidification. It serves as a hardener in lead and bearing alloys and in type metal. These uses account for about 90 percent of antimony consumption. Other uses are in semiconductors and as an alloying element in bismuth-telluride thermoelectric materials.

Cadmium's major use is for protective coatings on metals, chiefly by electroplating. It is easy to deposit, has excellent ductility, and good resistance to salt water and alkalies. Cadmium's high thermal neutron-capture cross section makes it useful in the nuclear field to control the rate of fission.

Indium is a silver-white metal, similar in color to platinum. It is one of the softest metals and can be deformed almost indefinitely by compression. The three main uses are in semiconductor devices, bearing metals, and fusible alloys. It is also one of the major ingredients in a glass-sealing alloy (50 percent indium and 50 percent tin).

Nickel and Cobalt

Nickel is one of the most important of our engineering metals, and ranks around tenth in the world's consumption of metals. It not only serves as an important alloying element in ferrous metals but is the base metal for a number of nickel alloys that possess combinations of properties unavailable in other materials.

7-27 *Major Characteristics and Properties*
Nickel has a face-centered cubic crystal structure. Its color is silvery white with a yellowish cast. It is magnetic up to 680°F and melts at about 2650°F. With a specific gravity of 8.85, it is somewhat heavier than steel.

In general, nickel alloys are stronger, tougher, and harder than most other nonferrous alloys and many steels. Their stiffness is about the same as steel, ranging from 26 to 31 million psi. They are about comparable in cost to the stainless steels and are therefore more costly than aluminum alloys and most ferrous alloys. The alloys have important electrical and magnetic properties. Some alloys have outstanding strength, toughness, and ductility at cryogenic temperatures, while others, the nickel-base superalloys, are suitable for high-strength applications at temperatures in the 2000 to 2200°F range. A majority of the alloys also have moderate to high resistance to most of the normal and special corroding media found in industry as well as resistance to oxidation and scaling at elevated temperatures.

In general, nickel alloys are formed and fabricated like steel. They can be readily hot and cold worked, cut, machined, and joined by the usual procedures. A general precautionary measure for all nickel alloys is to avoid heating them in the presence of sulfur to avoid the formation of nickel sulfide in the microstructure, which adversely affects mechanical properties. Nickel alloys are available in all commercial mill forms such as plate, sheet, strip, tube, bar, rod, and wire; and are also produced as sand, centrifugal, and precision castings.

Nickel and its alloys are usually classified into nine groups, which are discussed below.

7-28 Extra-high Nickel Alloys

The metals in this group contain down to 90 percent nickel. The highest purity wrought nickel commercially available has a nominal nickel content of 99.97 percent. It is used for plating bars, fluorescent lamp parts, resistance thermometers, and other components where high purity is required. At least seven other grades of commercially pure nickel have a nickel content of 99.5 percent and small amounts of other elements to meet specific application requirements. For example, extra-high nickel alloys are available with 2 to 4.75 percent manganese for such components as spark plug electrodes and electron-tube grid wires. Other alloys with around 4.5 percent cobalt are also used in electronic components.

Age-hardenable extra-high nickel alloys are capable of being hardened to very high strength while retaining good corrosion resistance. They are used for springs, glass molds, thermostat contact arms, dies and molds for glass and plastics, and chemical equipment. These alloys are also used to produce high-strength, wear- and corrosion-resistant castings.

7-29 Nickel-Copper Alloys

There are some 25 alloys based on the mutual solubility of nickel and copper in all proportions. Known as Monel alloys, they are the most important of the nickel alloys. They contain around 30 percent copper and have a combination of high strength, corrosion resistance, and good formability that makes them suitable for a general range of applications: oil refining and chemical plant equipment, pumps, valves, marine hardware, heat exchangers, and architectural trim. Free-machining grades of Monel are also available.

The highest strength nickel-copper alloys are the age-hardenable grades. Except for additions of titanium and aluminum, their composition is similar to the standard Monel alloys. They are nonmagnetic, highly corrosion resistant, and equivalent in strength to many heat-treated steels (up to around 200,000 psi).

Table 7-13 *Properties of Typical Wrought- and Cast-Nickel Alloys*

Type & Condition	Tensile Strength, 1,000 psi	Elongation, % in 2 in.	Modulus of Elasticity, 10^6 psi	Hardness Bhn
High-purity nickel (270)	50–95	4–50	30	80–210
Monel (400)	70–100	22–60	26	110–241
Monel (K500)	90–190	13–45	26	140–346
Inconel (600)	90	47	31	120–170
Hastelloy B	134	51	29	—
Illium B	65	1	—	225

Several Monel alloys, which have copper contents of 40 to 44 percent, are produced for special applications, such as where excellent brazing characteristics are required, or for pickling tanks, for seawater corrosion resistance, or for hot-water tanks.

7-30 Nickel-Chromium Alloys

There are two principal classes in this group. One consists of electrical-resistance alloys. The other covers high-temperature alloys, some of which are regarded as nickel-base superalloys. The electrical-resistance types, which contain from 60 to 80 percent nickel plus chromium and, in some cases, iron, form the bulk of materials used for heater elements. In the heat-resistant group are the original Inconel alloys, the standard composition being 80 percent nickel, 14 percent chromium, and 6 percent iron. They combine the inherent corrosion resistance, strength, and toughness of nickel with the extra resistance to atmospheric and high-temperature oxidation provided by chromium. Other nickel-chromium alloys with additions of copper (3 to 7 percent) and molybdenum (5 to 7 percent) have special corrosion-resistance properties.

7-31 Nickel-Chromium-Cobalt Alloys

This group, composed almost entirely of the nickel-base superalloys, has exceptional elevated-temperature strength, stress-rupture life, thermal-shock resistance, and good oxidation resistance. They are complex materials usually containing a minumum of four principal alloying elements. Many are rich in cobalt and chromium. Some of the other alloying elements are iron, aluminum, titanium, molybdenum, and tantalum. Most of the alloys are age-hardened for optimum strength. They are used primarily for aerospace parts and components in jet engines, missiles, and other high-temperature hardware.

7-32 Other Nickel Alloys
Nickel-molybdenum alloys have good high-temperature properties and are highly resistant to a range of corrosive media, especially salts and hot acids. Some also have excellent wear resistance. They are widely used in chemical equipment for valves, agitators, condensers, evaporators, and similar parts.

Nickel-chromium-molybdenum alloys, the Hastelloys, contain various amounts of chromium and molybdenum. Besides good high-temperature properties and corrosion resistance, they have exceptional resistance to certain chlorides.

Nickel-silicon alloys are chiefly represented by a sand-casting alloy containing about 9 percent silicon and small amounts of various other elements. The alloy is noted for its exceptional resistance to hot and cold sulfuric acid. It is hard, and therefore difficult to machine.

Nickel-chromium-iron alloys are distinguished from nickel-chromium alloys by their relatively high iron content. Although minimum iron content is set at 10 percent, the average content is about 23 percent. Most of the wrought alloys are age-hardenable and a few are considered to be in the superalloy category. A number of the alloys are used as electrical-resistance alloys.

7-33 Special Magnetic Alloys
Nickel and iron form a series of alloys with thermal-expansion and magnetic characteristics of special commercial importance. While a number of them are actually iron base, they are usually considered nickel alloys. Variations in nickel content and other elements provide a large number of alloys with particular coefficients for specific applications such as balance wheels in watches, tuning forks, and glass-to-metal seals. Permalloy, an alloy with 78 percent nickel, and many modifications of its composition, has high magnetic permeability at low field strengths. Nickel is also present in permanent magnets of the Alnico type, which contain 14 to 18 percent nickel plus aluminum, cobalt, and sometimes copper.

7-34 Cobalt
Cobalt is an alloying element in some 800 patented alloys. In addition, it serves as the base for a number of important special-purpose alloys. Like iron, cobalt is allotropic. Below 790°F it has a close-packed hexagonal structure. Above this temperature it is face-centered cubic.

Cobalt also has the distinction of being the metal with the highest curie temperature (2050°F)—the temperature at which it changes from a ferromagnetic to a paramagnetic metal.

It is a silvery white metal with a faint bluish tinge and closely resembles nickel in appearance and mechanical properties. However, it is rarer and more expensive. The commercially pure metal, which contains 99.9 percent cobalt plus nickel, iron, and copper, is used for X-ray-tube targets and other specialty components. So-called ductile cobalt contains about 5 percent iron to provide ductility for deep-drawn parts.

The best-known cobalt alloys are the cobalt-base superalloys. The desirable high-temperature properties of low creep, high stress rupture, and high thermal-shock resistance are attributed to cobalt's allotropic change to a face-centered cubic structure at high temperatures. Most of these superalloys contain around 20 percent chromium for good oxidation resistance. An important group of these alloys is known as Stellites. They contain chromium and various other elements such as tungsten, molybdenum, and silicon. These extremely hard alloy carbides in a fairly hard matrix give excellent abrasion and wear resistance and are used as hard facing alloys.

The interesting properties of cobalt-containing permanent, soft, and constant-permeability magnets are a result of the electronic configuration of cobalt and its high curie temperature. In addition, cobalt in the well-known Alnico-magnet alloys decreases grain size and increases coercive force and residual magnetism.

Cobalt is a significant element in many glass-metal seal and low-expansion alloys. One alloy with 54 percent cobalt and 36 percent nickel, known as Invar, has a thermal coefficient of expansion of nearly zero over a small temperature range. Another alloy, with 57 to 63 percent cobalt, named Co-Elinvar, is characterized by an invariable modulus of elasticity over a wide temperature range and low expansion.

Cobalt-chromium-base alloys are used in dental and surgical applications because they are not attacked by body fluids. Alloys named Vitallium are used as bone replacements and are ductile enough to permit anchoring of dentures on neighboring teeth.

Refractory Metals

This group is defined loosely as being composed of metals and their alloys that have melting temperatures above 3000°F and that can be used in load-bearing applications. The metals commonly assigned to this group are tungsten, molybdenum, tantalum, columbium (niobium), chromium, vanadium, hafnium, and rhenium. Of these, molybdenum, tungsten, columbium, and tantalum are the principal members; chromium and vanadium are minor members; and the other two, hafnium and rhenium, are too scarce and expensive for practical use.

As a group, refractory metals and alloys have high tensile strengths at room temperature and generally retain a greater percentage of this strength than other metals at elevated temperatures. Usually considered for structural applications at above around 1600°F, they also have good corrosion resistance. However, despite high-temperature strength, refractory metals are susceptible to rapid oxidation at elevated temperatures. Consequently, in most cases, they must be protected with high-temperature coatings.

7-35 *Molybdenum*

Besides its good elevated-temperature strength, some other attractive characteristics of molybdenum are a high modulus of elasticity at 2000°F (as high as steel at room temperature), excellent thermal-shock resistance, high thermal conductivity, low coefficient of expansion, and good corrosion resistance. On the minus side, it has low oxidation resistance at elevated temperatures and is brittle at around room temperature if the metal is not heavily cold worked.

Molybdenum retains good strength at temperatures above 900°F, and when alloyed with titanium and zirconium, it has even higher strengths. Because of its high rigidity and good thermal conductivity,

Table 7-14 *Properties of Some Refractory Metals*

Type & Condition	Tensile Strength, 1,000 psi	Reduction of Area, %	Modulus of Elasticity, 10^6 psi	Hardness VHN
Molybdenum				
Room temperature	95	60	47	250
2400°F	16	—	21	—
Tungsten				
Room temperature	220	0	59	480
2400°F	45	90	—	—
Tantalum				
Room temperature	60	95+	27	155
2400°F	13	95+	—	—

it is used in many airborne electronic devices. It is one of the few metals that has some resistance to hydrofluoric acid. Because it is not subject to hydrogen embrittlement, it is ideal for high-temperature parts where this could be a problem.

In recent years a number of molybdenum-base alloys have been developed. Three major ones are an alloy containing 0.5 percent titanium, a 0.5 percent titanium and 0.08 percent zirconium alloy (TZM), and one containing 30 percent tungsten. Of the three, TZM has the widest use because of its ready availability in a number of forms from forgings to foil and because of its good workability and machinability. Originally developed for high-temperature structural strength at about 1825 to 2200°F, its major uses are for parts operating in considerably lower temperatures, such as die casting and extrusion and forging dies.

7-36 Tungsten (Wolfram)

Tungsten has the highest melting point (6170°F) of all the metals, and, with about the same density as gold, it is one of the heaviest. Besides its good high-temperature strength, it has good electrical conductivity, and a long history of successful applications in the electrical and electronic fields for lamps, electron tubes, and electrical contacts.

Because tungsten has one of the highest ductile-to-brittle transition temperatures, its ductility and malleability at room temperature are low, and forming operations must be performed between 750 and 3000°F. The most effective method to lower the recrystallization temperature, and thereby improve ductility, is by cold work. This can lower the recrystallization temperature from about 600 to about 375°F. As is true of the other refractory metals, tungsten's oxidation resistance is poor and it must be protected from the atmosphere when used at high temperatures.

Tungsten alloys have been developed for a number of specialized applications. Tungsten's high density makes it ideal as a base for heavy-metal alloys produced by powder metallurgy to give desired combinations of density, corrosion resistance, and strength for parts such as counterweights, flywheels, gyroscope components, and radiation shielding. These alloys also have been found suitable for high-temperature tooling. Tungsten with up to about 2 percent thorium oxide has long been used for heliarc welding tips and for electron-tube filaments because of high emissivity. And rhenium-containing tungsten alloys are used as wire for thermocouples and in radio and cathode-ray valves.

7-37 Tantalum

Tantalum is the least abundant of the four principal metals. Like columbium and unlike tungsten and molybdenum, it is relatively ductile and can be worked at room temperature in much the same manner as fully annealed steel.

Its major uses are not in the high-temperature area. Its corrosion resistance and electrical properties have always directed most of its use into the chemical and electrical fields. Tantalum is one of the most—if not the most—corrosion-resistant metals available. It is only attacked by hydrofluoric acid, alkalies, and hot, concentrated sulfuric or phosphoric acid. However, it is susceptible to hydrogen embrittlement under certain conditions, and has poor oxidation resistance in air above 500°F.

Tantalum is used for acid-resistant heat exchangers, condensers, chemical lines, and other chemical-processing equipment. Because of its nontoxic properties and immunity to body chemicals, it is also used in the medical field for sutures, gauze, pins, and plates. The electrical characteristics of anodic films, which act as a dielectric on tantalum, make the metal suitable for use in capacitors and rectifiers. The combination of high melting point, low vapor pressure, and gettering properties is useful for parts in the vacuum-tube industry.

7-38 Columbium (niobium)

Columbium is primarily considered a nuclear and aerospace material. Its low neutron-absorption cross section and high-temperature strength make it promising as a fuel element and structural material in nuclear reactors.

Columbium's general characteristics include a high melting point, low vapor pressure, moderate density, excellent fabricability in the pure state, low cross section, poor oxidation resistance, rather low modulus of elasticity, and susceptibility to embrittlement by impurities such as carbon, oxygen, and nitrogen. Columbium, while having good wet-corrosion resistance at room temperature, is noted for its resistance to liquid metals—a factor which favors its use in nuclear reactors. For example, unalloyed columbium is resistant to lead in the liquid state up to about 1800°F.

A number of columbium alloys have been developed, but most are still experimental. Alloying can improve oxidation resistance more than 50-fold and increase high-temperature tensile strength by a factor of 4 or 5. However, alloying decreases ductility and workability.

Precious Metals

Eight metals make up the precious (or noble) metal family: gold, silver, and the platinum-group metals—platinum, palladium, rhodium, ruthenium, iridium, and osmium. As the family name implies, except for silver, these are the most expensive metals in the world. Rhodium, for example, costs around $3,000/lb, and platinum over $1,500/lb. Despite their high cost, these metals find considerable use because no other materials offer such a unique combination of properties including chemical inertness, stable thermoelectric behavior, high hardness and strength, and others. In many applications, by judicious design or by using them as cladding or plated coatings, the precious metals are competitive with less costly materials.

7-39 Silver

Silver is the most abundant and the least costly of the precious metals. Pure silver has the highest thermal and electrical conductivity (105.5 percent IACS) of any metal, as well as the highest optical reflectivity.

Table 7-15 *Properties of Some Precious Metals*

Type & Condition	Specific Gravity	Tensile Strength, 1,000 psi	Elongation,% in 2 in.	Hardness, Bhn	Electrical Resistivity, Microhm-cm
Gold					
Annealed	19.3	19	45	25	2.19
Cold rolled	—	32	4	58	—
Silver					
Annealed	10.5	22	48	30	1.47
Cold rolled	—	54	2.5	—	—
Platinum					
Annealed	21.5	20	35	52	9.83
Cold rolled	—	29	3.0	97	—
Palladium					
Annealed	12.0	25	32	46	10.0
Cold rolled	—	47	1.5	—	—
Iridium					
Annealed	22.5	80	—	170	5.3
Cold rolled	—	290	—	350	—
Rhodium					
Annealed	12.4	73	—	156	4.51
Cold rolled	—	300	—	390	—

Next to gold it is the most ductile and malleable of metals. It is one of the most corrosion-resistant metals, and under ordinary conditions is not affected by alkalies and most other chemicals unless hydrogen sulfide is present. Silver, however, dissolves in nitric acid and hot, concentrated sulfuric aicd. Although silver tarnishes quickly, it oxidizes slowly in air.

Silver is classifed by grades in parts of silver content per thousand. The commercial grades are fine silver and high fine silver. Pure silver has many uses in electronic and electrical equipment where maximum conductivity is needed. Typical uses are fluorescent lamp controls, refrigerator thermostats, and telephone relays.

Because of its softness, it is often alloyed with other metals, the most common being copper, which impart hardness and strength without appreciably degrading other desirable properties. Sterling silver (7.5 percent copper) is the best-known alloy. Others are coin silver (10 percent copper) and silver-copper eutectic (28 percent copper). The last alloy has the highest combination of strength, hardness, and electrical properties of any of the silver alloys. Silver-alloy filler metals, commonly called silver-brazing alloys, contain from 10 to 85 percent silver, and are widely used for joining all ferrous and nonferrous metals except aluminum and magnesium.

7-40 *Gold*
Gold is yellow, soft, and the most ductile of all metals. It has exceptionally high oxidation and chemical resistance. It is not attacked by common acids, but is soluble in aqua regia and cyanide solutions. It is attacked by chlorine at temperatures above 175°F.

For most uses, gold is alloyed to increase hardness. Copper is often used along with silver and small amounts of the platinum-group metals. Gold and its alloys are available in the usual forms of sheet wire, ribbon, and tubing. Precision casting is widely used to make gold jewelry. Economical use is enhanced by plating in such parts as reflectors, and in electrical components such as contacts and vacuum-tube grids.

Gold alloys are produced in a range of colors from white to many shades of yellow for jewelry and other decorative applications as well as for use in dentistry. The corrosion resistance and low melting point also make gold useful in brazing alloys.

7-41 *Platinum-group Metals*
These metals are grouped together in the periodic table (group VIII). Platinum and palladium are the most abundant of the group.

Platinum. This is the most important member of the group. It is a white, ductile metal that takes a high permanent polish. It has remarkable resistance to corrosion and chemical attack, a high melting point, retention of strength, and resistance to oxidation in air, even at very high temperatures. It is virtually nonoxidizable and is soluble only in liquids generating free chlorine, such as aqua regia. Platinum is alloyed with other precious metals to obtain improved properties. Iridium is added to enhance mechanical properties and further improve corrosion resistance. The addition of rhodium improves mechanical properties, particularly at high temperatures, and corrosion resistance. And gold is alloyed with platinum to provide special chemical and physical characteristics.

Palladium. This is a silvery white metal, very ductile, and slightly harder than platinum. It is not as corrosion resistant as platinum, and it is soluble in aqua regia and can be attacked by boiling nitric and sulfuric acids. Its electrical resistivity is the highest in the platinum group. Palladium is sometimes used as a substitute for platinum, but it begins to tarnish at temperatures above 750°F. It is used as a coating on printed circuit boards and various types of contacts. As the principal constituent in several contact alloys, it improves hardness and strength without lowering corrosion resistance.

Iridium. This is the most corrosion-resistant metallic element known. It is a very hard, brittle, tin-colored metal, and is soluble in aqua regia only when alloyed with sufficient platinum. It has high-temperature strength comparable to tungsten up to 3000°F. Because of high resistance to attack by lead, it is used for spark plug electrodes in aircraft and other engines where high reliability is needed.

Rhodium. This is the hardest metal and the highest in electrical and thermal conductivity in the platinum group. Its high polish and reflectivity make it ideal as a plated coating for special mirrors and reflectors. It is slightly soluble in aqua regia, and is used with platinum for thermocouple wire that can be used at temperatures up to 3000°F. Rhodium and its alloys are used in furnace windings and in crucibles at temperatures too high for platinum.

Ruthenium. This is a hard and brittle metal with a silver gray luster. Its tetraoxide is very volatile and poisonous. Unalloyed it is unworkable. It is added to platinum to increase resistivity and hardness.

Osmium. This is the heaviest known metal, having a density roughly three times that of steel. Its melting point (about 4900°F) is third highest of the metals, following that of tungsten and tantalum. It oxidizes readily when heated in air to form a very volatile and poisonous tetraoxide. As a metal it is practically unworkable, and it has been used predominantly as a catalyst.

Review Questions

1. List five nonferrous metals that have melting ranges below 800°F.
2. List four nonferrous metals that are commonly classified as light metals.
3. List five metals that are classified as refractory metals.
4. Compare the meaning of the term temper in reference to nonferrous and ferrous metals.
5. What method or methods are used to increase strength and hardness of:
    (a) Nonheat-treatable wrought-aluminum alloys?
    (b) Heat-treatable wrought-aluminum alloys?
6. Which series of wrought-aluminum alloys (or alloys) is identified with the following:
    (a) Lowest in strength of the heat-treatables?
    (b) Best electrical conductivity?
    (c) Highest in strength?
    (d) Most susceptible to corrosion?
7. Which group of aluminum-casting alloys would you choose to obtain:
    (a) Maximum castability?
    (b) Highest corrosion resistance?
    (c) Good bearing properties?
8. Compare magnesium alloys with aluminum alloys for the following property areas:
    (a) Density.
    (a) Strength-to-weight ratio.
    (c) Dimensional stability.
9. What safety hazard must be considered when using magnesium in a finely divided state?
10. Despite its relatively high cost, when should the use of titanium be considered?
11. What are the reasons for beryllium's limited use despite many exceptional properties?
12. What is the basic distinction between brasses and bronzes?

13. A metal is specified as being 75 percent IACS. How does its electrical conductivity compare with that of tough-pitch copper?
14. Explain the significance of dezincification which occurs in some brasses.
15. Use Tables 7-9 and 7-10 to list the hardened copper alloy or alloys with:
    (a) The greatest stiffness.
    (b) The lowest strength.
    (c) The highest ductility.
16. Name the copper alloy or alloys commonly used for:
    (a) Jewelry.
    (b) Heat exchangers and condenser tubing.
    (c) Thermocouple elements.
    (d) Screw machine parts.
    (e) Nonsparking tools.
17. Why are zinc die castings widely used for decorative and low-stress automotive parts?
18. What are some (name three) aging characteristics of zinc die castings?
19. Judging on the basis of most engineering properties, which other major group of metals do nickel alloys most closely resemble? Name two similarities.
20. Name three outstanding properties that cobalt contributes when it is used as an alloying element.
21. What is the major problem in the use of refractory metals at high temperatures?
22. Why are low-melting alloys generally unsuitable for structural parts?
23. Low-melting alloys are relatively ductile as a group. Name one exception.
24. What are some ways in which precious metals can be used economically?
25. Name the groups, classes, or series of nonferrous metals from which you would select candidate materials if the important or critical requirement(s) is:
    (a) High electrical conductivity (name three).
    (b) High strength-to-weight ratio (name two).
    (c) High strength-to-weight ratio and high stiffness (rigidity) (name one).
    (d) Excellent machinability (name two).
    (e) Die castings (name two).
    (f) High strength up to temperatures of 2000°F (name two).

Bibliography

American Society for Metals: *Metals Handbook*, 8th ed. Novelty, Ohio.
Clauser, H. R. (ed.): *Encyclopedia of Materials, Parts and Finishes*, Technomic Publishing Co., Westport, Conn., 1975.
Koves, G.: *Materials for Structural and Mechanical Functions*, Hayden Book Co., Inc., New York, 1970.
"Materials Selector Issue," *Materials Engineering*, Mid-September, 1972.
"Metals Reference Issue," *Machine Design*, Feb. 17, 1972.
"Wrought Aluminum and Its Alloys," *Materials Engineering*, June 1965.

PLASTICS MATERIALS

Plastics are relatively newcomers to the field of engineering materials. The first commercial plastic, celluloid, was developed in 1868 to replace ivory for billiard balls. Phenolic plastics, developed by Baekeland and named Bakelite after him, were introduced around the turn of the century. During World War II plastics were mainly a substitute for metals. Today, high-performance plastics are responsible for the successful operation of parts and products in the critical environments of electronics, communications, and aerospace systems. And, in recent years, thousands of plastics have been developed and have shown that in many industrial and consumer applications they can do an equal or better job at lower cost than other materials.

A look at plastics used in automobiles is proof of the great advances that plastics have made. The amount of plastics in the average car has increased from less than 50 lb in the 1960s to 100 lb or more in the 1970s. It is predicted that by 1980, automobiles will contain at least 200 lb of plastics. Similarly, in the building field, plastics application tripled from 1960 to 1970, going from about 0.7 to 2.1 billion lb. This increased use is typical of the rapid growth of plastic use. In the last few years, the total consumption of synthetic and plastic materials has been in the neighborhood of 25 billion lb.

Nature and Major Characteristics

8-1 *Definitions and Major Characteristics*
In discussing plastics, three terms—plastics, polymers, and resins—are often used interchangeably. However, in this book, *polymer* is used in the more general sense to encompass large molecular-chain hydrocarbon materials such as plastics, rubber, and wood. In a specific sense, polymer refers to the long-chainlike molecules themselves (see Chap. 3); *resin* refers to the basic chemical polymeric compounds; and *plastic* refers to the class of polymers covered in this chapter and defined in the following paragraph.

A plastic material, as defined by the Society of the Plastics Industry, is: "Any one of a large and varied group of materials consisting wholly or in part of combinations of carbon with oxygen, hydrogen, nitrogen and other organic and inorganic elements which, while solid in the finished state, at some stage in its manufacture is made liquid, and thus capable of being formed into various shapes, most usually through the application, either singly or together, of heat and pressure."

There are two basic types of plastics based on intermolecular bonding (Fig. 8-1). *Thermoplastics*, because of little or no cross-bonding between molecules, soften when heated and harden when cooled, no matter how often the process is repeated. *Thermosets*, on the other hand, have strong intermolecular bonding. Therefore, once the plastic is set into permanent shape under heat and pressure, reheating will not soften it.

Within these major classes, plastics are commonly classified on the basis of base monomers. There are over two dozen such monomer

Fig. 8-1 Schematic sketches showing molecular bonding in plastics: (a) weak secondary bonding found in thermoplastics, and (b) cross-link bonding between molecular chains found in thermosets.

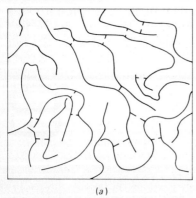
(a)

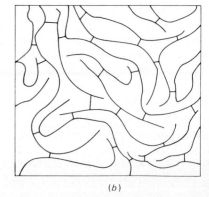
(b)

families or groups (see Chap. 9). Plastics are also sometimes classified roughly into three stiffness categories: rigid, flexible, and elastic. Another method of classification is by the "level" of performance or the general area of application, using such categories as engineering, general-purpose, and specialty plastics, or the two broad categories of engineering and commodity plastics.

Some of the major characteristics of plastics that distinguish them from other materials, particularly metals, are:

1. They are essentially noncrystalline in structure.
2. They are nonconductors of electricity and are relatively low in heat conductance.
3. They are, with some important exceptions, resistant to chemical and corrosive environments.
4. They have relatively low softening temperatures.
5. They are readily formed into complex shapes.
6. They exhibit viscoelastic behavior. That is, after an applied load is removed, plastics tend to continue to exhibit strain or deformation with time.

8-2 Types of Polymers

In Chap. 3 we learned that a plastic is composed of millions of giant chainlike molecules (polymers) constructed of smaller molecules called monomers. The atoms in both are held together by primary, covalent bonding—that is, by sharing electrons with each other.

Polymers can be built of one, two, or even three different monomers, and are termed homopolymers, copolymers, and terpolymers, respectively. Their geometrical form can be linear or branched. Linear or unbranched polymers are composed of monomers linked end-to-end to form a molecular chain that is like a simple string of beads or a piece of spaghetti (Fig. 8-2a). Branched polymers have side chains of molecules attached to the main linear polymer. These branches can be composed either of the basic linear monomer or of a different one. If the side molecules are arranged randomly, the polymer is atactic (b); if they branch out on one side of the linear chain in the same plane, the polymer is isotactic (c); and if they alternate from one side to the other, the polymer is syndiotactic (d).

8-3 Molecular Weight and Bonding

A plastic's properties and behavior are greatly influenced by the weight of its polymers. Molecular weight is simply the weight of the atoms per monomer times the number of monomers in the polymer molecule.

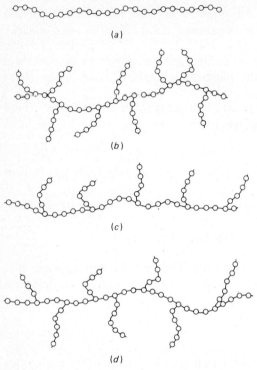

Fig. 8-2 Types of polymer molecules (small beads represent the monomers): (a) linear or unbranched; (b) atactic branched; (c) isotactic branched; (d) syndiotactic branched.

Because the polymer chains can vary widely in length, the molecular weight of a plastic is expressed as an average value (average molecular weight).

As we saw in Chap. 3, the forces (secondary van der Waals forces) holding the millions of polymer chains together are relatively weak in thermoplastics (see Fig. 8-1). In contrast, the molecular chains in thermoset plastics are cross-linked; that is, they are bonded together by stronger, primary covalent bonding forces, which is why thermosets are generally stronger and stiffer than thermoplastics.

In thermoplastics, the strength of the secondary bonding is dependent upon the geometrical form, the molecular weight (or length), and the various molecular-weight distributions of the polymers making up the plastic. In general, many of the mechanical properties and heat resistance increase with molecular weight. For example, if the chains are linear and the molecular-weight distribution is narrow, the chains will fit together tightly and promote more numerous secondary bonding, which in turn means relatively high strength and rigidity, but lower impact resistance. On the other hand, branched polymers cannot be packed as closely together, and therefore the bonding forces are weaker. Thus the resulting plastic is lower in strength, softer, and more flexible.

8-4 *Crystallinity*

Although plastics are not basically crystalline materials, they often contain regions in which the mixture of polymer molecules has an ordered or regular pattern. That is, plastic crystallinity generally consists of thousands of relatively small crystalline islands, called crystallites, surrounded by amorphous regions in which the polymers are randomly intermingled (Fig. 8-3). The crystalline regions can be either preferentially oriented in one direction or unoriented.

Crystallinity varies widely from one plastic to another. If a plastic is composed chiefly of linear polymers, the individual polymers tend to pack closely together and exhibit relatively high bonding forces, which in turn lock the chains together in an aligned pattern. In branched polymers, the bulky side groups keep the polymer chains further apart, thus decreasing the tendency to crystallize.

The density, or specific gravity, of a plastic is closely related to the degree of crystallinity present; that is, density increases with the percentage of crystallinity. Therefore, in practice, the degree of crystallinity is often expressed in terms of density. Because intermolecular-bonding forces are a function of crystallinity, both strength and elevated-temperature resistance increase with density, and ductility decreases. Also, since the space between the molecular chains is a

Fig. 8-3 Schematic sketches showing difference between (a) an amorphous plastic, and (b) a plastic with crystalline regions (*A*) in an amorphous matrix (*B*).

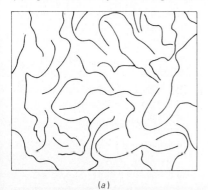

(a)   (b)

function of crystallinity, moisture and gas permeability decrease with a rise in density. Table 8-1 summarizes some of the effects that density and molecular weight have on plastic properties.

8-5 Glass Transition Temperature

As might be expected, the behavior of polymer molecules is temperature dependent. This is manifested clearly in the glass transition temperature (T_g). This point is the lower end of a narrow range above which a thermoplastic changes from a rigid, brittle, glasslike state to a more flexible substance. At least in part, the glass transition temperature is a function of secondary bonding. In general, the stiffer the polymer chains, the higher the T_g.

Many thermoplastics have glass transition temperatures above room temperature. Nevertheless, the crystallinity forces hold the molecules together and account for the presence of useful strength at normal temperatures. In contrast, plastics with little or no crystallinity must depend on the rigidity of their polymer chains for their strength above the T_g. Thus a plastic with a low T_g and high crystallinity has a combination of good strength, flexibility, and toughness.

8-6 Additives

Few plastics in use are totally composed of polymer resins. Nearly all contain one or more additive materials to modify or control properties, or to reduce costs. A recent survey by *Plastics Technology* maga-

Table 8-1 *Effects of Density and Molecular Weight on Plastic Properties*

Qualities	When Density (Crystallinity) Increases	When Average Molecular-Weight Increases	When Molecular-Weight Distribution Narrows
Softening temperature	Increases	Increases	Increases
Strength	Increases	Increases	Increases
Elongation	Decreases	Increases	
Stiffness	Increases	Increases	
Hardness	Increases	Increases	
Toughness	Decreases	Increases	
Creep resistance	Increases	Increases	Increases
Impermeability to gas and liquid	Increases	Increases	

zine showed that, on the average, plastics used in industry contain about 10 percent additives by weight, accounting for about 20 percent of the cost.

Fillers. These are probably the most common of the additives. They are usually used to either provide bulk or modify certain properties. Generally, they are inert and thus do not react chemically with the resin during processing. The fillers are often cheap and serve to reduce costs by increasing bulk. For example, wood flour, a common low-cost filler, sometimes makes up 50 percent of a plastic compound. Other typical fillers are chopped fabrics, asbestos, talc, mica, gypsum, and milled glass. Besides lowering costs, fillers can improve properties. For example, asbestos increases heat resistance, and cotton fibers improve toughness.

Plasticizers. These are added for either of two purposes: to improve flowability during processing by reducing the glass transition temperature, or to improve properties such as flexibility. Plasticizers are usually liquids that have high boiling points, such as certain phthalates. Substances which are themselves polymers of low molecular weight, such as polyesters, are also used as plasticizers.

Stabilizers. These are added to plastics to help prevent breakdown or deterioration during molding or when the polymer is exposed to sunlight, heat, oxygen, ozone, or combinations of these. Thus there is a wide range of compounds, each designed for a specific function. Stabilizers can be metal compounds, based on tin, lead, cadmium, barium, and others. And phenols and amines are added antioxidants that protect the plastic by diverting the oxidation reactions to themselves.

Catalysts. These, by controlling the rate and extent of the polymerization process in the resin, allow the curing cycle to be tailored to the processing requirements of the application. Catalysts also affect the shelf life of the plastics. Both metallic and organic chemical compounds are used as catalysts.

Colorants. These are widely used for decorative purposes. Colorants come in a great variety of pigments and dyestuffs. The traditional colorants are metal base pigments such as cadmium, lead, and selenium. More recently, liquid colorants, composed of dispersions of pigments in a liquid, have been developed.

Fire Retardants. These are added to plastic products that must meet fire-retardant requirements, for polymer resins are generally flammable, except for such notable exceptions as polyvinyl chloride. In general, the function of fire retardants is limited to the spread of fire. They do not normally increase heat resistance or prevent the plastic from charring or melting. Some fire-retardant additives include compounds containing chlorine or bromine, phosphate-ester compounds, antimony thrioxide, alumina trihydrate, and zinc borate.

Reinforcements. These are not normally considered additives. They are used in plastics primarily to improve mechanical properties, particularly strength. Although asbestos and some other materials are used, glass fibers are the predominant reinforcement for plastics (see Sec. 9-21).

Plastics Processing

Plastics are formed into final, usable shapes by processes that handle the material either as a liquid or as a solid. Plastics can also be formed by cutting and machining. And they can be assembled or fabricated by joining methods such as sewing, sealing, and adhesive bonding.

8-7 *Molding and Casting*

Injection Molding. This is the principal method of forming thermoplastic materials. (The process is sometimes modified and used for thermosetting plastics.) In injection molding, which is similar to metal die casting, plastic resin is fed into a heating chamber where the material is softened to a fluid state (Fig. 8-4). At the end of this chamber

Fig. 8-4 Injection molding.

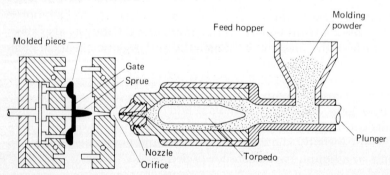

the fluid resin is forced at high pressure through a nozzle into a cool, closed mold. As soon as the plastic cools to a solid state, the mold opens and the finished piece is ejected from the press. Injection molding is a relatively fast, high production process suitable for parts weighing from around 2 to 50 oz.

Compression Molding. This is the most common method of forming thermosetting materials. (It is not generally used for thermoplastics.) Compression molding is simply squeezing a material into a desired shape by applying heat and pressure to the material in a mold (Fig. 8-5). Plastic molding powder, mixed with such fillers as wood flour, cellulose, and asbestos, is put directly into the open, heated mold cavity. The mold is then closed, pressing down on the plastic and causing it to flow throughout the mold. While the heated mold is closed, the thermosetting material undergoes a chemical change which permanently hardens it into the shape of the mold. The three compression-molding factors—pressure, temperature, and time the mold is closed—vary with the design of the finished article and the material being molded.

Transfer Molding. This method is a modification of compression molding, and is most generally used for thermosetting plastics. It differs in that the plastic is heated to a point of plasticity before it reaches the mold, and it is forced into a closed mold by means of a hydraulically operated plunger (Fig. 8-6). This form of molding was developed to facilitate molding intricate products with small, deep holes or

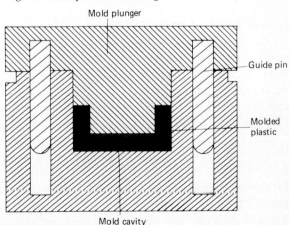

Fig. 8-5 Compression molding.

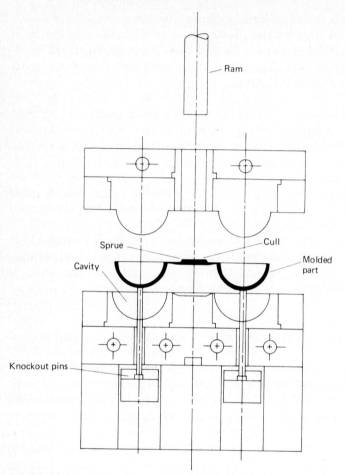

Fig. 8-6 Transfer molding.

numerous metal inserts. The dry molding compound used in compression molding sometimes disturbed the position of the metal inserts and the pins which formed the holes. The liquefied plastic material in transfer molding flows around these metal parts without causing them to shift position.

Rotational Molding. This technique is used to fabricate hollow, one-piece flexible parts from vinyl plastisols or from polyethylene powders. It consists of charging a measured amount of resin into a warm mold which is rotated about two axes in an oven. Centrifugal force distributes the resin evenly throughout the mold, and the heat melts and fuses the charge to the shape of the cavity. After removing and cooling the mold, the finished part is extracted. Rotational molding is particularly suited to the production of very large parts, since size is only limited by the capacity of the oven. The advantages of this type of molding are low mold cost, strain-free parts, and uniform wall thickness.

Solvent Molding. This method is applied to thermoplastics. A mold is immersed in a solution and withdrawn, or it is filled with a liquid resin and then emptied. The result in either case is that a layer of plastic film adheres to the sides of the mold. Some articles thus formed, such as bathing caps or vials, are removed from the molds. Other solvent moldings remain permanently on the form as, for example, a plastic coating on a metal tube.

Pulp Molding. This method is used with thermosetting plastics. A porous form, approximately the shape of the finished article, is lowered into a tank containing a mixture of pulp, plastic resins, and water. The water is drawn off through the porous form by a vacuum. This causes the pulp-and-resin mixture to be drawn to the form, where it adheres. When a sufficient thickness of pulp has been drawn onto the form, it is removed and then molded into the final shape.

Blow Molding. This method of forming thermoplastics consists of stretching and then hardening a plastic against a mold. There are two general ways this is done: the direct and the indirect method, each with several variations. In the direct method, a gob of molten thermoplastic is formed into the rough shape of the desired finished product. This shape is then inserted into a female mold, and air is blown into it, as into a balloon, to force it against the sides of the mold (Fig. 8-7). The formed material is then cooled before removal from the mold. In the

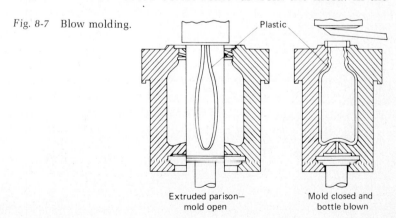

Fig. 8-7 Blow molding.

Extruded parison—
mold open

Mold closed and
bottle blown

indirect method, a thermoplastic sheet or special shape is first heated, and then clamped between a die and cover. Air pressure-forced between the plastic and the cover forces the material into contact with the die, which has the contour desired in the finished product. The plastic is cooled before removal from contact with the die.

Casting. This method is used for making special shapes, rigid sheets, film, sheeting, rods, and tubes of both thermoplastics and thermosets. The essential difference between casting and molding is that pressure is not used in casting. In casting, the resin material is heated to a fluid mass, poured into either open or closed molds, cured at varying temperatures depending on the resin used, and removed from the molds. Casting of film (and some sheeting) is done on a wheel or belt, or by precipitation in a chemical bath. In wheel or belt casting, the resin is spread to the desired thickness on the wheel or belt as it revolves and as the temperature is increased. The film is then dried and stripped off.

8-8 *Sheet, Extrusion, and Coating Processes*

Calendering. This method is used to process thermoplastics into film and sheeting, and to apply a plastic coating to textiles or other supporting materials. The term film is used for thicknesses up to and including 10 mils, while sheeting is used for thicknesses over 10 mils. In calendering film and sheeting, the resin compound is passed between a series of three or four large, heated, revolving rollers that squeeze the material between them into a sheet or film (Fig. 8-8). The thickness

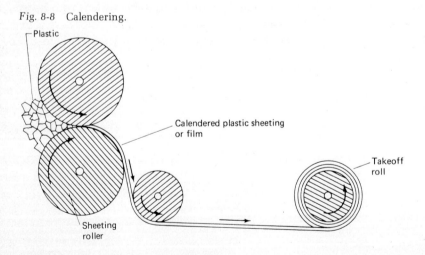

Fig. 8-8 Calendering.

of the finished material is controlled by the space between the rolls. The surface of the film or sheeting may be smooth or matted, depending on the surfacing on the rollers.

A plastic coating to a fabric or other material can be applied by the calendering process. The coating compound is first passed through two top horizontal rollers on a calender, while the uncoated material is passed through two bottom rollers. The smooth film of the coating material is anchored to the fabric when the fabric and film are passed between the same rolls.

Coating. Both thermosetting and thermoplastic materials are used as coatings on metal, wood, paper, fabric, leather, glass, concrete, ceramics, or other plastics. There are many coating processes, including knife or spread coating, spraying, roller coating, dipping, brushing, calendering, and the fluidized-bed process. In spread coating, the material to be coated passes over a roller and under a long blade or knife (Fig. 8-9). The plastic coating compound is placed on the material just in front of the knife, and the knife spreads it out over the material. In roller coating, two horizontal rollers are used. One roller picks up the plastic coating solution on its surface and deposits it on the second roller which, in turn, deposits the coating solution on the supporting material. A plastic coating may also be applied by spraying it through a spray gun or brushing it over the material to be coated, as in silk screen work. Articles to be coated may also be dipped into a plastic solution until the desired thickness of coating is achieved, and then dried.

High-pressure Laminating. Thermosetting plastics are generally used in this method to hold together the reinforcing materials that comprise

Fig. 8-9 Coating.

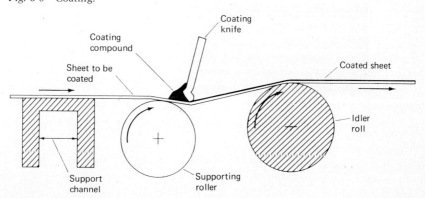

the body of the finished product. The reinforcing materials may be cloth, paper, wood, or glass fibers. The end product may be plain flat sheets, or decorative sheets as in counter tops, rods, tubes, or formed shapes. The first step is impregnating the reinforcing materials with resin. For flat surfaces, impregnated sheets are stacked between two highly polished steel plates and subjected to heat and high pressure in a hydraulic press which cures the resin and presses the plies of material into a single piece (Fig. 8-10). High-pressure tubing and resin-treated reinforcing sheets are wrapped, under tension and/or pressure, around a heated rod. The assembly is then cured in an oven. In producing formed shapes, the reinforcing material is cut into pieces that conform to the contour of the product; these are then fitted into the mold and cured under heat and pressure. (See Sec. 9-23 for low-pressure processing methods for glass-fiber-reinforced plastics.)

Thermoforming. This method involves heating a thermoplastic sheet to a formable plastic state and then applying air and/or mechanical assists to shape it to the contours of a mold (Fig. 8-11). The air pressure may range from almost zero to several hundred psi. Vacuum forming involves evacuating the space between the sheet and the mold in order to utilize the atmospheric pressure. This will satisfactorily reproduce the mold configuration in the majority of forming applications. When higher pressures are required, they are obtained by sealing a chamber above the top of the sheet and using compressed air to build pressure within. This system is known as pressure forming. Variations of thermoforming include straight, drape, plug and ring,

Fig. 8-10 High-pressure laminating.

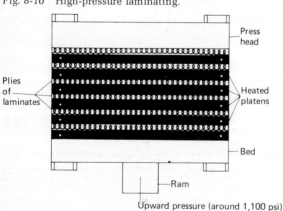

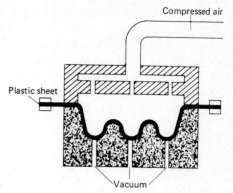

Fig. 8-11 Thermoforming.

slip, snapback, reverse-draw, air-slip, plug-assist, plug-assist and air-slip, and machine-die forming.

Extrusion. This molding method is used to form thermoplastic materials into continuous sheeting, film, tubes, rods, profile shapes, and filaments, and to coat wires, cables, and cord. Dry material is fed into a long heating chamber where it becomes molten. At the end of the chamber, the resin is forced out through a small opening or die with the shape desired in the finished product. As the plastic extrusion comes from the die, it is fed onto a conveyor belt where it is cooled, most frequently by blowers or by immersion in water (Fig. 8-12). When used to coat wires and cables, the thermoplastic is ex-

Fig. 8-12 Extrusion molding.

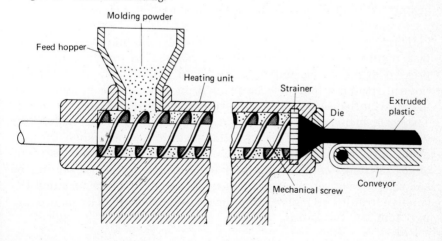

truded around a continuing length of the wire or cable which passes through the extruder die. In producing wide film or sheeting, the plastic is extruded in the form of a tube, which is split as it comes from the die and then stretched and thinned to the dimensions desired in the finished film. In a different process, the extruded tubing is inflated as it comes from the die, the degree of inflation regulating the thickness of the final film.

Stamping and Die Cutting. These processes are used to produce flat parts less than 0.060 in. thick, and thicker parts when molding is not economical. The two types of dies used are matched metal and steel rule. Matched-metal tool-steel dies are the same as those for stamping metals. Steel-rule dies are male dies in a plywood frame that stamp against a backup plate. Typical parts made in this manner are washers, gaskets, seals, cams, dials, panels, terminals, and windows.

8-9 *Machining, Finishing, and Joining*

Machining. Machining is used to make parts from stock shapes such as sheets, rods, and tubes, and for secondary operations to produce holes, slots, and other features impractical to mold. While all plastics can be machined, speeds and tool configurations vary from one type to another. Common machining operations include turning, milling, planing, drilling, threading, tapping, routing, sanding, sawing, shearing, and punching. Because plastics are relatively low heat conductors and do not readily dissipate the frictional heat generated by machining operations, scorching or warping of the workpiece must be guarded against. Also, dulling and gumming up of tools is a common problem.

Finishing. Film, sheeting, and coated materials may have their surface texture changed, either during processing or after, by being pressed against a heated roller or mold embossed with a pattern. Colorful patterns may also be printed on film, sheeting, and coated material surfaces either by letterpress, gravure, or silk-screening. There are many methods of adding decorative effects to the surface of rigid plastic parts. One is metal plating, accomplished through metal spraying, vacuum deposition, or metal dusting and painting. Other methods involve stamping the finished product, printing by silk-screen or offset processes, engraving, etching, or air-blasting. Decorative textures on molded products are often achieved by incorporating them into the molds.

Joining. Plastics can be joined by a variety of methods. A fusion type of bond can be obtained by melting the surfaces of plastic parts and then maintaining them in contact during cooling and solidification. In hot-plate welding, the surfaces to be joined are first held lightly against a heated metal surface to melt the surface layer, after which they are quickly brought together and held under pressure. In induction welding, an induction-heated metal insert provides the heat to melt the surfaces. In spin welding, the rotation of one part against another generates frictional heat which causes a melt film to develop. When the rotation is stopped, the weld film solidifies under pressure. Hot gas welding is used chiefly for assembling large parts or for short runs on small parts. The process resembles that of gas welding of metals. A gas flame melts a filler rod, and the melted plastic flows into the joint and fuses with the plastic pieces being joined. Filler rods are of the same or similar material as the plastics being joined.

In ultrasonic joining, mechanical vibrations are produced in the workpieces by a transducer that changes electrical energy into mechanical energy. The vibrations of one surface moving against the other generate frictional heat that produces the weld.

Plastics can be cemented to themselves or to dissimilar materials by adhesive bonding techniques. Cements for plastics are classified as solvent, dope, and chemical. With solvent cement, the surfaces to be joined go into solution. During cure, the solvent evaporates and produces a homogeneous joint. Dope cement consists of a solvent plus a plastic similar to the one being bonded. Chemical cements are resins that polymerize in the joint to form the bond. All cements can be used for bonding thermoplastics together. However, only chemical cements can be used with thermosets.

Review Questions

1. Which term, plastics or polymers, is the broader and encompasses several classes of hydrocarbon materials?
2. Explain the difference between thermoplastics and thermosets
   (a) In the bonding mechanism at the microstructural level.
   (b) When they are repeatedly heated and cooled.
3. Explain the difference between molecular weight and average molecular weight.
4. What type of polymer chain favors formation of crystalline areas in a plastic?

5. What is the difference in a plastic's physical state above and below its glass transition temperature?
6. List four properties of plastics that increase when density increases.
7. List four properties of plastics that increase when the average molecular weight increases.
8. What functions do the following additives perform in plastics?
   (a) Glass fibers.
   (b) Plasticizers.
   (c) Fillers.
9. Which plastics processing method (or methods) is used for
   (a) Forming thermosetting plastics shapes and parts.
   (b) Producing reinforced flat sheets laminated with thermosetting plastics.
   (c) Producing a large quantity of thermoplastic moldings.
   (d) Producing hollow objects such as bottles.
   (e) Producing tubes, rods, and profile shapes.
10. Explain three ways in which heat is generated for fusion bonding of plastics.

Bibliography

Kaufman, M.: *Giant Molecules*, Doubleday & Co., Inc., Garden City, N.Y., 1968.

Mark, H. F.: "The Nature of Polymeric Materials," *Scientific American*, September 1967.

"Modern Plastics Encyclopedia," *Modern Plastics*, October 1972.

Society of the Plastics Industry: *The Story of the Plastics Industry*, New York, 1966.

Stinson, S. C.: "Chemicals and Additives Today," *Plastics Technology*, July 1972.

THERMO-PLASTICS AND THERMOSETS

The preceding chapter described the nature, structure, and general characteristics of plastics materials, the additives used, and the methods of processing plastics into finished forms and shapes. This chapter broadly covers the properties and characteristics of the major groups of plastics. As a general guide, Table 9-1 provides a rough overall view of the comparative properties of the classes of plastics discussed.

Thermoplastics

Of the two basic types of plastics, the thermoplastic family is the larger, both in number and in volume used. With some 20 classes and with great diversity in each class, thermoplastics provide a range of properties that makes them useful in a wide range of consumer and industrial products.

As we have just seen, the basic distinguishing characteristic of thermoplastics is that they return to their original solid and firm condition even after being repeatedly softened by heating. With some exceptions (certain vinyls and nylons), thermoplastics do not become "true" liquids under the application of heat. Instead, they soften into a plastic state of high-melt viscosity and, in most cases, decompose before liquefying.

Most thermoplastics are furnished chemically complete, polymerized, and in the form of small pellets or cubes ready to mold. Some are also available as liquid resins for casting, laminating, and coating.

Table 9-1 *Comparative Properties of Plastics*

Property	Low		Medium		High	
	Thermoplastics	Thermosets	Thermoplastics	Thermosets	Thermoplastics	Thermosets
Density	Olefins ABS	Urethanes Polyesters	Nylons Polyphenylene Oxides Styrenes Carbonates Acrylics Cellulosics Sulfones	Allylics Aminos Silicones Epoxies	Fluoroplastics Vinyls Acetals Polyimides	Alkyds Phenolics
Tensile strength	Olefins Fluoroplastics	Silicones Urethanes	Styrenes ABS Cellulosics Vinyls	Phenolics Aminos Alkyds	Nylons Polyimides Carbonates Acrylics Acetals Polyphenylene Oxides	Epoxies Polyesters
Stiffness (Modulus of Elasticity)	Olefins Fluoroplastics	Urethanes	Carbonates Cellulosics Vinyls ABS Polyphenylene Oxides Sulfones	Polyesters Aminos Allylics Alkyds	Acrylics Styrenes Acetals Polyimides Nylons	Phenolics Silicones Epoxies

Thermoplastics and Thermosets

Impact strength	Styrenes Sulfones Acrylics Polyphenylene Oxides	Aminos	Cellulosics Polypropylenes ABS Styrene (high impact) Acetals Nylons	Alkyds Epoxies Silicones Polyesters	Carbonates Vinyl (PVC) Polyethylene (high density)	Phenolics
Electrical resistivity	Acetals Cellulosics Nylons Vinyls Acrylics	Phenolics Polyesters	Polyphenylene Oxides Sulfones Carbonates ABS	Silicones Aminos Alkyds	Styrenes Fluoroplastics Olefins	Epoxies Allylics
Dielectric strength	ABS Carbonates		Styrenes Fluoroplastics Cellulosics Polyphenylene Oxides Acrylics Nylons Acetals Polyethylenes	Allylics Phenolics Silicones Aminos Alkyds	Vinyls Propylenes	Urethanes Polyesters Epoxies

(continued)

Table 9-1 Comparative Properties of Plastics (continued)

Property	Low		Medium		High	
	Thermoplastics	Thermosets	Thermoplastics	Thermosets	Thermoplastics	Thermosets
Useful at high temperatures	Cellulosics Vinyls Acrylics Styrenes Polyethylenes Acetals ABS		Nylons Carbonates Propylenes	Polyesters Alkyds	Polyimides Fluoroplastics Polyphenylene Oxides Sulfones	Silicones Allylics Aminos Epoxies
Price per pound	Olefins Styrenes ABS Vinyls Acrylics	Polyesters Phenolics	Sulfones Carbonates Cellulosics Acetals	Alkyds Aminos	Fluoroplastics Polyimides Polyphenylene Oxides Nylons	Silicones Allylics Epoxies

Although they can be filled or reinforced, thermoplastics are most often used in the unfilled resin form.

9-1 Polyolefins

The polyolefins, based on the ethylene monomer, are the most widely used plastics today. Although they are generally thought of as being composed of linear polymer chains, most commercial grades have at least some branching.

There are two predominant groups of polyolefins: polyethylenes and polypropylenes. In addition, the polyolefin family includes a number of olefin copolymers.

Polyethylenes. Developed about 25 years ago, polyethylenes were the first of the olefins. Today, they are one of the lowest cost and one of the most widely used plastics. As a group, they are noted for toughness, excellent dielectric strength, and chemical resistance. Another outstanding characteristic is their low water absorption and permeability, which is the reason for their wide use in sheet form as moisture barriers. They are white in thick sections, but otherwise the range varies from translucent to opaque. They feel waxy.

The many available types of polyethylene, ranging from flexible to rigid materials, are classified by density (specific gravity) into three major groups:

Low density: 0.910 to 0.925.
Medium density: 0.926 to 0.940.
High density: 0.941 to 0.959.

The variations in properties among these three groups are directly related to density. As density increases, polymer cross-bonding or branching and crystallinity increase. Thus stiffness, tensile strength, hardness, and heat and chemical resistance increase with density in polyethylenes. Low-density polyethylenes are flexible, tough, and less translucent than high-density grades. High-density grades, often called linear polyethylene grades, are stronger, more rigid, and have high creep resistance under load, but they have lower impact resistance.

Typical uses of low-density polyethylenes include blow-molded bottles and containers, gaskets, paintbrush handles, and flexible-film packaging. High-density grades are used for wire insulation, beverage cases, dishpans, toys, and the film used for boil-in-bag packaging. In general, polyethylenes are not used in load-bearing applications because of their tendency to creep. However, a special type, high-

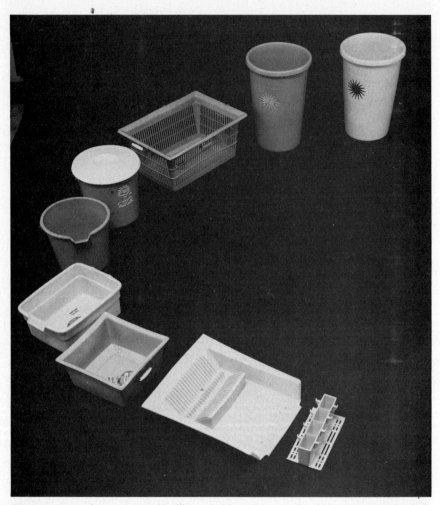

Fig. 9-1 Typical injection-molded household products made of low-density polyethylene. (Union Carbide Corp.)

molecular-weight polyethylene, is used for machine parts, bearings, bushing, and gears.

Polyethylenes can be processed, shaped, and formed by all thermoplastic processes. Injection and blow molding and film and sheet extrusion are widely used. Hollow parts, such as housings and containers, are commonly produced in low-density and some high-density grades by rotational molding.

Polyethylenes can be blended or combined with other monomers—propylene, ethyl acrylate, and vinyl acetate—to produce copolymers to improve such properties as stress-crack resistance and clarity and to increase flexibility. They can also be modified by exposure to high-energy radiation, which produces cross-linking and thereby increases heat resistance and stiffness.

Polypropylenes. These were developed in 1957 in Italy and Germany. They are similar in many respects to high-density polyethylenes. Their mechanical properties—particularly strength and stiffness—are at the high end of the polyethylene range. In color, they are milky white and semitranslucent.

Fig. 9-2 Many blow-molded containers, such as these bottles, are made of high-density polyethylene. (Union Carbide Corp.)

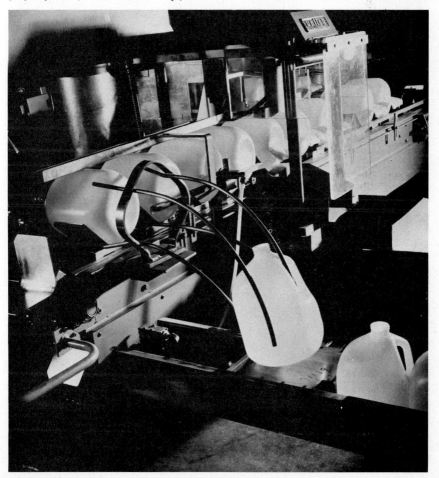

Table 9-2 *Typical Properties of Polyethylenes*

Property	Low Density	Medium Density	High Density	High Molecular Weight
Specific gravity	0.910–0.925	0.926–0.940	0.941–0.959	0.940
Modulus of elasticity in tension, 10^5 psi	0.21–0.27	0.35–0.75	0.6–1.5	1.0
Tensile strength, 1,000 psi	1.4–2.5	2.0–2.4	3.0–5.0	5.0
Elongation, %	500–725	200–425	100–1,000	400
Hardness, Shore	D41–D46	D55–D56	D60–D70	D60–D65
Impact strength, ft-lb/in. (notch)	No break	—	1.5–20	No break
Maximum service temperature, °F	175–200	200–235	250–266	200
Volume resistance, ohm-cm	10^{17}–10^{19}	10^{15}	10^{15}	10^{15}
Dielectric strength, V/mil	480	480	480	480

The many different grades of polypropylenes fall into three basic groups: homopolymers, copolymers, and reinforced and polymer blends. Properties of the homopolymers vary with molecular-weight distribution and the degree of crystallinity. Commonly, copolymers are produced by adding other types of olefin monomers to the propylene monomers to improve properties such as low-temperature toughness. Copolymers are also made by radiation grafting. Polypropylenes are frequently reinforced with glass or asbestos fibers to improve mechanical properties and increase resistance to deformation at elevated temperatures.

As a class, the wide usefulness of polypropylenes can be attributed to their rather nice balance of a number of desirable properties. They are light in weight, fairly rigid, and have relatively good resistance to heat and a wide range of chemicals. They also have negligible water absorption, excellent electric resistivity, and are easily processed.

A unique property is their ability in thin sections to withstand prolonged flexing. This characteristic has made polypropylenes popular for "living hinge" applications. In tests, they have been flexed over 70 million times without failure.

Polypropylenes have relatively low long-term creep resistance, and the homopolymers lose most of their impact strength at low temperatures. Reinforced with asbestos, they have good strength and stiffness at temperatures up to 250°F.

About 50 percent of the polypropylenes produced is processed into finished parts by injection molding, thus indicating their good processability. Moldings are noted for their high surface gloss and hardness. Other applicable processing methods are extrusion, blow molding, laminating, sintering, and rotational molding. However, because of their low melt strength, polypropylenes are not readily thermoformed. Also, blow-molded parts are limited in size.

This good combination of properties makes polypropylenes suitable for many appliance and automotive products, such as washing machine agitators and automotive interior parts and closure panels. The high-impact and reinforced grades are used for kick panels, fender skirts, lamp housings, and battery cases. And polypropylenes' good electrical properties qualify them for TV yokes, coil bobbins, cable covers, insulators, and fuse housings.

Olefin Copolymers. The principal olefin copolymers are the polyallomers, ionomers, and ethylene copolymers. The polyallomers, which are highly crystalline, can be formulated to provide high stiffness and medium impact strength; moderately high stiffness and high impact strength; or extra-high impact strength (Table 9-4). Polyallomers, with their unusually high resistance to flexing fatigue, have "hinge" properties better than those of polypropylenes. They have the characteristic

Table 9-3 *Typical Properties of Polypropylenes*

Property	General Purpose	High Impact	Asbestos Filled
Specific gravity	0.900–0.910	0.900–0.910	1.11–1.36
Modulus of elasticity in tension, 10^5 psi	1.6–2.2	1.3	—
Tensile strength, 1,000 psi	4.8–5.5	—	—
Elongation, %	30>200	30>200	3–20
Hardness, Rockwell R	80–100	28–95	90–100
Impact strength, ft-lb/in. (notch)	0.4–2.2	1.5–12	0.5–1.5
Maximum service temperature, °F	240	240	250
Volume resistance, ohm-cm	$>10^{17}$	$>10^{17}$	1.5×10^{15}
Dielectric strength, V/mil	650	450–650	450

Table 9-4 *Typical Properties of Some Olefin Copolymers*

	Polyallomer	Ionomer	EVA
Specific gravity	0.898–0.905	0.94	0.94
Tensile strength, 1,000 psi	3–4.5	3–5	0.5–1.0
Elongation, %	350	450	650
Hardness, shore D	50–85 R(R)	60	35
Impact strength, ft-lb/in. (notch)	1.5	9–14	—
Softening point, Vicat, °F	250–275	162	147
Dielectric strength, V/mil	500–650	1000	525
Volume resistivity, ohm-cm	10^{15}	10^{15}	—

milky color of polyolefins; they are softer than polypropylene, but have greater abrasion resistance. Polyallomers are commonly injection-molded, extruded, and thermoformed, and they are used for such items as typewriter cases, snap clasps, threaded container closures, embossed luggage shells, and food containers.

Ionomers are nonrigid plastics characterized by low density, transparency, and toughness. Unlike polyethylenes, density and other properties are not crystalline-dependent. Their flexibility, resilience, and high molecular weight combine to provide high abrasion resistance. They have outstanding low-temperature flexural properties, but should not be used at temperatures above 160°F. Resistance to attack from organic solvents and stress cracking chemicals is high. Ionomers have high melt strength for thermoforming and extrusion coating, and a broad temperature range for blow molding and injection molding. Representative ionomer parts include injection-molded containers, housewares, tool handles, and closures; extruded film, sheet, electrical insulation, and tubing; and blow-molded containers and packaging.

There are four commercial ethylene copolymers, of which ethylene vinyl acetate (EVA) and ethylene ethyl acrylate (EEA) are the most common.

Ethylene vinyl acetate (EVA) copolymers approach elastomers in flexibility and softness, although they are processed like other thermoplastics. Many of their properties are density-dependent, but in a different way from polyethylenes. Softening temperature and modulus of elasticity decrease as density increases, which is contrary to the behavior of polyethylene. Likewise, the transparency of EVA increases with density to a maximum that is higher than that of polyethylenes, which become opaque when density increases above around 0.935 g/cm³. Although EVA's electrical properties are not as good as those of low-density polyethylene, they are competitive with vinyl and elastomers normally used for electrical products. The major limitation of EVA plastics is their relatively low resistance to heat and solvents, the Vicat softening point being 147°F. EVA copolymers can be injection-, blow-, compression-, transfer-, and rotationally molded; they can also be extruded. Molded parts include appliance bumpers and a variety of seals, gaskets, and bushings. Extruded tubing is used in beverage vending machines and for hoses for air-operated tools and paint spray equipment.

Ethylene ethyl acrylate (EEA) is similar to EVA in its density-property relationships. It is also generally similar to EVA in high-temperature resistance, and like EVA it is not resistant to aliphatic and aromatic hydrocarbons as well as chlorinated versions thereof. However, EEA is superior to EVA in environmental stress cracking and resistance to ultraviolet radiation. As with EVA, most of EEA's applications are related to its outstanding flexibility and toughness. Typical uses are household products such as trash cans, dishwasher trays, flexible hose and water pipe, and film packaging.

Two other ethylene copolymers are ethylene hexene (EH) and ethylene butene (EB). Compared with the other two, these copolymers have greater high-temperature resistance, their useful service range being between 150 and 190°F. They are also stronger and stiffer, and therefore less flexible, than EVA and EEA. In general, EH and EB are more resistant to chemicals and solvents than the other two, but their resistance to environmental stress cracking is not as good.

9-2 *Polystyrenes*

Polystyrenes are second only to polyethylenes in volume of use. They are a large family of thermoplastics based on the styrene monomer. Styrene resins are basically rigid, with low crystallinity and considerable branching. However, various modifications create a wide range of structures and properties. As a class, polystyrenes are low in cost, exceptional in electrical-insulating properties, range in appearance from clear to almost any color, have high hardness and gloss, and are readily molded.

Although they have relatively high tensile strengths, they are subject to creep. Therefore the long-term bearing strength (over 2 weeks) is only about one-third the short-time tensile strength. Also, since their maximum useful service temperature is about 160°F, their use is restricted chiefly to room-temperature applications.

Because of their low cost and ease of processing, polystyrenes are widely used for consumer products. The impact grades and glass-

Table 9-5 *Typical Properties of Polystyrenes*

Property	General Purpose	Impact Grades	Styrene Acrylonitrile (SAN)
Specific gravity	1.04	1.04–1.07	1.04–1.07
Modulus of elasticity in tension, 10^5 psi	4.6–5.0	2.0–4.7	4.0–5.5
Tensile strength, 1,000 psi	5–10	3.3–6.0	9.5–13
Elongation, %	1.0–2.3	3.0–40	0.5–3.7
Hardness, Rockwell M	65–80	20–65	75–90
Impact strength, ft-lb/in. (notch)	0.2–0.4	0.5–10	0.30–0.50
Maximum service temperature, °F	160–205	125–165	175–190
Volume resistance, ohm-cm	$>10^{16}$	$>10^{16}$	$>10^{16}$
Dielectric strength, V/mil	400–500	300–600	400–500

filled types are used quite widely for engineering parts and semi-structural applications. Also, polystyrene foams are highest in volume use of all the plastics.

Polystyrenes can be divided into the following major types: general-purpose grades, the lowest in cost, are characterized by clarity, colorability, and rigidity. They are applicable where appearance and rigidity, but not toughness, are required. Common uses are wall tiles, compact cases, knobs, brush backs, and container lids.

Impact grades of polystyrenes are produced by physically blending styrene and rubber. Grades are generally specified as medium, high, and extra-high. As impact strength increases, rigidity decreases. Medium-impact grades are used where a combination of moderate toughness and translucency is desired, for example, in such products as containers, closures, and small radio cabinets. High-impact polystyrenes have improved heat resistance and surface gloss. They are used for refrigerator door liners and crisper trays, containers, toys, and heater ducts in automobiles. The extra-high-impact grades are quite low in stiffness, and their use is limited to parts subject to high-speed loading.

Styrene-acrylonitrile (SAN) polystyrene, sometimes considered the chemical-resistant grade, has excellent resistance to acids, bases, salts, and some solvents. It also is among the stiffest of the thermoplastics, with a tensile modulus of 400,000 to 550,000 psi.

Because of good processing characteristics, polystyrenes are produced in a wide range of forms. They can be extruded, injection compressed, blow molded, and thermoformed. They are also available as film sheet and foam.

9-3 Vinyls

Vinyls are the third most highly used plastic. Like polyethylenes and polystyrenes, they are low in cost and have wide versatility. Vinyls, as a class, are largely amorphous, branched polymers, and thus generally have low crystallinity, and are basically stiff and rigid. However, by various modifications and the use of plasticizers, flexible, rubber-like grades are produced.

Vinyls are noted for their excellent chemical resistance, and thus are used mainly in chemical and weathering environments. They also have good electrical resistivity and high abrasion resistance. They are self-extinguishing, but are not useful at temperatures above 165°F for long periods of time.

The most widely used vinyls, by far, are polyvinyl chloride (PVC) and PVC-acetate copolymer. They are produced in both rigid and flexible forms. Rigid PVC is composed of unplasticized vinyl homopolymers, and has a combination of high tensile-strength modulus, hardness, and toughness. However, relatively low creep strength limits its use for load-bearing applications. Rigid PVC is used extensively in sheet form for chemical tanks, ducts, and hoods, and as

Table 9-6 *Typical Properties of Vinyls*

Property	Nonrigid Polyvinyl Chloride (PVC) and PVC-Acetate Copolymer	Rigid Polyvinyl Chloride (PVC) and PVC-Acetate Copolymer	Vinylidene Chloride
Specific gravity	1.20–1.55	1.32–1.44	1.68–1.75
Modulus of elasticity in tension, 10^5 psi	Flexible	3.5–6.0	0.5–0.8
Tensile strength, 1,000 psi	1–3.5	5.5–8.0	4–8
Elongation, %	200–450	1–10	15–25
Hardness, Rockwell	—	R110–120	M50–65
Impact strength, ft-lb/in. (notch)	Flexible	0.4–20	1–5
Maximum service temperature, °F	150–220	150–165	170–200
Dielectric strength, V/mil	250–1,000	500–1,300	400–600

architectural shapes such as moldings, gutters, down spouts, and window frame parts. Rigid-PVC pipe is used in waterworks, and in the oil, chemical, and food processing industries.

Flexible PVC is composed of either homopolymers or copolymers with additions of plasticizers. Varying the percentage of plasticizer and adding fillers makes the range of flexible vinyls almost unlimited. Flexible vinyl film and sheeting are used in packaging, pipe wraps, and upholstery. Flexible extrusions in many different shapes and forms are used as electrical wire and cable insulation, gaskets, and weather stripping.

Vinylidene chloride, better known as saran, and polyvinyl dichloride are two other important vinyl plastics. Vinylidene chloride is flexible, transparent, and has greater chemical and solvent resistance than PVC, along with heat resistance up to 200°F. Polyvinyl dichloride is rigid, has improved solvent resistance, and retains strength and stiffness up to 212°F. It is being used for rigid components such as hot-water pipes and fittings.

Other members of the vinyl family are, in brief, polyvinyl butyral, which is used mainly for safety- and decorative-glass interlayers, fabric coatings, and in metal primers; polyvinyl formal, which is used for magnetic wire insulation and in coatings; and polyvinyl acetate, which is used in adhesive formulations.

9-4 *Acrylics*

Most acrylic plastics are based on polymers of methyl methacrylate which may be modified by copolymerizing or blending with other monomers. Acrylics are well known by their trade names Lucite and Plexiglas. They are noted for excellent optical properties, having a light transmission of about 92 percent. Besides the transparent grades, they can be obtained in translucent or opaque colors as well as the natural color of water white. Acrylic moldings have a deep luster and high surface gloss, and, for this reason, are widely used for decorative parts. Painted or vacuum metallized acrylics, for example, are used for automotive, appliance, and other product medallions.

Acrylics have excellent weathering characteristics. Because they are only slightly affected by sunlight, rain, and corrosive atmospheres, they are well suited for outdoor applications. In general, the majority of grades can be used up to about 200°F. Thermal expansion is relatively high.

Acrylics are hard, stiff, and strong. Their tensile strength ranges from 5,000 to about 11,000 psi, which is relatively high for plastics. However, regular grades are somewhat brittle. High-impact grades are

Table 9-7 *Typical Properties of Acrylics*

Property	Cast Sheets & Rods	General-Purpose Moldings	High-Impact Moldings
Specific gravity	1.17–1.20	1.18–1.19	1.12–1.16
Modulus of elasticity in tension, 10^5 psi	3.5–5.0	3.5–5.0	2.3–3.3
Tensile strength, 1,000 psi	6–10	9.5–10.5	5.5–8.0
Elongation, %	2–7	3–5	>25
Hardness, Rockwell	M80–100	M80–103	R100–120
Impact strength, ft-lb/in. (notch)	0.4	0.2–0.4	0.8–3
Maximum service temperature, °F	150–225	165–225	170–200
Volume resistance, ohm-cm	$>10^{15}$	$>10^{14}$	2×10^{16}
Dielectric resistance, V/mil	450–525	400	400–500

produced by blending with rubber stock. However, the high strength is useful only for short-term loadings. For long-term service, to avoid crazing or surface cracking, tensile stresses must be limited to about 1,500 psi.

About half of the acrylic-resin production is processed into sheets by casting against polished glass plates. Cast rods, tubes, and blocks are also common forms. Sheets are produced in thicknesses from ⅛ to ⅜ in. and in sizes to 10 by 12 ft. A special process that produces molecular orientation in the cast product is used to make crack-resistant aircraft cabin windows and fighter-plane canopies.

Acrylic molding powders are used for conventional injection moldings and extrusions. Moldings as large as 36 by 36 in. have been produced. Typical parts include knobs, handles, escutcheons, vending machine parts, and a wide variety of lenses for light control, signal lamps, and the like.

9-5 Cellulosics

The first plastic, celluloid, was developed in the last quarter of the nineteenth century. It is a cellulose nitrate polymer, but is not used much today because of its flammability.

Cellulosics are modified natural polymers produced from refined cellulose of wood pulp and cotton linters. Some of their major distinguishing characteristics are very high toughness, low strength, transparency, excellent colorability, and high surface gloss. They are easily processed with the addition of proper amounts of plasticizer. Although they are tough materials, they are not generally recom-

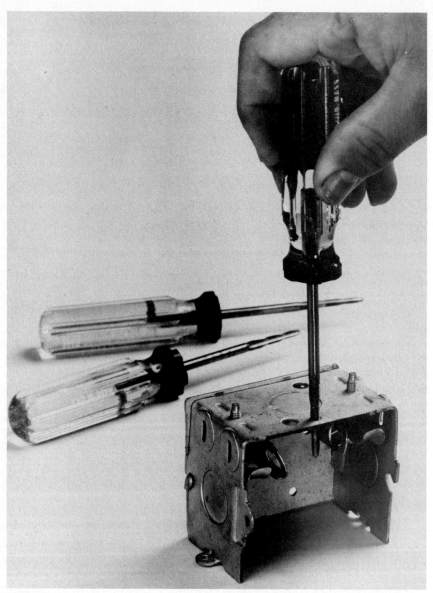

Fig. 9-3 Cellulose acetates are widely used for tool handles. (Celanese Plastics Co.)

mended for load-bearing applications because of their relatively low strength. As a class, they have good dielectric properties.

There are four major cellulosic plastics: Cellulose acetate, often simply called acetate, is the lowest in cost. It has good toughness and rigidity and is available in a variety of grades from soft to hard. It

Table 9-8 *Typical Properties of Cellulosics*

Property	Cellulose Acetate	Cellulose Acetate Butyrate	Cellulose Acetate Propionate
Specific gravity	1.29–1.31	1.15–1.20	1.18–1.22
Modulus of elasticity in tension, 10^5 psi	0.6–4	0.5–2	0.6–2.0
Tensile strength, 1,000 psi	2–8	3–7	4–6.5
Hardness, Rockwell R	35–120	20–114	60–100
Impact strength, ft-lb/in. (notch)	1.0–6.5	3–10	2–9.5
Maximum service temperature, °F	140–175	140–200	150–200
Volume resistance, ohm-cm	10^{10}–10^{13}	10^{11}–10^{14}	10^{11}–10^{14}
Dielectric strength, V/mil	250–600	250–400	300–450

softens at 175°F and has relatively high water absorption. Acetate is used in optical frames, pen barrels, toys, transparent machine covers, and small housings.

Cellulose acetate butyrate, commonly called butyrate or CAB, is somewhat tougher, and has lower moisture absorption and a higher softening point (190°F) than acetate. Special formulations with good weathering characteristics plus transparency are used for outdoor applications such as signs, light globes, and lawn sprinklers. Clear sheets of butyrate are available for vacuum-forming applications. Other typical uses include transparent dial covers, television screen shields, tool handles, and typewriter keys. Extruded pipe is used for electrical conduits, pneumatic tubing, and low-pressure waste lines.

Cellulose propionate is similar to butyrate in both cost and properties. Some grades have slightly higher strength and modulus of elasticity. Propionate has better molding characteristics but lower weatherability than butyrate. Molded parts include steering wheels, fuel filter bowls, and appliance housings. Transparent sheeting is used for blister packaging and food containers.

Ethyl cellulose, considered the toughest of the cellulosics, has high impact strength over a wide temperature range. It has good impact resistance down to −40°F. It is also the lightest and has the lowest water absorption of the cellulosics. However, it is less abrasion and scratch resistant than the others. Ethyl cellulose is used for helmets, rollers, slides, gears, flashlight housings, and refrigerator parts.

Cellulosics are formed by all the conventional processes. Molded parts usually have a fine, lustrous surface. The harder grades are favored for extrusion because of their greater melt strength. Thermoforming is commonly used for shaping sheeting.

9-6 Polyamides (Nylon)

Nylon is probably the best known of all synthetic materials. Although its extensive popular recognition stems from its use as a fiber in consumer products, especially stockings, it is also an important engineering plastic. Since it has good mechanical and friction properties and high chemical resistance, it was the first load-bearing plastic, and today is widely used for machine parts.

Nylon basically has a linear polymer structure, with a relatively high degree of crystallinity. This is mainly why all nylons have excellent mechanical properties, including high tensile and fatigue strengths and good impact resistance. They also possess very high abrasion resistance and good frictional characteristics, and are resistant to most chemicals and solvents with the exception of some strong acids and certain polar solvents. Their major disadvantage is their relatively large moisture absorption and resulting dimensional changes, which can be more than 2 percent in a 100 percent relative humidity.

The most common types of nylon, designated 6/6, 6, and 6/10, have the same structure, differing only in the length of the polymer molecules. This structural difference accounts for the variations among them in melting points, stiffness, and water absorption.

Type 6/6 is the most widely used, perhaps because it is the strongest over the widest range of moisture and temperature. Unlike most plastics, it has a sharp melting point and exhibits no appreciable creep below that point. Type 6 is quite similar in properties to type 6/6, but has greater water absorption and retains less strength with increases in moisture and temperature. It has a somewhat lower melting point and slightly lower modulus. However, it has better processability and permits closer tolerances in molded parts.

Type 6/10 (also 6/11 and 6/12) has substantially less moisture absorption than the other types, and therefore has a three- to fourfold improvement in dimensional stability. Although more flexible, it is about the same as the other types in most other respects.

In addition to the above homopolymer types, a number of copolymers have been developed for specific uses. Also, nylon plastics with a wide variety of fillers and reinforcements are available. For example, the solid lubricant molybdenum disulfide is used as a filler to further improve wear, and frictional and mechanical properties. And glass fibers can be added to increase tensile strengths up to as high as 30,000 psi.

Nylons are processed mainly by molding and extruding. Sintering of nylon in powder form is often used for small- to medium-sized parts. In this process, oil, graphite, or molybdenum disulfide fillers

Table 9-9 *Typical Properties of Nylons*

Property	Type 6	Type 6/6	Type 6/6, Glass Filled, 30 Percent	Type 6/10
Specific gravity	1.14	1.13–1.15	1.37	1.08
Modulus of elasticity in tension, 10^5 psi	—	4.75	14	2.8
Tensile strength, 1,000 psi	9.5–12.5	11.8	25	8.5
Elongation, %	30–220	60	1.8	85
Hardness, Rockwell	R120	R118	E60	R111
Impact strength, ft-lb/in. (notch)	0.8–1.2	1.0	2.5	1.6
Maximum service temperature, °F	250–300	250–300	250–300	250–300
Volume resistance, ohm-cm	4.5×10^{13}	10^{14}–10^{15}	5.5×10^{15}	10^{15}
Dielectric strength, V/mil	385	385	400	470

can be added to enhance bearing properties. For parts over 3 lb, casting the liquid nylon polymer into molds is often more economical than injection molding.

The exceptional range of properties makes nylons suitable for many machine parts. Their most common industrial use is as a bearing material. Typical applications are gears, cams, sleeve bearings, slide fasteners, door hinges, wire insulation, high-pressure flexible tubing, and a wide variety of parts in textile- and food-handling machinery.

9-7 Acetals

Acetals are highly crystalline plastics that were specifically developed to compete with zinc and aluminum die castings. The natural acetal resin is translucent white and can be readily colored. There are two basic types: a homopolymer (Delrin) and a copolymer (Celcon). In general, the homopolymers are harder, more rigid, and have higher tensile flexural and fatigue strength, but lower elongation. The copolymers are more stable in long-term high-temperature service and have better resistance to hot water. Special types of acetals are glass filled, providing higher strengths and stiffness, and tetrafluoroethylene (TFE) filled, providing exceptional frictional and wear properties.

Acetals are among the strongest and stiffest of the thermoplastics. Their tensile strength ranges from 8,000 to about 13,000 psi, the tensile modulus of elasticity is about 500,000 psi, and fatigue strength at room temperature is about 5,000 psi. Their excellent creep resistance and low moisture absorption (less than 0.4%) give them excellent dimensional stability. They are useful for continuous service up to about 220°F (Table 9-10).

Acetals' low friction and high abrasion resistance, though not as good as nylon's, rates them high among thermoplastics. Their impact resistance is good and remains almost constant over a wide temperature range. Acetals are attacked by some acids and bases, but have excellent resistance to all common solvents.

They are processed mainly by molding or extruding. Some parts are also made by blow and rotational molding. Typical parts and products made of acetal include pump impellers, conveyor links, drive sprockets, automobile instrument clusters, spinning reel housings, gear valve components, bearings, and other machine parts.

9-8 *Polycarbonate*

Polycarbonate is a linear, low-crystalline, transparent, high-molecular-weight plastic. It is generally considered to be the toughest of all plastics. In thin sections, up to about 3/16 in., its impact strength is as high as 16 ft-lb/in. (notch) (see Table 9-10). In addition, polycarbonate is one of the hardest plastics. It also has good strength and rigidity, and, because of its high modulus of elasticity, is resistant to creep. These properties, along with its excellent electrical resistivity, are maintained over a temperature range of about −275 to 250°F. It has negligible moisture absorption, but it also has poor solvent resistance, and, in a stressed condition, will craze or crack when exposed to some chemicals. It is generally unaffected by greases, oils, and acids.

Polycarbonate is easily processed by extrusion, by injection, blow, and rotational molding, and by vacuum forming. It has very low and uniform mold shrinkage. With a white light transmission of almost 90 percent and high impact resistance, it is a good glazing material. It has more than 30 times the impact resistance of safety glass. Other typical applications are safety shields and lenses. Besides glazing, polycarbonate's high impact strength makes it useful for air-conditioner housings, filter bowls, portable tool housings, marine propellers, and housings for small appliances and food-dispensing machines.

Table 9-10 Typical Properties of Acetal, Polyphenylene Oxide, and Carbonate Plastics

Property	Acetal Homopolymer		Acetal Copolymer		Phenylene Oxide		Poly-carbonate
	Standard	20% Glass	Standard	25% Glass	PPO	Noryl	
Specific gravity	1.43	1.56	1.41	1.61	1.06	1.06	1.51
Modulus of elasticity in tension, 10^5	5.2	6–10	4.1	12.5	3.8	3.6	17
Tensile strength, 1,000 psi	12	8.5	8.8	18.5	11	9.6	18
Elongation, %	25	7	65	3	60	60	0–5
Hardness, Rockwell	M94	M90	M80	M79	R120	R120	M97
Impact strength, ft-lb/in. (notch)	1.4	0.8	1.3	1.8	1.5–1.9	1.3	5–16
Maximum service temperature, °F	195	195	220	220	212	212	250
Volume resistance, ohm-cm	10^{15}	5×10^{14}	10^{14}	1.2×10^{17}	10^{17}	10^{17}	1.4×10^{15}
Dielectric strength, V/mil	500	500	500	500	—	—	475

242   Industrial and Engineering Materials

Fig. 9-4   Polycarbonate, one of the toughest and unbreakable plastics, is used as glazing. (General Electric Co.)

9-9 *ABS Plastics*
The letters ABS identify a family of terpolymer plastics composed of the monomers acrylonitrile, butadiene, and styrene. Common trade names for these materials are Cycolac, Kralastic, and Lustran. They are opaque and distinguished by a good balance of properties, including high impact strength, rigidity, and hardness over a wide temperature range (−40 to 230°F). Compared with other structural or engineering plastics, they are generally considered to fall at the lower end of the scale in strength.

Medium-impact grades are hard, rigid, and tough, and are used for appearance parts that require high strength, good fatigue resistance, and surface hardness and gloss. The high-impact grades are formulated for similar products where additional impact strength is gained at some sacrifice in rigidity and hardness. The extra-high-impact grades provide the higher room-temperature impact resistance with further decreases in strength, rigidity, and hardness. The low-temperature impact grades have high impact strength down to −40°F. Again, some sacrifice is made in strength, rigidity, and heat resistance.

Heat-resistant, high-strength grades provide the best heat resistance (continuous use up to about 200°F, and a 264-psi heat-distortion temperature of around 215°F). Impact strength is about comparable

Table 9-11 *Typical Properties of ABS Plastics*

Property	Medium Impact	Extra-high Impact	Low-Temperature Impact	Heat Resistant
Specific gravity	1.05–1.07	1.01–1.06	1.02–1.04	1.06–1.08
Modulus of elasticity in tension, 10^5 psi	3.3–4.0	2.0–3.1	2.0–3.1	3.5–4.2
Tensile strength, 1,000 psi	6.3–8.0	4.5–6.0	4–6	7–8
Elongation, %	5–20	20–50	30–200	20
Hardness, Rockwell R	108–115	85–105	75–95	107–116
Impact strength, ft-lb/in. (notch)	2–4	5–7	6–10	2–4
Maximum service temperature, °F	180	180	180	220

to that of medium-impact grades, but strength, modulus of elasticity, and hardness are higher.

At stresses above their tensile strength, ABS plastics usually yield plastically rather than rupturing, and impact failures are ductile. Because of their relatively low creep, they have good long-term load-carrying ability for stresses up to 1,000 psi. The low creep plus low water absorption and relatively high heat resistance provide ABS plastics with good dimensional stability.

ABS plastics are readily processed by extrusion, injection and blow molding, calendering, and vacuum forming. Special resins have been developed for cold forming or stamping from extruded sheet. Typical applications are helmets, refrigerator liners, luggage, tote trays, housings, grills for hot-air systems, and pump impellers. Extruded shapes include tubing and pipe. ABS-plated parts are now in wide use, replacing metal parts in the automotive and appliance fields. Typical plated applications include mirror housings, knobs, decorative trim, insignia, and some plumbing fixtures.

9-10 Fluoroplastics

The fluoroplastics are a group of high-performance, high-price, engineering plastics. They are composed basically of linear polymers in which some or all of the hydrogen atoms are replaced with fluorine, and are characterized by relatively high crystallinity and molecular weight.

All fluoroplastics are natural white and have a waxy feel. They

Fig. 9-5 This automobile shift console is made of vinyl laminated to an ABS plastic. The composite sheet is vacuum formed. (Uniroyal Inc.)

range from semirigid to flexible. As a class, they rank among the best of the plastics in chemical resistance and elevated-temperature performance. Their maximum service temperature ranges up to about 500°F. They also have excellent frictional properties and cannot be wet by many liquids. Their dielectric strength is high and is relatively insensitive to temperature and power frequency. Mechanical properties, including tensile creep and fatigue strength, are only fair, although impact strength is relatively high. Their mechanical properties are considerably improved by adding fillers and reinforcements.

There are three major classes of fluoroplastics. In order of decreasing fluorine replacement of hydrogen, they are fluorocarbons, chlorotrifluoroethylene, and fluohydrocarbons.

There are two fluorocarbon types: tetrafluoroethylene (PTFE or TFE) and fluorinated ethylene propylene (FEP). They are similar in most respects. PTFE is the most widely used fluoroplastic. It has the highest useful service temperature (500°F) and chemical resistance. FEP's chief advantage is its low-melt viscosity, which permits it to be conventionally molded.

Chlorotrifluoroethylene (CTFE or CFE) is stronger and stiffer than the fluorocarbons and has better creep resistance. Like FEP and unlike PTFE, it can be molded by conventional methods.

The fluohydrocarbons are of two kinds: polyvinylidene fluoride (PVF_2) and polyvinyl fluoride (PVF). While similar to the other fluoroplastics, they have somewhat lower heat resistance, and considerably higher tensile and compressive strength.

Except for PTFE, the fluoroplastics can be formed by molding, extruding, and other conventional methods. However, processing must be carefully controlled. Because PTFE cannot exist in a true molten state, it cannot be conventionally molded. The common method of fabrication is by compacting the resin in powder form and then sintering.

Typical uses of fluoroplastics are self-lubricating bearings and components, chemical-resistant pipes and parts, high-temperature electronic components, tank linings, gaskets, packings, and seals.

9-11 Chlorinated Polyether

This plastic has a good combination of mechanical and chemical-resistant properties. It has a tensile strength of around 6,000 psi and good creep resistance, and it maintains its mechanical properties in a wide range of chemical and corrosive environments (see Table 9-12). However, its high price (approximately $4/lb) limits it to special high-performance parts chiefly in chemical-processing equipment.

9-12 Polyphenylene Oxide

This is a relatively new plastic that is notable for its high strength and broad temperature resistance. There are two major types: phenylene oxide (PPO) and modified phenylene oxide (Noryl) (see Table 9-10). These materials have a deflection temperature ranging from 212 to 345°F at 264 psi. Their coefficients of linear thermal expansion are among the lowest for engineering thermoplastics. Room-temperature strength and modulus of elasticity are high and creep is low. In addition, they have good electrical resistivity. Their ability to withstand steam sterilization and their hydrolytic stability make them suitable for medical instruments, electric dishwashers, and food dispensers. They are also used in the electrical and electronic fields and for business-machine housings.

Table 9-12 *Typical Properties of Fluoroplastics and Chlorinated Polyether*

Property	Tetra-fluoro-ethylene (PTFE)	Ethylene Propylene (FEP)	Poly-vinylidine Fluoride (PVF$_2$)	Chlorinated Polyether
Specific gravity	2.1–2.3	2.14–2.17	1.77	1.4
Modulus of elasticity in tension, 10^5 psi	0.40–0.65	0.5–0.7	1.7–2.0	1.5
Tensile strength, 1,000 psi	2.5–6.5	2.5–3.5	7.2–8.6	6
Elongation, %	250–350	250–325	200–300	130
Hardness, Rockwell	D52	D58	R110	R100
Impact strength, ft-lb/in. (notch)	2.5–4.0	No break	3.8	0.4
Maximum service temperature, °F	500	400	340	275
Volume resistance, ohm-cm	$>10^{18}$	$>2 \times 10^{18}$	5×10^{14}	1.5×10^{16}
Dielectric strength, V/mil	400–500	500–600	260	400

9-13 Superpolymers

Superpolymers are several plastics developed in recent years that maintain mechanical and chemical integrity above 400°F for extended periods. They are polyimide, polysulfone, polyphenylene sulfide, polyarylsulfone, and aromatic polyester.

In addition to their high-temperature resistance, superpolymers have in common high strength and modulus of elasticity, and excellent resistance to solvents, oils, and corrosive environments. They are also high in cost, ranging from $1 to over $20 per lb (at this writing). Their major disadvantage is processing difficulty. Molding temperatures and pressures are extremely high compared to conventional plastics. Some of them, including polyimide and aromatic polyester, are not molded conventionally. Because they do not melt, the molding process is more of a sintering operation. Their high price largely limits their use to specialized applications in the aerospace and nuclear-energy fields.

One indication of the high-temperature resistance of the superpolymers is their glass transition temperature of well over 500°F, as compared with less than 350°F for most conventional plastics. In the case of polyimides, the glass temperature is greater than 800°F, and the material decomposes rather than softens when heated excessively.

Polysulfone has the highest service temperature of any melt-processable thermoplastic. Its flexural modulus stays above 300,000 psi at up to 320°F. At such temperatures it does not discolor or degrade.

Aromatic polyester, a homopolymer, does not melt, but at 800°F can be made to flow in a nonviscous manner similar to metals. Thus, filled and unfilled forms and parts can be made by hot sintering, high-

Fig. 9-6 Polysulfone is used for several parts in Polaroid's SX-70 camera. (Union Carbide Corp.)

Table 9-13 *Typical Properties of Heat-Resistant Plastic (Superpolymers)*

Property	Polyimide	Polysulfone	Polyphenylene Sulfide	Polyarylsulfone	Aromatic Copolyester
Specific gravity	1.43–1.47	1.24	1.34	1.36	1.40
Modulus of elasticity in tension, 10^5 psi	5.4–7.5	3.6	4.8	3.7	3.5
Tensile strength, 1,000 psi	10.5	10.2	—	13	14
Elongation, %	1.2	50–100	3	15–20	8
Hardness, Rockwell	H85	R120	R124	M110	R88
Impact strength, ft-lb/in. (notch)	0.5	1.3	—	5.0	1.0
Maximum service temperature, °F	500	340	500	500	550
Volume resistance, ohm-cm	4×10^{15}	5×10^{16}	—	3.2×10^{16}	10^{15}
Dielectric strength, V/mil	310	425	595	350	350

velocity forging, and plasma spraying. Notable properties are high thermal stability, good strength at 600°F, high thermal conductivity, good wear resistance, and extra-high compressive strength. Aromatic copolyesters have also been developed for injection and compression molding. They have long-term thermal stability and a strength of 3,000 psi at 550°F.

Plastics Alloys. Plastics, like metals, can be alloyed. And like metal alloys, the resulting materials have different, and often better, properties than the base materials making up the alloys.

There are about a half-dozen plastic alloys commercially available. However, a number of others are in use that have not been announced for various reasons. Some have been developed by users who do not want to reveal any information about them. And others have been "tailor-made" by resin suppliers for large-volume special applications.

The plastics most widely used in alloys today are polyvinyl chloride

(PVC), ABS, and polycarbonate. These three plastics can be combined with each other or with other types of polymers.

ABS, besides being used with polycarbonate, can also be alloyed with polyurethane. Commercially available in two grades, these alloys combine the excellent toughness and abrasion resistance of the urethanes with the lower cost and rigidity of ABS. The materials can be injection molded into large parts but cannot be extruded. Typical applications for which they are suitable include such parts as wheel treads, pulleys, low-load gears, gaskets, automotive grilles, and bumper assemblies.

ABS is also being successfully combined with polyvinyl chloride; this is available commercially in several grades. One of the established grades provides self-extinguishing properties, thus eliminating the need for intumescent (nonburning) coatings in present ABS applications, such as power tool housings, where self-extinguishing materials are required. A second grade possesses an impact strength about 30 percent higher than general-purpose ABS. This improvement, combined with this grade's ability to be readily molded, has resulted in its use for automobile grilles.

ABS-PVC alloys also can be produced in sheet form. The sheet materials have improved hot strength, which allows deeper draws than are possible with standard rubber-modified PVC base sheet. They also are nonfogging when exposed to the heat of sunlight. This is an important advantage where transparent materials are needed. Some properties of ABS-PVC alloys are lower than those of the base resins. Rigidity, in general, is somewhat lower, and tensile strength is more or less dependent on the type and amount of ABS in the alloy.

Another sheet material, an alloy of about 80 percent PVC and the rest acrylic plastic, combines the nonburning properties, chemical re-

Table 9-14 *Properties of Typical Plastics Alloys*

	ABS-Poly-carbonate	PVC-Acrylic
Specific gravity	1.04	1.35
Tensile strength, psi	6,200	6,500
Tensile modulus, 10^5 psi	3.7	3.4
Impact strength, ft-lb/in. (notch)	10	15
Heat deflection temperature, °F at 255 psi	250	160

sistance, and toughness of vinyl plastics with the rigidity and deep drawing merits of the acrylics. The PVC-acrylic alloy approaches some metals in its ability to withstand repeated blows. Because of its unusually high rigidity, sheets ranging in thickness from 0.60 to 0.187 in. can be formed into thin-walled, deeply drawn parts. Typical thermoformed products include luggage, truck cargo liners, and a wide variety of machine and equipment housings.

PVC is also alloyed with chlorinated polyethylene (CPE) by end-users to gain materials with improved outdoor weathering or to obtain better low-temperature flexibility. Applications include wire and cable jacketing, extruded and molded shapes, and film sheeting.

Acrylic-base alloys with a polybutadiene additive have also been developed, chiefly for blow-molded products. The acrylic content can range from 50 to 95 percent, depending on the application. Besides blow-molded bottles, the alloys are suitable for thermoformed products such as tubs, trays, and blister pods. The material is rigid and tough and has good heat-distortion resistance up to 180°F.

Another group of plastics, polyphenylene oxide (PPO), can be blended with polystyrene to produce an alloy with improved processing traits and lower cost than nonalloyed PPO. The addition of polystyrene reduces tensile strength and heat deflection temperature somewhat and increases thermal expansion.

Thermosetting Plastics

As already discussed, the major structural characteristic of thermosets is that the polymer chains are bonded to each other by strong covalent bonds in contrast to the weaker secondary bonds that prevail in thermoplastics. This primary cross-linking accounts for the major attributes of thermosets and their differences in behavior from the thermoplastics.

Thermosetting plastics are hard and relatively brittle substances. They have greater thermal stability than thermoplastics and greater resistance to creep. However, fillers or reinforcements must almost always be added to the resins to make them useful as moldings. They are generally more difficult to process than thermoplastics, and require longer molding cycles. But, once molded or formed by heat and polymerized, they remain stable and will not return to their original state; that is, they are infusible, will not soften or flow on reheating, and are not chemically soluble.

Thermosetting plastics in the raw-material stage are composed of two major components: a resin system and fillers and/or reinforcements. The resin component usually consists of a polymer, curing agents, hardeners, inhibitors, and plasticizers. It has a major influence on the finished plastic's dimensional stability, heat and chemical resistance, electrical properties, and flammability. The other component consists of one or more of the following: mineral or organic particles, inorganic or organic fibers, inorganic or organic chopped cloth or paper. This component usually has a major effect on the plastic's mechanical properties.

The mixture of components is usually supplied in the form of dry pellets or granules, as soft putty, or as slugs or ropes. The processing cycle usually involves several steps:

1. Preforming, in which the compound is compressed at ambient temperature into a shape that fits into the mold cavity.
2. Preheating, in which the preformed shape is heated to shorten the molding time and to promote flow of the material in the mold.
3. Molding, in which the compound is shaped and polymerized (cured) under heat and pressure to produce the finished plastic part.

9-14 Phenolics

Phenolics, the oldest of the thermosetting plastics, are known chemically as phenol formaldehyde. Their natural color is brown, darkening with age. As a group, they are among the least costly of the thermosets. They are readily molded, and rank high in stiffness and impact resistance. Although their electrical resistivity is low compared to the other thermosets, they are still widely used for their electrical-insulating qualities.

The hundreds of different phenolic molding compounds can be divided into the following six groups on the basis of major performance characteristics:

1. General-purpose phenolics are low-cost compounds with fillers such as wood flour and flock, and are formulated for noncritical functional requirements. They provide a balance of moderately good mechanical and electrical properties, and are generally suitable in temperatures up to 300°F.
2. Impact-resistant grades are higher in cost. They are designed for use in electrical and structural components subject to impact loads. The fillers are usually either paper, chopped fabric, or glass fibers.
3. Electrical grades, with mineral fillers, have high electrical resistivity plus good arc resistance, and they retain their resistivity under high-temperature and high-humidity conditions.

Fig. 9-7 The insulating qualities of phenolics make them suitable for many electrical parts, such as these plugs. (Union Carbide Corp.)

4. Heat-resistant grades are usually mineral- or glass-filled compounds that retain their mechanical properties in the 375 to 500°F temperature range. Special grades, such as phenyl silanes, provide long-term stability at temperatures up to 550°F.
5. Special-purpose grades are formulated for service applications requiring exceptional resistance to chemicals or water, or combinations of conditions such as impact loading and a chemical environment. The chemical-resistant grades are inert to most common solvents and weak acids, and their alkali resistance is good.
6. Nonbleeding grades compounded specially for use in container closures and for cosmetic cases.

About one-third of all phenolic resins are processed into parts by molding. Compression and transfer molding are the principal processes used, but phenolics can also be extruded and injection molded. Typical molded parts are washing machine agitators, wheels, motor housings, electrical coil forms, ignition parts, fuse blocks, photographic development tanks, and radio and TV cabinets. Other major

Table 9-15 *Typical Properties of Phenolics*

Property	No Filler	General Purpose (Wood flour & Flock)	Heat Resistant (Mineral)	Impact Resistant (Glass Fiber)	Electrical (Mineral)
Specific gravity	1.28	1.34–1.46	1.36–1.43	1.75–1.90	1.6–3.0
Modulus of elasticity in tension, 10^5 psi	7.5–10	8–13	14–15	30–33	10–30
Tensile strength, 1,000 psi	7.5	5.0–8.5	5–7	5–10	6
Elongation, %	—	—	0.40–0.55	0.2	—
Hardness	M126	E85–100	E80–94	E50–70	E80–90
Impact strength, ft-lb/in. (notch)	Nil	0.24–0.50	0.6–0.8	10–33	0.32
Maximum service temperature, °F	—	300–350	250–300	350–450	400
Volume resistance, ohm-cm	10^{11}	10^9–10^{12}	$>10^{10}$	10^{13}	6×10^{12}
Dielectric strength, V/mil	—	200–425	200–350	200–370	380

uses are as the binder or adhesive in laminated materials and as binders in grinding wheels, friction parts, and particle board.

9-15 Polyesters and Alkyds

Polyesters. The name polyester in a general sense covers three classes of commercial polymeric materials, all similar in chemical structure:

1. The saturated polyesters are usually produced in the form of films or fibers, and are thermoplastic products.
2. The unsaturated polyesters are thermosets composed of linear polyester polymers cross-linked with another monomer such as styrene or diallyl phthalate. These are generally formulated as solid molding compounds and are called alkyds.
3. The liquid polyester resins, called simply polyesters, are converted to solid plastics simply by the addition of an organic peroxide catalyst that triggers curing or polymerization. During the polymerization process and before hardening, fillers and reinforcements are added. Curing can be done at room temperature, but heat accelerates the reaction. The liquid resins vary in color from nearly water white to amber. They can be colored with common pigments.

About three-quarters of all polyester production goes into glass-reinforced plastics, either as moldings or laminates. (Their properties in these uses are discussed in the section on Fiber-reinforced Plastics.) Unreinforced polyesters have only limited use. They are produced as castings in both rigid and flexible forms, but they have excessive mold shrinkage. Their tensile strength ranges from 7,000 to 10,000 psi, and their maximum service temperature is about 300°F. Among the thermosets, they are second only to the urethanes in dielectric strength. Typical cast-polyester products include electrical components, buttons, and decorative architectural shapes.

Alkyds. Alkyd plastics are composed of a polyester resin and, usually, a diallyl phthalate monomer plus various inorganic fillers depending on the desired properties. The raw material is produced in three forms: rope, granules, and putty and glass reinforced. As a class, the alkyds have excellent heat resistance up to about 300°F, high stiffness, and moderate tensile and impact strength. Their low moisture absorption combined with good dielectric strength makes them particularly suitable for electronic and electrical hardware such as switchgear, insulators, and parts for motor controllers and automotive igni-

Table 9-16 *Typical Properties of Polyesters and Alkyds*

Property	Polyesters				Alkyds	
	Cast Rigid	Molding (Asbestos)	Rope		Granular	Glass Reinforced
Specific gravity	1.12–1.46	1.5–1.75	2.20–2.22		2.21–2.24	2.02–2.10
Modulus of elasticity in tension, 10^5 psi	1.5–6.5	12–15	19–20		24–29	20–25
Tensile strength, 1,000 psi	4–10	4–6	7–8		3–4	5–9
Elongation, %	1.7–2.6	—	—		—	—
Hardness	—	—	—		—	—
Impact strength, ft-lb/in. (notch)	0.18–0.40	10–13	2.2		0.30–0.35	8–12
Maximum service temperature, °F	250–300	375–400	300		300	300
Volume resistance, ohm-cm	10^{13}	10^{12}–10^{13}	10^{14}		10^{14}–10^{15}	10^{14}
Dielectric strength, V/mil	300–400	350	290		300–350	300–350

tion systems. They are easily molded at low pressures and cure rapidly.

9-16 Allylics

These are premium, high-priced plastics chiefly used in electrical and electronic hardware because of their exceptionally high insulation and dielectric values plus their very low moisture absorption. The three principal types are diallyl phthalate (DAP), diallyl isophthalate (DAIP), and allyl diglycol carbonate.

As a class, besides good electrical properties, they have good solvent and chemical resistance, moderate strength, and fair impact strength. Maximum service temperatures are 350°F for DAP, 450°F for DAIP, and 250°F for allyl carbonate. The allyl carbonate, which is a hard, water-white material, has exceptionally high transparency and high stability of optical properties under load and heat, and in many chemical environments. It is available only in cast stock shapes for machining or in finished machined parts.

9-17 Epoxies

Epoxies, perhaps best known as adhesives, are premium thermosets, and are generally employed in high-performance uses where their high cost is justified. They are available in a wide variety of forms, both liquid and solid, and are cured into the finished plastic by a catalyst or with hardeners containing active hydrogens. Depending on the type, they are cured at either room temperature or at elevated temperatures.

Table 9-17 *Typical Properties of Allylics*

Property	Diallyl Phthalate (DAP)	Diallyl Isophthalate (DAIP)	Allyl Diglycol Carbonate (Cast)
Specific gravity	1.27	1.26	1.32
Modulus of elasticity in tension, 10^5 psi	—	—	3.0
Tensile strength, 1,000 psi	4	4.3	5–6
Elongation, %	—	—	—
Hardness, Rockwell M	114–116	119–121	95–100
Impact strength, ft-lb/in. (notch)	0.2–0.3	0.2–0.3	0.2–0.4
Maximum service temperature, °F	350	450	212–250
Volume resistance, ohm-cm	1.7×10^{16}	3.9×10^{17}	4×10^{14}
Dielectric strength, V/mil	450	420	290

Table 9-18 *Typical Properties of Epoxies*

Property	Bisphenol A		Novalac		Cyclo-aliphatic, Cast, Rigid
	Cast, Rigid	Molded, Mineral	Cast, Rigid	Molded, Mineral	
Specific gravity	1.15	1.6–2.1	1.24	1.7	1.22
Modulus of elasticity in tension, 10^5 psi	4.5	—	4–5	—	5
Tensile strength, 1,000 psi	9.5–11.5	5–7	8–12	5.3	9.5–12
Elongation, %	4.4	—	2–5	—	2.2–4.8
Hardness, Rockwell M	106	101	107–112	—	—
Impact strength, ft-lb/in. (notch)	0.2–0.5	0.25–0.45	0.5	0.3–0.5	—
Maximum service temperature, °F	175–190	300–500	450	450–500	450–500
Volume resistance, ohm-cm	6×10^{15}	9×10^{15}	2×10^{14}	3×10^{14}	$>10^{16}$
Dielectric strength, V/mil	>400	350–400	—	280–400	444

The major characteristics of epoxies as a class include excellent electrical and mechanical properties, low moisture absorption, excellent adhesive properties, excellent chemical resistance, and ease of processing. Their maximum service temperature is about 525°F. When used with glass-fiber reinforcement, they can achieve the highest strengths possible with plastic materials—from 160,000 to 250,000 psi in filament-wound structures.

Liquid epoxies are used for casting, for potting or encapsulation, and for laminating. They are used unfilled or with any of a number of different mineral or metallic powders. Molding compounds are available as liquids, and also as powders with various types of filters and reinforcements. Typical uses include potting and encapsulation of electronic parts, electrical moldings, laminated tooling, and aircraft structural parts.

9-18 Aminos

There are two major groups of amino plastics: melamines and ureas. The aminos are generally rated as the hardest of all plastics. They are relatively low in cost and are noted for their good chemical resistance and unlimited colorability. These latter two characteristics explain their wide use for dishes and tableware.

The base resin is water white and transparent. Where color is not a consideration, wood, flour, asbestos, chopped fabrics, and glass fibers are added to melamines for engineering or industrial uses. Wood flour added to urea provides a low-cost industrial material.

Table 9-19 *Typical Properties of Aminos*

Property	Melamine Cellulose (Electrical)	Melamine Glass Fiber	Urea Cellulose
Specific gravity	1.43–1.50	2.0	1.52
Modulus of elasticity in tension, 10^5 psi	10.5 x 10	24	10–15
Tensile strength, 1,000 psi	5–9	5–10	5.5–13
Elongation, %	0.6	—	—
Hardness, Rockwell M	110	120	120
Impact strength, ft-lb/in. (notch)	0.25–0.35	0.5–10	0.25–0.35
Maximum service temperature, °F	280	300–350	175
Volume resistance, ohm-cm	$10^{12} \times 10^{13}$	$2\text{–}7 \times 10^{11}$	—
Dielectric strength, V/mil	350–400	250–300	300–400

Amino molded products are hard, rigid, and abrasion resistant, and do not become brittle down to temperatures as low as −70°F. They are flame resistant but do not have exceptional heat resistance. Unfilled ureas are limited to temperatures of 170°F and the unfilled melamines to temperatures of 210°F. They have good electrical-insulation values and are used in automotive ignition systems. As a class, they do not have good impact strength, and are only fair in dimensional stability. Shrinkage after molding can be as high as 2 percent over a long period of time. Besides being used for dishes, melamines are used in the electrical and mechanical fields. Ureas are used for housings, toilet seats, and low-cost electric switch plates, wiring devices, and parts requiring high arc resistance.

9-19 *Silicones*
Unlike all the other plastic polymers, silicones are not hydrocarbons. Because the polymers are composed of monomers in which oxygen atoms are attached to silicon atoms, they can be classed as semiorganic polymers. The polymers are bonded to each other by silicon-oxygen-silicon cross-linking similar to that found in quartz and glass. Various organic groups (radicals) are also attached to the silicon atoms, thus producing a variety of materials including liquids, elastomers, and rigid solids.

Silicone molding compounds consist of silicone resin, inorganic filler, and catalyst. Because of the quartzlike structure, molded parts have exceptional thermal stability. Their maximum continuous-use service temperature is about 500°F. Special grades exceed this and go

Table 9-20 *Typical Properties of Silicones*

Property	Molding		Casting Flexible
	Glass Fiber	Mineral	
Specific gravity	1.88	1.86–2.0	1.15
Modulus of elasticity in tension, 10^5 psi	—	—	—
Tensile strength, 1,000 psi	6.5	4–6	8–10
Elongation, %	<3	<3	100
Hardness, Rockwell M	87	70–90	—
Impact strength, ft-lb/in. (notch)	3–5	0.34	—
Maximum service temperature, °F	>500	>500	400
Volume resistance, ohm-cm	9×10^{14}	5×10^{14}	1.4×10^{14}
Dielectric strength, V/mil	280 (in oil)	380 (in oil)	550

as high as 700 and 900°F. Their heat-deflection temperature for 265 psi is 900°F. They also have high dielectric strength. Their moisture absorption is low, and reistance to petroleum products and acids is good. Nonreinforced silicones have only moderate tensile and impact strength, but fillers and reinforcements provide substantial improvement.

Because silicones are high in cost, they are premium plastics and are generally limited to critical or high-performance products such as high-temperature components in the aircraft, aerospace, and electronics fields. Silicone liquids serve as damping fluids, and lubricants are available for applications where heat and chemical resistance are required.

9-20 *Urethanes*
Polyurethane polymers have only limited use in solid plastic form. They are produced commercially as rigid and flexible foams, elastomers, and coatings. They are best known as plastic foam. In this form, they are widely used for insulation, cushioning, and packaging. Flexible foams range from about 1 lb/ft^3 up to 4 or 5 lb/ft^3. Rigid foams have densities ranging from about 5 to 50 lb/ft^3.

Fiber-reinforced Plastics

In recent years, the fastest growing engineering material has been fiber-reinforced plastics (FRP). These materials are extremely versatile composites with relatively high strength-to-weight ratios and excellent corrosion resistance. In both these respects they outperform most metals. In addition, they can be formed economically into virtually any shape and size. In size, FRP products range from tiny electronic components up to boat hulls of 70 ft and longer. In between these extremes, there are a wide variety of FRP gears, bearings, housings, and other parts used in all the product-manufacturing industries.

The basic nature and structure of FRPs are covered in Chap. 14. Here we will note that FRPs are composed of three major components: matrix, fiber, and bonding agent. The plastic resin serves as the matrix in which the fibers are embedded. Adherence between matrix and fibers is achieved by a bonding agent or binder, sometimes called the coupling agent. Most plastics, both thermosets and thermoplastics, can be used for the matrix. In addition to these three major components, a wide variety of additives—fillers, catalysts, inhibitors, stabilizers, pigments, and fire retardants—can be used to fit specific application needs.

Fig. 9-8 This ventilating fan for cooling explosion-proof motors is molded of glass-reinforced nylon with an aluminum antistatic additive. (Fiberfil Division, Dart Industries, Inc.)

9-21 *Fibers*

Glass is by far the most used fiber in FRPs. Plastics composites reinforced with glass are referred to as GFRP or GRP. Asbestos has some use, but is largely limited to applications where maximum thermal insulation or fire resistance is required. Other fibrous materials in limited use are paper, sisal, cotton, and nylon. Metals can be used as wire mesh or cloth for special applications. High-performance and costly fibers, such as boron and graphite, and metal fibers are covered in Chap. 14. In recent years, other fibers for FRPs have been developed that will find increasing use in the years ahead. One of these is polyvinyl-alcohol fiber (PVA) developed in Japan.

The standard glass fiber used in GRP is a borosilicate type known as E-glass. The fibers are spun as single glass filaments with diameters ranging from 0.002 to 0.001 in. These filaments are collected into strands, usually around 200 per strand, and are manufactured into many forms of reinforcement. The E-glass fibers have a tensile strength of 500,000 psi. A relatively new glass fiber known as S-glass is higher in strength but, because of its higher cost, its use is limited to advanced, high-performance applications. In general, glass content runs between 20 and 40 percent in reinforced thermoplastics; in thermosets, it runs as high as 80 percent in the case of filament-wound structures.

There are a number of standard forms in which glass fibers are produced and applied in GRP (Fig. 9-9):

1. Continuous strands of glass supplied either as twisted, single-end strands (yarn), or as untwisted multistrands (continuous) roving.
2. Fabrics woven from yarns in a variety of types, weights, and widths.
3. Woven rovings, which are continuous rovings woven into a coarse, heavy, drapable fabric.
4. Chopped strands made from either continuous or spun roving cut into ⅛ to ½ in. lengths.
5. Reinforcing mats made of either chopped strands or continuous strands laid down in a random pattern.
6. Surfacing mats composed of continuous glass filaments in random patterns.

9-22 *Resins*

Although a number of different plastic resins are used for reinforced plastics, thermosetting polyester resins are the most common. The combination of polyester and glass provides a good balance of mechanical properties as well as corrosion resistance along with low cost and

Fig. 9-9 Some forms of glass reinforcements for plastics: continuous strand mat, woven roving, yarn, continuous roving, and chopped strands. (Owens-Corning Fiberglas Corp.)

good dimensional stability. In addition, curing can be done at room temperature without pressure, thus making for low processing-equipment costs. For high-volume production, special sheet-molding compounds are available in continuous sheet form for use in the matched die process (see Sec. 9-23). (In recent years, resin mixtures of thermoplastics with polyesters have been developed to produce high-quality surfaces in the finished molding. The common thermoplastics used are acrylics, polyethylenes, and styrenes.)

Other glass-reinforced thermosets include phenolics and epoxies. GRP phenolics are noted for their low cost and good overall performance in low-strength applications. Because of their good electrical resistivity and low water absorption, they are widely used for electrical housings, circuit boards, and gears. Epoxies, being more expensive than polyesters and phenolics, are limited to high-performance parts where their excellent strength, thermal stability, chemical resistance, and dielectric strength are required.

Up until about a decade ago, GRP materials were largely limited to thermosetting plastics. Today, however, more than 1,000 different types and grades of reinforced thermoplastics are commercially available. Leaders in volume use are nylon and the styrenes. Others include sulfones and ABS. Unlike thermosetting resins, GRP-thermoplastic parts can be made in standard injection-molding machines. The resin can be supplied as pellets containing chopped glass fibers $\frac{1}{8}$ to $\frac{1}{2}$ in. long. As a general rule, a GRP thermoplastic with chopped fibers at least doubles the plastic's tensile strength and stiffness.

Glass-reinforced thermoplastics are also produced as sheet materials for forming on metal stamping equipment. Produced in sheet and coil form, the materials are preheated at about 400°F and then formed and trimmed. The mechanical properties are comparable to GRP parts made by other methods.

9-23 Processing Methods

Matched Metal Die Molding. This is the most efficient and economical method for mass producing high-strength parts. Parts are press-molded in matched male and female molds at pressures of 200 to 300 psi and at temperatures of 235 to 260°F (Fig 9-10a).

Four main forms of thermosetting-resin reinforcements are used:

1. Chopped-fiber preforms, shaped like the part, are saturated with resin at the mold. They are best for deep-draw, compound-curvature parts.
2. Flat mats, saturated with resin at the mold, are used for shallow parts with simple curvatures.
3. Sheet molding compound, a preimpregnated material, has advantages for parts with varying thickness.
4. Bulk molding compound is a premix made up of short fibers preimpregnated with resin. It is used for parts similar to castings.

Injection Molding. In this high-volume process, a mix of short fibers and resin is forced by a screw or plunger through an orifice into the heated cavity of a closed matched metal mold (Fig 9-10b). It is the major method for forming reinforced thermoplastics, and is beginning to be used for thermoplastic-modified, thermosetting bulk molding compounds.

Hand Lay-up. This is the simplest of all the methods of forming thermosetting composites. It is the best employed for quantities under 1,000 for prototypes and sample runs, for extremely large parts, and for larger volume where model changes are frequent, as in boats.

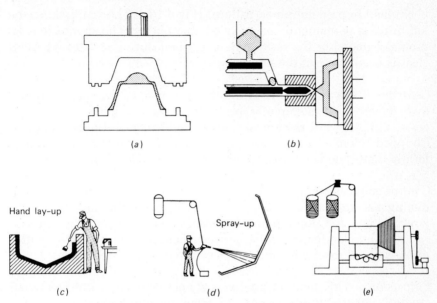

Fig. 9-10 Some methods of producing fiber-reinforced plastics: (a) matched die molding, (b) injection molding, (c) hand lay-up, (d) spray-up, and (e) filament winding.

In hand lay-up, only one mold, usually female, is used. It can be made of low-cost wood or plaster (Fig 9-10c), and duplicate molds are inexpensive. The reinforcing mat or fabric is cut to fit, laid in the mold, and saturated with resin by hand, using a brush, roller, or spray gun. Layers are built up to the required thickness, and then the laminate is cured to permanent hardness, generally at room temperature.

Spray-up. Like hand lay-up, the spray-up method uses a single mold, but it can introduce a degree of automation (Fig 9-10d). This method works well for complex thermosetting moldings, and its portable equipment eases on-site fabrication and repair. Short lengths of reinforcement and resin are projected by a specially designed spray gun so they are deposited simultaneously on the surface of the mold. Curing is usually accomplished by a catalyst in the resin at room temperature.

Filament Winding. Filament-wound parts, because of their high glass-to-weight ratio, have the highest strength-to-weight ratio of any reinforced thermoset. The method is generally limited to revolving round, oval, tapered or rectangular surfaces, but a high degree of automation can be achieved to offset this limitation. Continuous fiber strands are wound on a suitably shaped mandrel or core and precisely positioned in predetermined patterns (Fig 9-10e). The mandrel may be left in place permanently or removed after curing. The strands may be preimpregnated or the resin may be applied during or after winding. Heat is used to effect the final cure.

Centrifugal Casting. This is another method of producing round, oval, tapered, or rectangular parts. It offers low labor and tooling costs, uniform wall thicknesses, and good inner and outer surfaces. Chopped fibers and resin are placed inside a mandrel and are uniformly distributed as the mandrel is rotated inside an oven.

Continuous Laminating. This is the most economical method of producing flat and corrugated panels, glazing, and similar products in large volume. Reinforcing mats or fabrics are impregnated with resin, run through laminating rolls between cellophane sheets to control thickness and resin content, and then cured in a heated zone.

Pultrusion. This method produces shapes with high unidirectional strength, for example, I beams, flat stock for building siding, fishing rods, and shafts for golf clubs. Continuous fiber strands combined with mat or woven fibers for cross strength, are impregnated with resin and pulled through a long heated steel die. The die shapes the product and controls the resin content.

9-24 Properties

The mechanical properties of fiber composites depend on a number of complex factors (see Chap. 14). Two of the dominating factors in glass-reinforced plastics are the length of the fibers and the glass content by weight. In general, strength increases with fiber length. For example, reinforcing a thermoplastic with chopped glass fibers at least doubles the plastic's strength, and long-glass-fiber-reinforced thermoplastics exhibit increases of 300 and 400 percent. Also, heat-distortion temperatures usually increase by about 100°F, and impact strengths are raised appreciably. Similarly, as a general rule, an increase in glass content results in the following property changes:

Tensile and impact strength increase.

Modulus of elasticity increases.

Heat-deflection temperature increases, sometimes as much as 300°F.

Creep decreases, and dimensional stability increases.

Thermal expansion decreases.

Table 9-21 *General Properties of Glass-Fiber-reinforced Polyester Plastics*

Property	No Reinforcement	Glass Cloth	Mat or Preform	Premix
Glass content, % by weight	—	60–70	35–45	10–40
Specific gravity	1.20–1.30	1.7–1.9	1.5–1.6	1.6–1.9
Hardness, Rockwell M	100–110	100–110	90–100	55–75
Flexible strength, 1,000 psi	13–17	40–85	25–35	5–20
Tensile strength, 1,000 psi	8–12	30–55	15–25	3–6
Modulus of elasticity in tension, 10^6 psi	0.45–0.55	1.8–3.0	0.8–1.6	0.8–1.2
Impact strength, ft-lb/in. (notch)	0.17–0.25	15–30	10–20	1.5–3.0
Shear strength, 1,000 psi	—	15–25	12–18	—
Water absorption (24 hr), %	0.15–0.25	0.10–0.20	0.2–0.5	0.3–1.0

More specifically, moldings of glass-reinforced polyesters range in strength from around 15,000 to 55,000 psi, while nonreinforced polyesters have strengths of 8,000 to 12,000 psi (Table 9-21). GFR-thermoplastic moldings, such as nylon, have strengths up to about 30,000 psi, while unreinforced parts have strengths of 10,000 psi (Table 9-22). In the high-performance area, GFR epoxies using glass cloth laminates have strengths as high as 60,000 psi. Filament-wound structures can exceed 240,000 psi (Table 9-23). Typical GRP-epoxy applications of this kind are rocket motor cases, chemical tanks, and pressure bottles.

There are similar improvements in impact strength and stiffness. For example, the impact strengths of GRP polyesters run from 10 to 30 ft-lb compared to less that 1 ft-lb for unreinforced moldings; and the modulus of elasticity ranges up to 2.5 and 3.0 million psi in contrast to 500,000 psi. And the modulus of elasticity of most thermoplastics is usually doubled or tripled by adding glass fibers.

Another outstanding characteristic of GRP materials is their chemical stability, which has been used to advantage in a large variety of tanks, containers, and piping used in corrosive environments.

Table 9-22 Typical Properties of Glass-Fiber-reinforced Thermoplastics

Resin	Specific gravity	Tensile Strength, 1,000 psi	Elongation, %	Modulus of Elasticity in Tension, 10^5 psi	Impact Strength, ft-lb/in.	Deflection Temperature at 264 psi, °F
ABS	1.19–1.36	14.5–18	1.5–2.0	8.0–10.0	0.4–3.0	210–230
Acetal	1.54–1.69	9–18	1.0–3.0	8.0–14.5	0.8–3.0	310–335
Nylon	1.32–1.52	15–30	1.8–3.0	8.0–20.0	1.0–4.5	390–500
Polycarbonate	1.34–1.58	13–21	1.2–4.0	8.1–17.0	1.5–4.0	285–295
Polyethylene (linear)	1.09–1.28	7–11	1.5–3.0	7.0–10.5	1.2–3.5	240–260
Polypropylene	1.04–1.22	6–9	2.0–3.0	4.5–9.0	1.2–3.0	270–300
Polystyrene	1.20–1.34	10–15	1.0–1.6	8.0–19.0	0.4–2.2	210–220
Polysulfone	1.31–1.47	11–17	1.5–4.0	6.5–10.0	0.9–2.5	340–350
Styrene acrylonitrile	1.22–1.35	13.5–18	1.1–2.0	9.5–18.0	0.6–3.0	210–220
Vinyl chloride	1.49	12	1.5	12.0	1.0	170

NOTE: Ranges shown represent various amounts of glass content, mostly from 20 to 40 percent.
SOURCE: "Plastics/Elastomers Issue," *Machine Design*, Feb. 15, 1973.

Table 9-23 *Typical Properties of Glass-Fiber-reinforced Thermosets*

Property	Phenolics (Fibers)	Melamines (Fibers)	Alkyds (Fibers)	Epoxies (Laminates)	Epoxies (Filament Wound)	Silicones (Woven Fabrics)
Specific gravity	1.75–1.90	1.9–2.0	2.0–2.1	1.8	2.17	1.75–1.8
Modulus of elasticity in tension, 10^5 psi	30–33	24	20–25	33–36	64–72	28
Tensile strength, 1,000 psi	5–10	5–10	5–9	50–60	230–240	30–35
Elongation, %	0.2	—	—	—	—	—
Hardness, Rockwell	E50–70	—	—	M115–117	M98–120	—
Impact strength, ft-lb/in. (notch)	10–33	5–10	8–12	12–15	—	10–25
Maximum service temperature, °F	350–450	300–400	300	250–350	250–350	450–500
Volume resistance, ohm-cm	7–10×10^{12}	1–7×10^{11}	10^{14}	—	—	2–5×10^{14}
Dielectric strength, V/mil	200–370	250–300	300–350	450–550	—	725

Improved GRP properties are usually accompanied by increases in material cost. The added cost varies widely depending on the plastic, the glass content, the form of the glass, and so on, but it is generally at least 15 percent, and can be several times higher. However, material cost is only one consideration. Processing costs and other considerations may make the reinforced plastic more economical than unreinforced plastic.

Review Questions

1. Which of the two basic families of plastics, thermoplastics and thermosets,
    (a) Requires the longer molding cycles?
    (b) Once polymerized, will not soften or flow on reheating?
    (c) Is relatively hard and brittle?
    (d) Is most widely used in terms of volume and range of applications?
    (e) Has many members that range from flexible to rigid?
2. Which of the three density classes of polyethylene has the highest degree of crystallinity?
3. List four types or groups of polyolefins in order of decreasing stiffness.
4. Which polyolefin is noted for its ability to withstand prolonged flexing?
5. What three groups of thermoplastics are lowest in cost and most widely used in consumer products?
6. Which plastics would you consider if you wanted to replace transparent glass with a relatively "unbreakable" plastic?
7. Which plastic has large volume use as moisture-barrier sheeting?
8. Which group of plastics is most often considered competitive with zinc and aluminum die castings? Why?
9. What characteristic of nylon is a disadvantage when the service environment includes high humidity?
10. What is the major reason for the exceptional corrosion resistance of fluoroplastics?
11. Which of the following have useful service temperature above 400°F?
    (a) Polyethylene.
    (b) Polyimide.
    (c) Nylon.
    (d) Tetrafluoroethylene (PTFE).
    (e) Silicone.
    (f) Epoxy.
    (g) Vinyl.
    (h) Polyphenylene sulfide.
    (i) Chlorinated polyether.
12. What thermoset plastic or plastics are prominently identified with
    (a) Adhesives?

(b) Glass-reinforced plastics?
(c) Exceptionally high dielectric values?
(d) Dishes?
(e) High-strength filament-wound structures?
13. Arrange the following plastics in order of increasing heat resistance:
   (a) Polyethylene.
   (b) ABS.
   (c) Nylon.
   (d) Polystyrene.
   (e) Phenolics (filled).
   (f) Epoxy.
14. What distinguishes silicones from other polymeric materials?
15. Arrange the following plastics—low-density polyethylene, rigid PVC, type-6 nylon, medium-impact ABS, Noryl, acetal homopolymer (standard), general-purpose polypropylene, and polycarbonate—into two equal groups on the basis of high strength or low strength.
16. Which of the following plastics would you consider for an application in which colorability is important?
   (a) Polyethylene.
   (b) Urea.
   (c) Polycarbonate.
   (d) Fluoroplastic.
   (e) Melamine.
   (f) Polystyrene.
17. List the three major components of fiber-reinforced plastics.
18. Give three reasons why glass-fiber-reinforced polyester is the most widely used FRP.
19. When chopped glass fibers are used to reinforce thermoplastics, what is the effect on
   (a) Stiffness?
   (b) Tensile strength?
   (c) Elongation?
20. What FRP processing method produces
   (a) Thermoplastic parts?
   (b) The highest strength-to-weight cylindrical vessels?
   (c) High-volume production-run parts?
   (d) Corrugated panels for patio roofing?
   (e) Structural shapes?
21. What is the general relationship of the length and content by weight of fibers in a FRP to the strength of the composite material?

Bibliography

Brady, G. S.: *Materials Handbook*, 10th ed., McGraw-Hill Book Co., New York, 1971.

Clauser, H. R. (ed.): *Encyclopedia of Materials, Parts and Finishes*, Technomic Publishing Co., Westport, Conn., 1975.

Lubin. G. (ed.): *Handbook of Fiberglass and Advanced Plastics Composites*, Van Nostrand-Reinhold Publishing Corp., New York, 1969.
"Materials Selector Issue," *Materials Engineering*, Mid-September, 1972.
Milby, R. V.: *Plastics Technology*, McGraw-Hill Book Co., New York, 1973.
"Plastics/Elastomers Reference Issue," *Machine Design*, Feb. 15, 1973.

ELASTOMERS

Elastomers, commonly referred to as rubbers, are hydrocarbon, polymeric materials similar in structure to plastic resins. The difference between plastics and elastomers is largely one of definition based on the property of extensibility, or stretching. The American Society for Testing and Materials defines an elastomer as "a polymeric material which at room temperature can be stretched to at least twice its original length and upon immediate release of the stress will return quickly to approximately its original length." Some grades of plastics approach this rubberlike state, for example, certain of the polyethylenes. Also, a number of plastics have elastomer grades, such as the olefins, styrenes, fluoroplastics, and silicones.

As indicated above, the major distinguishing characteristic of elastomers is their great extensibility and high-energy storing capacity. Unlike many metals, for example, which cannot be strained more than a fraction of 1 percent without exceeding their elastic limit, elastomers have usable elongations up to several hundred percent. Also, because of their capacity for storing energy, even after they are strained several hundred percent, virtually complete recovery is achieved once the stress is removed.

As a family, elastomers share many characteristics with plastics. For example:

1. They are essentially noncrystalline in structure.
2. They are nonconductors of electricity, and are relatively low heat conductors.

3. They are high in resistance to chemical and corrosive environments.
4. They have relatively low softening temperatures.
5. They exhibit viscoelastic behavior generally to a greater extent than plastics.
6. They oxidize, or age, causing deterioration and changes in properties, generally more so than plastics.

Up until World War II, almost all rubber was natural. During the war, synthetic elastomers began to replace the scarce natural rubber, and, since that time, production of synthetics has increased until now their use far surpasses that of natural rubber.

Like plastics, there are thousands of different elastomer compounds. Not only are there many different classes of elastomers, but individual types can be modified with a variety of additives, fillers, and reinforcements. In addition, curing temperatures, pressures, and processing methods can be varied to produce elastomers tailored to the needs of specific applications.

Structure and Properties

10-1 *Structure*

As noted, elastomers are hydrocarbon polymers, and thus are similar in structure to plastic resins. In the raw-material or crude stage, elastomers are thermoplastic. The individual polymers which in general are longer than in plastics, are linear and only weakly joined to each other by secondary bonds (Fig. 10-1a). Thus crude rubber has little resiliency and practically no strength. By a vulcanization process in which sulfur and/or other additives are added to the heated crude rubber, the polymers are cross-linked by means of covalent bonds to each other, producing a thermoset-like material (Fig. 10-1b). The amount of cross-linking which occurs between the sulfur (or other additive) and the carbon atoms determines many of the elastomer's properties. As cross-linking increases, resistance to slippage of the polymers over each other increases, resilience and extensibility decreases, and the elastomer approaches the nature of a thermosetting plastic. For example, hard rubbers, which have the highest cross-linking of the elastomers, in many respects are similar to phenolics.

In the unstretched state, elastomers are essentially amorphous because the polymers are randomly entangled and there is no special preferred geometrical pattern present. However, when stretched, the polymer chains tend to straighten and become aligned, thus increasing in crystallinity (Fig. 10-1c). This tendency to crystallize when stretched is related to an elastomer's strength. Thus, as crystallinity increases, strength also tends to increase.

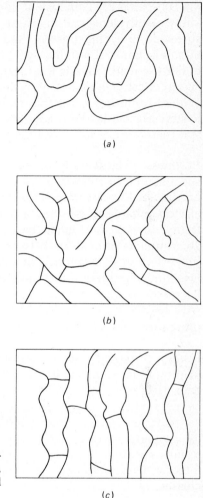

Fig 10-1 Schematic drawing of polymer arrangements in elastomers: (a) unvulcanized, (b) vulcanized, and (c) stretched vulcanized rubber.

10-2 Property Definitions

Because of elastomers' nature, in particular their extensibility and viscoelastic behavior, some of the properties and tests used to evaluate them differ from those employed for plastics and other materials. These properties are discussed briefly below.

Resilience and Hysteresis. Resilience is a key characteristic of elastomeric materials. It is a measure of an elastomer's capacity to store energy, and is defined as the ratio of energy output to energy input expressed as a percentage. Because some input energy is always lost through internal friction, resilience values must always be less than 1 (or 100 percent). This loss of energy in the form of frictional heat is known as hysteresis. It manifests itself as heat built up in elastomers subjected to cyclic stresses. Figure 10-2 shows the stress-strain curve of an elastomer through a cycle of loading and unloading. Area *ACD* under the load curve is the input energy, and area *BCD* is the output energy. Heat-energy loss, or hysteresis, is represented by area *ABD*. It is evident from the curves that resilience has an inverse relationship to hysteresis; that is, the higher the resilience, the lower the energy loss, and therefore the lower the hysteresis.

Hardness. Hardness is one of the most widely used properties in the specification of rubbers. Hardness can also be varied to meet specific needs. The Shore durometer test is the standard method for measuring elastomer hardness (see Sec. 4-11). On the durometer scale, which runs from 0 to 100, the higher the number the harder the material. There are rough relationships between hardness and several mechanical properties of elastomers. For example, the lower the hardness, the less tendency there is for the elastomer to creep or flow. Elongation decreases as hardness increases. Tensile strength generally increases with hardness up to about 50 on the durometer scale, and then falls off if hardness continues to increase.

Tensile Strength. Since elastomers are only infrequently used for loading in tension, tensile strength is not directly used for application data. However, it serves as an overall performance indicator. Values below 1,000 psi usually mean that most other mechanical properties are poor, and values above 2,000 psi point to generally good mechanical properties. For dynamic applications, a minimum of 1,000 psi is usually specified. Tensile strength is also used as a measure of an elastomer's deterioration in service, for, as aging sets in, strength decreases.

Compression Set or Creep. This characteristic refers to the percentage of deflection or distortion remaining in an elastomer after a load (usually compressive) is removed. It is comparable to creep or cold flow in plastics, and is dependent on many factors, such as strain rate, temperature, and type and size of the loading. In general, compression-set values are about 20 percent less than shear, and tension creep is about 30 percent more than shear.

Tear Resistance. Elastomers as a class have low tear strength. If tear strength is extremely low, small nicks or cuts can cause catastrophic

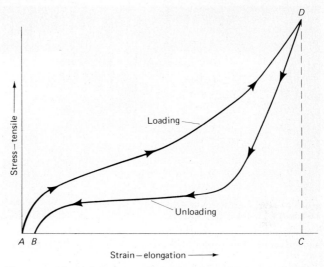

Fig 10-2 Typical stress-strain curve of an elastomer, showing behavior during loading and unloading, and hysteresis.

failure, and, under flexing, failure will quickly occur. Also, elastomers with poor tear strength generally have low abrasion resistance.

Abrasion Resistance. Because many elastomer applications involve friction and wear, abrasion resistance is usually an important consideration. As a general rule, abrasion resistance improves with an increase in hardness. However, for some kinds of abrasion, for example, where small particles are involved, relatively soft elastomers perform better.

Deterioration Resistance. All elastomers undergo changes in properties with time because of oxidation. This deterioration, which is referred to as aging, is affected by many different environmental factors such as sunlight, heat, and ozone, all of which accelerate oxidation. Aging results in a loss of resilience, a decrease of hardness, and eventual cracking. The classification system for elastomers, a joint specification of the ASTM (ASTM D2000) and the Society of Automotive Engineers (SAE J200), is based in part on their dry-heat resistance (aging).

Oil Resistance. This is an important criteria in the selection of elastomers because many of their applications involve an environment of hydrocarbon fluids such as oil and gasoline. Resistance to oils is based

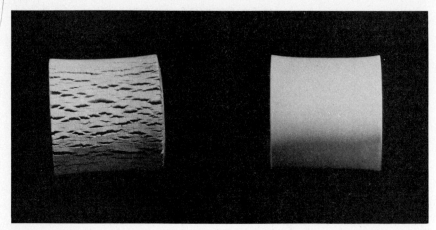

Fig 10-3 Left: Appearance of an elastomer that has deteriorated by aging. Right: Appearance of an elastomer with high resistance to aging. (B. F. Goodrich Chemical Co.)

on how much the elastomer swells in the presence of oil. This criterion is also used as part of the classification system for elastomers.

Types of Elastomers

There are roughly 20 major classes of elastomers; we cannot do much more here than identify them and highlight the major characteristics of each group.

There are two basic specifications that provide a standard nomenclature and classification system for elastomers. The ASTM standard D1418 categorizes rubbers into compositional classes (see Table 10-3). A joint ASTM-SAE specification (ASTM D2000/SAE J200) provides a classification system based on material properties. The first letter indicates specific resistance to heat aging (Table 10-1), and the second letter denotes resistance to swelling in oil (Table 10-2).

Table 10-3 lists typical properties of each of the compositional groups of elastomers. It also lists the combinations of heat aging and swelling resistance available in grades within that elastomer group. Table 10-4 provides a guide for the use of various elastomer groups for off-the-shelf and custom-made parts.

10-3 *Natural Rubber (NR)*

Natural rubber is processed from rubber latex, a milky liquid obtained from certain tropical trees. The latex is chemically coagulated and smoked to produce a spongy mass. This raw material is then rolled

Table 10-1 Types of Elastomers Classified by Resistance to Heat Aging (ASTM D2000/SAE J200)

Type (First Digit)	Test Temperature, °F
A	158
B	212
C	257
D	302
E	347
F	392
G	437
H	482
J	527

into sheets of crude rubber from which the usable elastomer is eventually produced.

Natural rubber is a homopolymer composed of polymers of the isoprene monomer. Although it is classified as non-oil resistant, natural rubber is among the low-cost elastomers and still meets the needs of many applications. A number of different types are produced by adding fillers, particularly carbon black, silica, and silicates. Antioxidants decrease the loss of strength and resilience under atmospheric conditions. Carbon black added in relatively large amounts produces hard rubbers (for example, Ebonite) with mechanical proper-

Table 10-2 Classes of Elastomers Classified by Resistance to Swelling in Oil (ASTM D2000/SAE J200)

Class (Second Digit)	Maximum Volume of Swelling, %
A	No requirement
B	140
C	120
D	100
E	80
F	60
G	40
H	30
J	20
K	10

Table 10-3 Typical Properties of Major Groups of Elastomers

	Natural Rubber	Polyisoprene	Styrene Butadiene	Butadiene	Isobutylene Isoprene	Chlorinated Isobutene Isoprene
ASTM D1418 Designation	NR	IR	SBR	BR	IIR	CIIR
ASTM D2000/SAE J200 Type and Class	AA	AA	AA, BA	AA	AA, BA	AA, BA
Specific gravity, base polymer	0.92	0.91	0.94	0.91	0.92	0.92
Hardness, Shore A	30–90	30–90	40–90	40–80	40–80	40–80
Tensile strength, maximum, reinforced, psi	4,000	4,000	3,500	3,000	3,000	3,000
Elongation, maximum, reinforced, %	700	700	600	600	800	700
Resilience, at room temperature	A	A	B	A	D	C
Electrical resistance	A	A	A	A	A	A
Tear resistance	A	A	C	C	B	B
Abrasion resistance	A	A-B	A	A	C	C
Flame resistance	D	D	D	D	D	D
Compression set	A	A	A	B	C-B	B-A

Property					
Brittle point, °F	−80	−80	−80	−100	−80	−80
Ozone resistance	NR	NR	NR	NR	A	A
Weather resistance	D	D	D	D	A	A
Adhesion to metal	A	A	A	A	C-A	C-A
Oxidation resistance	B	B	C	A	A	A
Water resistance	A	A	B-A	B	C-A	B-A
Steam resistance	C	C	C	C	C-A	B-A
Acid resistance, dilute/concentrated	A/C-B	A/C-B	C-B/C-B	C-B/C-B	A/A	A/A
Alkali resistance, dilute/concentrated	A/C-B	A/C-B	C-B/C-B	C-B/C-B	A/A	A/A
Synthetic-lubricant resistance	NR	NR	NR	NR	NR	NR
Animal- and vegetable-oil resistance	D-B	D-B	D-B	D-B	B-A	B-A
Lubricating-oil resistance	NR	NR	NR	NR	NR	NR
High aniline point	NR	NR	NR	NR	NR	NR
Low aniline point	NR	NR	NR	NR	NR	NR
Aliphatic hydrocarbon resistance	NR	NR	NR	NR	NR	NR
Aromatic hydrocarbon resistance	NR	NR	NR	NR	NR	NR
Gas-permeability resistance	C	C	C	C	A	A

A = excellent; B = good; C = fair; D = use with caution; NR = not recommended.
SOURCE: "Plastics/Elastomers Reference Issue," *Machine Design*, Feb. 15, 1973.

(Continued)

Table 10-3 (Continued)

	Ethylene Propylene Copolymer EPM	Ethylene Propylene Terpolymer EPDM	Chloro-sulfonated Polyethylene CSM	Chloroprene CR	Nitrile Butadiene (High Nitrile) NBR	Nitrile Butadiene (Low Nitrile) NBR
ASTM D1418 Designation						
ASTM D2000/SAE J200 Type and Class	AA, BA, CA	AA, BA, CA	CE	BC, BE	BF, BG, BK, CH	BF, BG
Specific gravity, base polymer	0.86	0.86	1.18	1.24	1.00	1.00
Hardness, Shore A	30–90	30–90	50–90	30–90	40–90	40–90
Tensile strength, maximum, reinforced, psi	3,000	3,000	3,000	4,000	4,000	3,500
Elongation, maximum, reinforced, %	600	600	500	600	600	600
Resilience, at room temperature	B	B	C	B-A	B	B
Electrical resistance	A	A	A	A	D-C	D-C
Tear resistance	C	C	B	C-B	B	B
Abrasion resistance	B	B	B-A	A	A	A
Flame resistance	D	D	B-A	B-A	D	D
Compression set	A	B-A	D-C	B-A	B-A	B-A
Brittle point, °F	−90	−90	−70	−80	−40	−90
Ozone resistance	A	A	A	A	D-C	D-C

Elastomers 283

Weather resistance	A	A	A	A	D	D
Adhesion to metal	C-B	C-B	B-A	B-A	B-A	B-A
Oxidation resistance	A	A	A	A	B	B
Water resistance	A	A	B-A	B	A	B-A
Steam resistance	B-A	B-A	B	C	C-B	C-B
Acid resistance, dilute/concentrated	A/A	A/A	A/A	A/A	B/B	B/B
Alkali resistance, dilute/concentrated	A/A	A/A	A/A	A/A	B/B	B/B
Synthetic lubricant resistance	NR	NR	D	D	B-A	D
Animal- and vegetable-oil resistance	B-A	B-A	B	B	B	B
Lubricating-oil resistance	NR	NR	A	A	A	A
High aniline point	NR	NR	B	B	A	A
Low aniline point	NR	NR	C	C	A	B
Aliphatic hydrocarbon resistance	NR	NR	D-C	D	B	D
Aromatic hydrocarbon resistance	C	C	B	B	B-A	B
Gas-permeability resistance						

A = excellent; B = good; C = fair; D = use with caution; NR = not recommended.
SOURCE: "Plastics/Elastomers Reference Issue," *Machine Design*, Feb. 15, 1973.

(Continued)

Table 10-3 (Continued)

	Epichloro-hydrin	Polyacrylate	Silicone	Urethane	Fluoro-silicone	Fluorocarbon
ASTM D1418 Designation	CO, ECO	ACM	MQ, PMQ, VMQ, PVMQ	AU, EU	FVMQ	FKM
ASTM D2000/SAE J200 Type and Class	CH	DF, DH	FC, FE, GE	BG	FK	HK
Specific gravity, base polymer	1.36–1.27	1.09	0.98	1.05–1.30	0.98	1.85
Hardness, Shore A	40–90	40–85	30–85	40–100	60–80	60–95
Tensile strength, maximum, reinforced, psi	2,500	2,500	1,200	10,000	1,200	2,500
Elongation, maximum, reinforced, %	400	400	700	700	400	300
Resilience, at room temperature	C-B	C	D-A	C-A	C	C
Electrical resistance	B	C	B-A	B	A	B
Tear resistance	C-B	D-C	D-C	A	D	C-B
Abrasion resistance	C-B	C	D-C	A	D	B
Flame resistance	B-D	D	C-A	D-A	A	A
Compression set	D-C	B	C-A	D-A	B	B-A
Brittle point, °F	−10 to −50	−40	−90 to −180	−60 to −90	−85	−40
Ozone resistance	B	A	A	A	A	A

284 Industrial and Engineering Materials

Elastomers

Property						
Weather resistance	B	A	A	A	A	A
Adhesion to metal	C/B	B	B	C-B	C	C
Oxidation resistance	B	B	A	B	A	A
Water resistance	B	D	B-A	D-C	A	A
Steam resistance	C-B	NR	C-B	D	C-B	B
Acid resistance, dilute/concentrated	B-C	D-C/D-C	B/C	C/D	A/B	A/A
Alkali resistance, dilute/concentrated	B-D	D-C/D-C	A/A	C/D	A/B	A/A
Synthetic-lubricant resistance	C-B	D	NR	D	A	A
Animal- and vegetable-oil resistance	A	B	A	A	A	A
Lubricating-oil resistance	A	A	B	A	A	A
High aniline point	A	A	C	B	A	A
Low aniline point	A	B	NR	B	A	A
Aliphatic hydrocarbon resistance	B	D	NR	C	A	A
Aromatic hydrocarbon resistance	B	D	NR	D	A	A
Gas-permeability resistance	A	B	D	D	D	A

A = excellent; B = good; C = fair; D = use with caution; NR = not recommended.
SOURCE: "Plastics/Elastomers Reference Issue," *Machine Design*, Feb. 15, 1973.

Table 10-4 *Typical Applications of Major Types of Elastomers*

Part	Natural Rubbers	Polyisoprene Rubbers	Hard Rubbers	Butyls	Fluorocarbons	Chlorosulfonated Polyethylenes	Ethylene Propylenes	Epichlorohydrins	Neoprenes	Nitriles	Polyacrylates	Polybutadienes	Styrene Butadienes (SBR)	Sulfides	Silicones	Fluorosilicones	Polyurethanes
OFF-THE-SHELF																	
Air springs	●								●								
Ball seals			●							●							●
Bumper and cushions	●	●	●						●								●
Cable coverings	●				●	●	●	●	●						●		
Conveyor belts	●	●			●	●	●	●	●	●	●		●	●	●		●
Drive belts	●	●					●	●	●	●	●						
Flexible bearings	●	●					●		●	●	●					●	●
Flexible couplings																	●
Gas masks			●		●				●						●		
Hose, coolant and water	●	●	●				●	●	●	●	●	●					●
Hose, hydraulic fluid					●	●			●	●	●			●	●	●	
Hose, oil and gasoline					●				●					●			
Metalforming dies																	●
Molded diaphragms					●	●		●	●	●	●				●	●	
Molded shaft seals						●			●	●					●		
Mountings	●		●				●		●	●	●		●		●		●
O rings				●	●	●			●	●	●				●	●	●

(continued)

Table 10-4 (continued)

Part	Natural Rubbers	Polyisoprene Rubbers	Hard Rubbers	Butyls	Fluorocarbons	Chlorosulfonated Polyethylenes	Ethylene Propylenes	Epichlorohydrins	Neoprenes	Nitriles	Polyacrylates	Polybutadienes	Styrene Butadienes (SBR)	Sulfides	Silicones	Fluorosilicones	Polyurethanes
OFF-THE-SHELF																	
Rubber covered rolls	●	●		●	●	●	●		●	●	●				●	●	●
Sheet packings	●	●		●	●	●	●		●	●	●	●	●	●	●		
Solid tires	●	●															●
Suction cups	●	●		●					●	●					●		
Valve liners				●	●	●			●					●	●		●
V belts						●	●		●	●							
Wear pads						●				●							●
CUSTOM-MADE																	
Air ducts				●	●		●		●	●							
Bellows	●	●							●	●							
Door seals	●	●		●	●	●	●		●	●	●	●		●			
Drive wheels	●	●															●
Electrical connectors				●	●	●	●		●						●		
Fuel cells										●							
Handles and grips	●	●				●	●		●				●				●
Housings			●			●		●									●
Hydroclave bags				●					●								

(continued)

Table 10-4 (continued)

Part	Natural Rubbers	Polyisoprene Rubbers	Hard Rubbers	Butyls	Fluorocarbons	Chlorosulfonated Polyethylenes	Ethylene Propylenes	Epichlorohydrins	Neoprenes	Nitriles	Polyacrylates	Polybutadienes	Styrene Butadienes (SBR)	Sulfides	Silicones	Fluorosilicones	Polyurethanes
CUSTOM-MADE																	
Knobs	●	●	●														
Pedals	●	●					●		●								
Protective boots	●	●			●	●		●	●	●				●			
Pump impellers			●	●													●
Radome covers					●		●			●							
Sprockets																	●
Tubing					●		●	●		●	●	●			●	●	
Weather strips						●	●		●	●				●	●		
Window channels	●	●			●	●		●		●			●				

SOURCE: "Rubber Parts," *Materials Engineering*, May 1967.

ties which approach those of the thermosets. Their tensile strength runs about 10,000 psi, and elongation is as low as 2 percent. Also, chemical resistance is better than that of soft natural rubbers. The soft grades, however, have excellent resilience and good abrasion resistance. Compared to the synthetics, they have low hysteresis, or heat buildup, under dynamic loading such as repeated flexing.

These properties make natural rubber attractive for conveyor belts and some large or special tire products. Their low cost makes them useful for such consumer products as rubber bands, hot-water bottles, and carpet backing, and for mechanical goods such as sound-damping equipment and gaskets.

Several types of modified natural rubber are used in the production of coatings, protective films, and adhesives. These types are chlorinated rubber, rubber hydrochloride, and cyclized or isomerized rubber. Chlorinated rubber, for example, modified with any one of a number of plastic resins, provides maximum protection against a wide range of chemicals, and the coatings are widely used in chemical plants, in gas works, and as tank-car linings.

10-4 *Styrene Butadiene (SBR)*
These elastomers, sometimes also called Buna S or GR-S, are copolymers of butadiene and styrene. They are similar in many ways to the natural rubbers, and were the first widely used synthetics. They top all elastomers in volume of use chiefly because of their low cost and use in auto tires. A wide range of property grades are produced by varying the relative amounts of styrene and butadiene. For example, styrene content varies from as low as 9 percent in low-temperature resistant rubbers to 44 percent in rubbers with excellent flow characteristics. Those grades with styrene content above 50 percent are by definition considered plastics. Carbon black is sometimes added also as it substantially improves processing and abrasion resistance.

SBR elastomers are similar in properties to natural rubber. They are non-oil resistant and are generally poor in chemical resistance. Although they have excellent impact and abrasion resistance, they are somewhat below natural rubber in tensile strength, resilience, hysteresis, and some other mechanical properties.

The largest single use is in tires. Other applications are similar to those of natural rubber.

10-5 *Chloroprene (Neoprene) (CR)*
Neoprene was developed in the 1930s, and has the distinction of being the first commercial synthetic rubber. It is chemically and structurally similar to natural rubber, and its mechanical properties are also similar. Its resistance to oils, chemicals, sunlight, weathering, aging, and ozone is outstanding. Also, it retains its properties at temperatures up to 250°F, and is one of the few elastomers that does not support combustion, although it is consumed by fire. In addition, it has excellent resistance to permeability by gases, having about one-fourth to one-tenth the permeability of natural rubber, depending on the gas. Although it is slightly inferior to natural rubber in most mechanical properties, Neoprene has superior resistance to compression set,

290   Industrial and Engineering Materials

Fig 10-4 Printing rolls and other industrial rolls which must withstand ink, chemicals, and metal chips are made of ethylene propylene rubber. (B. F. Goodrich Chemical Co.)

particularly at elevated temperatures. It can be used for low-voltage insulation, but is relatively low in dielectric strength.

Typical products made of chloroprene elastomers are heavy-duty conveyor belts, V-belts, hose covers, footwear, brake diaphragms, motor mounts, rolls, and gaskets.

10-6 *Isobutylene-Isoprene (Butyl) (IIR)*
Butyl rubbers are copolymers of isobutylene and about 1 to 3 percent isoprene. They are similar in many ways to natural rubber, and are one of the lowest-priced synthetics. They have excellent resistance to abrasion, tearing, and flexing. They are noted for low gas and air permeability (about 10 times better than natural rubber), and, for this reason, make a good material for tire inner tubes, hose, tubing, and diaphragms. Although butyls are non-oil resistant, they have excellent resistance to sunlight and weathering, and generally have good chemical resistance. They also have good low-temperature flexibility and heat resistance up to around 300°F; however, they are not flame re-

sistant. They generally have lower mechanical properties such as tensile strength, resilience, abrasion resistance, and compression set than the other elastomers.

Because of their excellent dielectric strength, they are widely used for cable insulation, encapsulating compounds, and a variety of electrical applications. Other typical uses include weather stripping, coated fabrics, curtain wall gaskets, high-pressure steam hoses, machinery mounts, and seals for food jars and medicine bottles.

10-7 *Isoprene (IR)*
This elastomer is synthetic natural rubber. It is processed like natural rubber, and its properties are quite similar although isoprene has somewhat higher extensibility. Like natural rubber, its notable characteristics are very low hysteresis, low heat buildup, and high tear resistance. It also has excellent flow characteristics, and is easily injection molded. Its uses complement those of natural rubber. And its good electrical properties plus low moisture absorption make it suitable for electrical insulation.

10-8 *Polyacrylate (ACM)*
Polyacrylate elastomers, also called acrylics, are based on polymers of butyl or ethyl acrylate. They are low-volume use, specialty elastomers, chiefly used in parts involving oils (especially sulfur-bearing) and elevated temperatures up to 300°F and even as high as 400°F. A major use is for automobile transmission seals. Other oil-resistant uses are gaskets and O rings. Mechanical properties such as tensile strength and resilience are low. And, except for recent new formulations, they lose much of their flexibility below $-10°F$. The new grades extend low-temperature service to $-40°F$. Polyacrylates have only fair dielectric strength, which improves, however, at elevated temperatures.

10-9 *Nitrile Butadiene (NBR)*
Nitrile elastomers, known originally as Buna N, are copolymers of acrylonitrile and butadiene. They are principally known for their outstanding resistance to oil and fuels at both normal and elevated temperatures. Their properties can be altered by varying the ratio of the two monomers. In general, as the acrylonitrile content increases, oil resistance, tensile strength, and processability improve while resilience, compression set, low-temperature flexibility, and hysteresis characteristics deteriorate. Most commercial grades range from 20 to 50 percent acrylonitrile. Those at the high end of the range are used where maximum resistance to fuels and oils is required, such as in

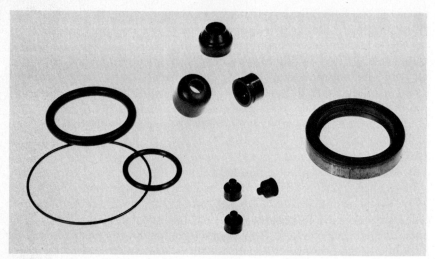

Fig 10-5 Typical synthetic rubber O rings, seals, and other automotive parts. (B. F. Goodrich Chemical Co.)

oil-well parts and fuel hose. Low-acrylonitrile grades are used where good flexibility at low temperatures is of primary importance. Medium-range types, which are the most widely used, find applications between these extremes. Typical products are flexible couplings, printing blankets, rubber rollers, and washing machine parts.

Nitriles as a group are low in most mechanical properties. Because they do not crystallize appreciably when stretched, their tensile strength is low, and resilience is roughly one-third to one-half that of natural rubber. Depending on acrylonitrile content, low-temperature brittleness occurs at from −15 to −75°F. Their electrical-insulation quality varies from fair to poor.

10-10 *Polybutadiene (BR)*
Among the newer types, these elastomers are notable for their low-temperature performance. With the exception of silicone, they have the lowest brittle or glass transition temperature (−100°F) of all the elastomers. They are also one of the most resilient, and have excellent abrasion resistance. However, resistance to chemicals, sunlight, weathering, and permeability by gases is poor. Some uses are shoe heels, soles, gaskets, and belting. They are also often used in blends with other rubbers to provide improvements in resilience, abrasion resistance, and low-temperature flexibility.

10-11 *Polysulfide (Thiokol)*
Thiokol, a specialty elastomer, is rated highest in resistance to oil and gasoline. It also has excellent solvent resistance, extremely low gas permeability, and good aging characteristics. Thus it is used for such products as oil and gasoline hoses, gaskets, washers, and diaphragms. Its major use is for equipment and parts in the coating production and application field. It is also widely applied in liquid form in sealants for the aircraft and marine industries. Thiokol's mechanical properties, including strength, compression set, and resilience, are poor. Although Thiokol is poor in flame resistance, it can be used in temperatures up to 250°F.

10-12 *Ethylene Propylene (EPM and EPDM)*
These relatively new elastomers are available as copolymers (EPM) and terpolymers (EPDM). They offer good resilience, flexing characteristics, compression-set resistance, and hysteresis resistance, along with excellent resistance to weathering, oxidation, and sunlight. Although fair to poor in oil resistance, their resistance to chemicals is good. Their maximum continuous service temperature is around 350°F. Typical applications are electrical insulation, footwear, auto hose, and belts.

10-13 *Urethane (AU and EU)*
These elastomers are copolymers of diisocyanate with their polyester (AU) or polyether (EU). Both are produced in solid gum form and viscous liquid. With tensile strengths above 5,000 psi, and some grades approaching 7,000 psi, urethanes are the strongest available elastomers. They are also the hardest, and have extremely good abrasion resistance. Other notable properties are low compression set, and good aging characteristics and oil and fuel resistance. The maximum temperature for continuous use is under 200°F, and their brittle point ranges from −60 to −90°F. Their largest field of application is for parts requiring high wear resistance and/or strength. Typical products are forklift truck wheels, airplane tail wheels, shoe heels, bumpers on earth-moving machinery, and typewriter damping pads.

10-14 *Chlorosulfonyl Polyethylene (Hypalon) (CSM)*
This specialty elastomer contains about one-third chlorine and 1 to 2 percent sulfur. It can be used by itself or blended with other elastomers. Hypalon is noted for its excellent resistance to oxidation, sunlight, weathering, ozone, and many chemicals. Some grades are satis-

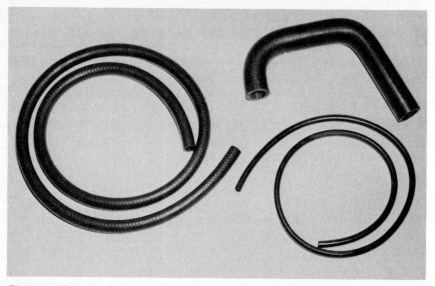

Fig 10-6   Elastomers used in hose applications must have low-temperature flex, oil or fuel resistance, and gas impermeability. (B. F. Goodrich Chemical Co.)

factory for continuous service at temperatures up to 350°F. It has moderate oil resistance. It also has unlimited colorability. Its mechanical properties are good but not outstanding, although abrasion resistance is excellent. Hypalon is frequently used in blends to improve oxidation and ozone resistance. Typical uses are tank linings, high-temperature conveyor belts, shoe soles and heels, seals, gaskets, and spark plug boots.

10-15   *Epichlorohydrin (CO and ECO)*
These recently developed specialty elastomers are noted for their good resistance to oils, and excellent resistance to ozone, weathering and intermediate heat. The homopolymer (CO) has extremely low permeability to gases. The copolymer (ECO) has excellent resilience at low temperatures. Both have low heat buildup, making them attractive for parts subjected to repeated shocks and vibrations.

10-16   *Fluorocarbon (FKM)*
These fluorine-containing elastomers, like their plastic counterparts, are highest of all the elastomers in resistance to oxidation, chemicals, oils, solvents, and heat—and they are also the highest in price. They can be used continuously at temperatures over 500°F and do not support combustion. Their brittle temperature, however, is only −10°F. Their mechanical and electrical properties are only moderate. Unrein-

forced types have tensile strengths of less than 2,000 psi and only fair resilience. Some recent developments have improved compression-set characteristics. Typical applications are brake seals, O rings, diaphragms, and hose.

10-17 Silicone (MQ, PMQ, VMQ, FVMQ, and PVMQ)

Silicones are not hydrocarbons, but polymers composed basically of silicon and oxygen atoms. There are four major elastomer composition groups, as indicated by the above letter designations. In terms of application, silicone elastomers can be divided roughly into the following types: general-purpose, low-temperature, high-temperature, low-compression-set, high-tensile-high-tear, fluid-resistant, and room-temperature vulcanizing. All silicone elastomers are high-performance, high-price materials. The general-purpose grades, however, are competitive with some of the other specialty rubbers, and are less costly than the fluorocarbon elastomers.

Silicone elastomers are the most stable group of all the elastomers. They are outstanding in resistance to high and low temperatures, oils, and chemicals. High-temperature grades have maximum continuous service temperatures up to 600°F; low-temperature grades have glass transition temperatures of -180°F. Electrical properties, which are comparable to the best of the other elastomers, are maintained over a temperature range from -100 to over 500°F. However, most grades have relatively poor mechanical properties. Tensile strength runs only around 1,200 psi. However, grades have been developed with much improved strength, tear resistance, and compression set.

Fluorosilicone elastomers (FVMQ) have been developed which combine the outstanding characteristics of the fluorocarbons and silicones. However, they are expensive and require special precautions during processing. A unique characteristic of one of these elastomers is its relatively uniform modulus of elasticity over a wide temperature range and under a variety of conditions.

Silicone elastomers are used extensively in products and components where high performance is required. Typical uses are seals, gaskets, O rings, insulation for wire and cable, and encapsulation of electronic components.

Review Questions

1. What is the major characteristic that distinguishes elastomers from plastics?
2. Explain the difference in molecular structure between crude and vulcanized rubber.

3. Which type of rubber most resembles phenolic plastic?
4. When a piece of rubber is stretched, what happens to the arrangement of the molecular, or polymer, chains?
5. In Fig. 10-2, if the energy input ACD is 12 ft-lb and the energy output BCD is 8 ft-lb, what is the resilience value?
6. If elastomer A has a resilience of 75 percent and elastomer B has a resilience of 60 percent,
   (a) Which of the two has the higher (or more) hysteresis?
   (b) When the two are subjected to the same cyclic stresses, which one has the greater heat buildup?
7. Elastomers A, B, and C have hardnesses 70, 45, and 35, respectively, on the durometer scale. Which of the three is likely to be
   (a) Highest in tensile strength?
   (b) Lowest in elongation?
   (c) Highest in wear or abrasion resistance?
   (d) Lowest in cold flow or creep?
8. Three elastomers have the following ASTM type (first letter) and class (second letter) designations: BC, EB, and CH. Which of the three
   (a) Has the best resistance to aging?
   (b) Has the best oil resistance?
9. What effect does the addition of carbon black to rubber have on the following properties?
   (a) Hardness and abrasion resistance.
   (b) Tensile strength.
10. Name the elastomer or elastomers which
    (a) Are most similar to natural rubber.
    (b) Are highest in resistance to oil and gasoline.
    (c) Are lowest in glass transition, or brittle, temperature (name two).
    (d) Have the highest tensile strength.
    (e) Have the best resistance to chemicals, oxidation, oils, and heat.
11. Which type of rubber is most widely used for automobile tires?

Bibliography

King, W. H.: "A Fresh Look at Elastomers," *Machine Design*, Jan. 25, 1973.
"Plastics/Elastomers Reference Issue," *Machine Design*, Feb. 15, 1973, p. 164.
"Rubber Parts," *Materials Engineering*, May 1967, p. 99.

WOOD AND PAPER

Wood was probably one of the first materials ever used. Today, it is still the most extensively used of all materials. Annual world consumption is about 1,000 million tons as compared to 400 million tons of steel, and, in the United States, consumption per person is about double that of steel—roughly 2,400 lb and 1,100 lb, respectively. The major reasons for wood's prolific use are its availability, low cost, and easy workability. In addition, its light weight and relatively high strength add up to a good strength-to-weight ratio.

Nature and Structure

11-1 *Microstructure*
Wood is vegetable matter, of course, composed of two principal ingredients: about 70 percent (by volume) cellulose, and 20 to 30 percent lignin, which is nature's glue for holding the cellulose together. A small residue is made up of minerals, waxes, tannins, and oils, which often gives wood unusual properties. For example, the oils in cypresses provide decay resistance, and the aromatic oils in cedars give a distinctive scent that makes them desirable for use in storage chests.

Wood, like all living vegetable matter, is composed of long, thin, hollow tubes, or cells. The cell walls consist of cellulose fibers bonded together by lignin and aligned parallel to the axis of the cell. The cellulose fibers are made up of cellulose-chain polymers that in turn are composed of glucose monomers joined end-to-end. The glucose (simple sugar) monomer is a molecule of carbon, oxygen, and hydrogen atoms. The cellulose polymers are joined to each other by strong intermolecular forces, giving high rigidity and crystallinity.

The wood cells, sometimes referred to as fibers, are aligned predominantly parallel to each other and to the axis of the tree's trunk or limb. A second system of cells, called rays, runs perpendicular to the axis of the trunk or limb. Because the cells are hollow, wood is not a solid material like most other polymeric materials. While the density of cellulose and lignin in the cell walls is about the same for all kinds of wood, the bulk density (weight per unit of volume) varies considerably, depending on cell wall thickness and cell cavity size. The thicker the tubular cell walls, the higher the bulk density. And, as we shall see, wood's mechanical properties are closely related to density.

Because wood cells are highly hygroscopic (taking up and retaining moisture), the moisture content significantly influences wood's properties. Rigidity and dimensional stability, in particular, decrease as moisture content increases.

11-2 *Macrostructure*
Wood's macrostructure is directly related to the structural characteristics of the tree's (or shrub's) growth cycle (Fig. 11-1). Trees grow in thickness in concentric layers that are called annual, or growth, rings. The rings are divided into two parts. The inner part, called springwood, which is formed first in the growing season, is usually lighter in weight, denser, and stronger than the outer part of the ring, which grows later and is called summerwood. Wood's grain, texture, and coloring are largely determined by the thickness of the growth rings and the relative amounts of spring- and summerwood. In general, the thicker the growth ring, the coarser the texture.

A wood's structure and properties also vary across the cross section of the tree trunk. The older, central growth rings, termed the heartwood, contain greater amounts of minerals and other chemical residues than do outer rings, called the sapwood. Some trees, such as maples and yellow and ponderosa pines, are mainly composed of sapwood. In some species, such as firs and spruces, there is little distinction. In general heartwood is usually darker than sapwood, although there are a number of exceptions, for example, in hemlocks, spruces, basswoods, and firs. Heartwood is also generally more decay resistant. Contrary to popular belief, however, it is not necessarily stronger than sapwood, although it is usually denser because of the presence of chemical residues.

The manner in which lumber is cut out of logs also determines cer-

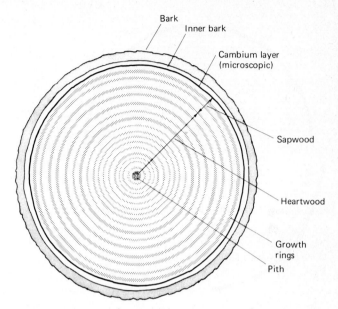

Fig 11-1   Drawing of tree trunk cross section showing macro-structural characteristics of wood.

tain structural characteristics (Fig. 11-2). Lumber cut parallel with the rays is called quarter-sawed, and, when cut tangent to the growth rings, it is called plain sawed. Softwoods cut the latter way are often referred to as flat or slash grained.

In plain-sawed lumber, the surface grain appears as ellipses, parabolas, and arrows of regular or irregular width and outline. The surface of quarter-sawed lumber has longitudinal stripes or ribbons with the wood rays appearing as flecks or flakes. Besides these differences in appearance, plain-sawed wood compared to quarter-sawed is usually cheaper, shrinks and swells less in thickness, and does not collapse as easily in drying. On the other hand, quarter-sawed wood shrinks and swells less in width, warps and twists less, splits and checks less, and wears more evenly.

Kinds and Grades

All woods are divided into two major classes on the basis of the type of tree from which they are cut: *Hardwoods* are from broad-leaved deciduous trees. *Softwoods* are from conifers, which have needle- or scalelike leaves, and which are, with few exceptions, evergreens. The terms hard- and softwoods do not refer to the relative hardnesses of

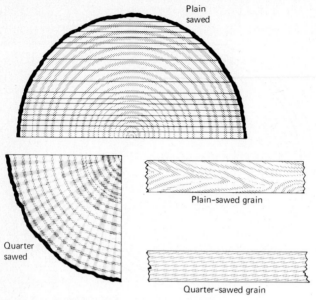

Fig 11-2 Plain-sawed and quarter-sawed wood.

the woods in these two classes. For example, some hardwoods, such as cottonwoods and aspens, are softer than many softwoods.

Some of the main distinctions between the two classes are:

1. Softwoods are more plentiful, accounting for about 75 percent of lumber production.
2. Hardwoods are generally higher in cost.
3. Hardwoods are mainly used for furniture, paneling, flooring, and finished and semifinished factory-made wood products.
4. Softwoods are mainly used for construction and building purposes, particularly in the home-construction field.

Besides these differences, the lumber industry is divided quite distinctly into hardwood and softwood producers, whose methods of manufacture and grading differ from each other.

11-3 *Softwood Grades*

There are three classes of softwood lumber: yard lumber, factory and shop lumber, and structural lumber, or timbers.

Yard lumber is less than 5 in. thick, and is chiefly used for general building purposes (Table 11-1). It is produced in three grades:

1. Finish grades, which are less than 4 in. thick and under 16 in. wide, are graded A, B, C, and D. They are intended for finishing work, as the name implies.
2. Common boards, which are less than 2 in. thick, are graded No. 1, No. 2, No. 3, No. 4, and sometimes No. 5. They are intended for general use where appearance is not important.
3. Dimension lumber, which is over 2 in. thick and 5 in. wide, is graded No. 1, No. 2, and No. 3. It is intended for general-utility purposes such as framing.

Factory and shop lumber is graded by the number of usable cuttings that can be made from a given board. Much of this type lumber is used for general millwork products, patterns and models, and for factory-made softwood items.

Structural lumber is never less than 2 in. thick, and is intended for use where definite strength requirements are specified. The allowable stresses specified for structural lumber depend on the size, number, and placement of defects. Because the location of defects is important, the piece must be used in its entirety for the specified strength to be realized.

11-4 *Hardwood Grades*

Hardwood grading is based on the amount of usable lumber that can be obtained from a given board. The size and number of defects are not significant except when the size of the affected area reduces the number of cuttings below that required for a given grade. Higher-grade boards usually have greater average lengths and widths than the lower-grade boards. The hardwood grades, from the highest to the lowest quality, are:

1. Firsts (the top quality) and seconds. These two grades are usually marketed as one grade called Firsts and Seconds (FAS).
2. Selects.
3. Common grades, rated No. 1, No. 2, No. 3A, and No. 3B.

Sometimes a grade is further specified, such as FAS One Face, which means that only one face is of FAS quality, or WHND, which means that wormholes are not considered defects in determining the grade.

Rough, ungraded hardwood lumber is also available. It is known as hardware dimension, and is usually kiln dried, cut to length and

Table 11-1 *Grades of Yard Lumber*

Finish or Select	Common Board and Strip	Dimension
Building lumber of good appearance and finish. Less than 4 in. thick, and 16 in. or less in width.	Contain certain defects and blemishes which detract from appearance, but are suitable for general-utility and construction purposes. Boards are less than 2 in. thick and 8 in. or more in width. Strips are less than 2 in. thick and less than 8 in. wide.	Graded principally on requirements for framing uses, such as joists, rafters, and studding. Strength, stiffness, and size uniformity are essential requirements.
GRADES A AND B These are usually combined and sold as "B and Better." Grade A pieces are practically clear. Grade B pieces have a few small defects or blemishes, such as small checks, stains, pitch areas, or pin knots. Grades B and Better are commonly used for high-class interior and exterior trim and where wood is to receive a natural finish.	CONSTRUCTION (NO. 1) This grade is sound and tight knotted. The size of knots and blemishes is limited. Used as siding and exterior trim. It is inferior for interior trim. It is difficult to conceal knots entirely with paint.	NO. 1 While sound, this grade can have small knots, depending on the size of the piece, and can have pitch and checks, if strength is not materially affected. Used for joists, rafters, scaffolding.
GRADE C This grade has a limited number of small defects or blemishes that can be covered with paint, so it is especially suited for high-class paint finish. It is used for exteriors of houses, porch flooring, and painted interior trim.	STANDARD (NO. 2) This grade has large and coarse features, such as knots that may be considered grain tight. The popular type of knotty finish is usually this grade. Used in construction as subfloors, sheathing, and concrete forms.	NO. 2 This grade has large, coarse, unsound knots, and may be warped. Used for joists, rafters, plates, sills, and studding. NO. 3 This is the lowest grade. It has the defects of No. 2, but without limitations on size and number.

GRADE D

This grade can have any number of surface defects or blemishes provided they can be covered by paint. It has the same uses as grade C.

UTILITY (NO. 3)

This grade has larger and coarser knots than No. 2, occasional knotholes, and season checking. Used like No. 2, where a cheaper material can be used.

ECONOMY (NO. 4)

This grade is of low quality, with the coarsest defects, such as decay and holes.

width, and planed to the needed dimensions by the wood-product manufacturer or by producers specializing in this form of hardwood.

11-5 *Green and Seasoned Lumber*
Freshly cut lumber contains large amounts of water, ranging from 30 to more than 200 percent of the dry weight of the wood. If at least some of the water is not removed, the wood will shrink and warp excessively in service. There are three general trade classifications to indicate the seasoned condition of the lumber.

1. *Shipping dry* indicates that the lumber has been partially dried to prevent stain and mold during shipping.
2. *Air-dried* indicates that the lumber has been dried in air either outdoors or in an unheated shed. In general, moisture content ranges between 10 and 20 percent.
3. *Kiln-dried* lumber has been dried in a kiln to reduce moisture content to between about 6 and 10 percent.

Most finish grades as well as wood for factory-made wood products are kiln dried. However, lumber used in exterior service can have 10 to 15 percent moisture content without causing excessive warping or shrinkage.

11-6 *Imported Woods*
A number of woods produced outside of the United States have properties not found in native species. For example, lignum vitae from the West Indies and tropical America is a dense, tough, resinous, dark reddish-brown heartwood. It is hard, strong, and the heaviest of commercial woods. It is used for such products as bearings, pulley sheaves, and mallet heads.

Balsa, one of the lightest woods, is from South America. It is nearly white, and is used for floats, model building, and other products where light weight is required. Ebony, from Africa, is a hard, heavy, black heartwood, and is used chiefly for inlay work, knife handles, and piano keys. Rosewood, from India, is dense, dark brown to nearly black, oily in nature, and has a roselike odor. It is used for knife handles and scientific instruments. Lauan or tanguile is the name for a group of Philippine woods sold as Philippine mahogany. They are dark reddish in color, but have a more open grain pattern than true mahoganies.

Properties
Because wood is a natural material subject to many relatively uncontrollable influences, such as speed of growth, grain nature, and defects,

the variability in its properties is greater than that encountered in most man-made materials. Therefore, properties of average wood in service can vary more widely from reported property values than happens with most other materials. To eliminate the influence of defects, property values are given here for "clear" wood samples.

11-7 Mechanical

Density, or Specific Gravity. This is the major factor related to the strength properties of wood. As a general rule, strength increases with density. The other factors that influence most mechanical properties are moisture content and defects in the wood. In endwise compression, for example, wood with a 12 percent moisture content is twice as strong as green wood. However, moisture generally doesn't affect toughness and shock resistance, although they sometimes decrease as moisture content decreases.

Strength. Strength is measured by the following tests:

1. The modulus of rupture measures the ability of a beam to support a slowly applied load for a short time. It is a widely accepted criterion of wood strength.
2. Static bending, or flexure stress at proportional limit, is another strength indicator.
3. The maximum crushing strength is the maximum stress sustained by a compression load which is slowly applied parallel to the grain. In general, tensile strength parallel to the grain is 40 times that perpendicular to the grain. Compressive strength parallel to the grain is 3 to 10 times greater than that perpendicular to the grain.

Wood strength tends to decrease at temperatures above normal. Air-dried wood can be exposed to temperatures of up to about 150°F for a year or more without significant loss in most strength properties, but at such elevated temperatures strength is temporarily reduced. Exposure to temperatures of 150°F or more for extended periods of time permanently weakens wood, the extent depending chiefly on moisture content, temperature, and length of exposure.

Stiffness or Rigidity. Stiffness is expressed in terms of the modulus of elasticity in bending, and thus is a measure of resistance to deflection. The modulus is high compared to most polymeric materials, ranging from around 1.2 to 2.5×10^6 psi. The modulus of elasticity in compression parallel to the grain is usually about 10 percent higher than the bending modulus.

Table 11-2 Strength Properties of Some American Woods

Species	Specific Gravity	Modulus of Rupture in Bending, 1,000 psi	Bending Stress at Proportional Limit, 1,000 psi	Modulus of Elasticity in Bending, 10^6 psi	Maximum Crushing Strength, Parallel to Grain, 1,000 psi
		Softwoods			
Cedar:					
Port Orford	0.42	11.3	7.7	1.3	6.5
Eastern red	0.47	8.8	3.8	0.9	6.0
Cypress, southern bold	0.46	10.6	7.2	1.4	6.4
Douglas fir, coast	0.48	12.2	8.1	2.0	7.4
Fir, balsam	0.36	7.6	5.2	1.2	4.5
Hemlock:					
Eastern	0.40	8.9	6.1	1.2	5.4
Western	0.42	10.1	6.8	1.5	6.2
Pine:					
Eastern white	0.35	8.6	6.0	1.2	4.8
Longleaf	0.58	14.7	9.3	2.0	8.4
Ponderosa	0.40	9.2	6.3	1.3	5.3
Redwood, virgin	0.40	10.0	6.9	1.3	6.2
Spruce, litka	0.40	10.2	6.7	1.6	5.6

Hardwoods

Ash, white	0.60	15.4	8.9	1.8	7.4
Basswood	0.37	8.7	5.9	1.5	4.7
Beech:	0.64	14.9	8.7	1.7	7.3
Yellow	0.62	16.6	10.1	2.0	8.2
Cottonwood, black	0.35	8.3	5.3	1.3	4.4
Elm, rock	0.63	14.8	8.0	1.5	7.1
Hickory-Shag Bark	0.72	20.2	8.9	2.2	9.2
Locust, black	0.69	19.4	12.8	2.1	10.2
Maple, sugar	0.63	15.8	9.5	1.8	7.8
Oak:					
Red	0.63	14.3	8.4	1.8	6.8
White	0.68	15.2	8.2	1.8	7.4
Poplar, yellow	0.42	10.1	6.1	1.6	5.5
Walnut, black	0.55	14.6	10.5	1.6	7.6

Table 11-3 *Strength Properties of Some Imported Woods*

Species	Specific Gravity	Modulus of Rupture in Bending, 1,000 psi	Modulus of Elasticity in Bending, 10^6 psi	Maximum Crushing Strength, Parallel to Grain, 1,000 psi
Balsa	0.14	2.2	0.46	1.0
Rosewood, eastern India	0.79	17	1.77	9.1
Lignum vitae	1.09	—	—	11.4
Mahogany				
Tropical America	0.50	11.1	1.43	6.4
African	0.47	10.7	1.48	5.7
Philippine	0.57	12.3	1.69	6.8
Ebony, African	—	26	2.73	12.9
Teak	—	11.4	1.67	5.9
Lemonwood, tropical America	0.78	22.3	2.27	9.7

Toughness. The ability of wood to absorb shock or impact loads is chiefly dependent on strength, and therefore on density, flexibility, and moisture content. As a class, woods compare favorably with other polymeric materials. Because of moisture content, green wood is generally tougher than seasoned or dry wood. Hardwoods are tougher than softwoods, although some species, such as basswoods and sycamores, are quite brittle. Among the softwoods, longleaf pine is about the toughest.

Fatigue Resistance. Because it is a fibrous material, wood is less sensitive to repeated loads than are the more crystalline structural materials, especially metals. Endurance limits are also usually higher in proportion to ultimate strength than is the case for some metals. For tension in fatigue parallel to the grain, the endurance load for 30 million cycles of stress can be 40 percent of the strength of air-dried wood.

Hardness. Hardness is a measure of wood's resistance to scratching, indentation, and wear. It is measured by the load required to embed into the wood a 0.444 in. ball to one-half its diameter. End-grain wood surfaces are harder than side-grain surfaces, sometimes by more than 50 percent. Exceptions are lignum vitae and teak, where side-grain hardness is higher. (As previously stated, the names softwoods and hardwoods are not indicative of the relative hardness of these woods.) Among hardwoods, basewoods, poplars, aspens, cottonwoods, and willows are quite soft, whereas yews and some cedars have high hardness.

11-8 *Thermal*

Thermal conductivity varies with wood density. Thus the lighter-weight woods are better heat insulators, with balsa, the lightest, the best. For a given density, thermal conductivity can still vary considerably, depending on moisture content and its distribution. In high-density woods, for example, conductivity can increase 20 to 25 percent when the moisture content increases from 12 to 30 percent. Other influencing factors on thermal conductivity are the direction of the grain, the proportion of springwood to summerwood, and defects such as checks and knots.

Thermal expansion differs in the longitudinal and in the radial and tangential directions of wood. Along the grain, values are independent of density and vary from about 1.7×10^{-6} to $2.5 \times 10^{-6}/1°F$. Thus the coefficient of linear thermal expansion along the grain is from one-tenth to one-third that of common metals, concrete, and glass. Across the grain, thermal expansion generally varies with density, and although it is usually higher than in the grain direction, it is still usually less than that of other common structural materials.

11-9 *Chemical and Decay Resistance*

Wood has good resistance to common organic acids and solutions of acidic salts, but poor resistance to most alkaline solutions and inorganic acids. All oxidizing solutions of salts, acids, and alkalies rapidly attack wood.

Chemicals can affect wood's strength by (1) producing swelling, (2) hydrolysis of the cellulose by acids or salts, and (3) attacking the lignin. Of the three, swelling is almost completely reversible, so that if the swelling liquid is removed from the wood, the original dimensions and strength are regained.

The combination of moisture, air, and microorganisms is the major cause of wood decay. If kept dry or submerged in water, wood does not decay. Decay also does not take place unless the moisture content is above 20 percent. The rate of decay varies greatly depending on local conditions and temperature. For example, wood in warm, humid climates deteriorates more rapidly than in cool, dry areas. The sapwood of most species has much lower decay resistance than the heartwood.

Woods with low decay resistance can be improved by a number of different preservative treatments (see Sec. 11-12).

11-10 Electrical
Wood is essentially a nonconductor, but its electric resistance and dielectric properties vary considerably with moisture content. The specific resistance (resistance of a cubic centimeter of wood) of some common woods varies from 3×10^{17} to 3×10^{18} ohm-cm for kiln-dried wood to 10^8 ohm-cm for wood with 16 percent moisture content. The dielectric constant of kiln-dried wood is about 4.2, but it increases with moisture content, since the dielectric constant of water is approximately 81.

11-11 Working and Forming
One of the major advantages of wood is its good working qualities, particularly with hand tools. Although there is no test for quantitatively evaluating workability, a qualitative classification based on experience at the Forest Products Laboratory is given in Table 11-4.

The machinability of various woods often can be rated according to the quality of the surface produced when machine worked. Table 11-5 roughly rates various hardwoods for their machinability in planing and shaping operations.

Although wood cannot be hot or cold formed in the same way as metals or other polymers, it can be bent or shaped to a considerable extent by soaking it in hot water or steam and then shaping and clamping it in position until it dries. Table 11-5 also roughly rates various hardwoods for their ability to be steam bent.

Modified Woods

There are three major ways in which woods can be modified to change or improve properties. They can be (1) chemically treated, (2) impregnated with another material, or (3) broken into chips or particles and reconstituted.

11-12 Wood Treatments
Woods can be treated with various types of chemicals to increase their resistance to decay by fungi, insects, and marine borers. Wood preservatives are of two major types: oils and water solutions of salts. The oldest oil preservative is creosote. Although it is still one of the best

Table 11-4 *Classification of Certain Hardwood and Softwood Species According to Ease of Working with Hand Tools**

Group 1 Easy to Work	Group 2 Relatively Easy to Work	Group 3 Least Easy to Work
HARDWOODS		
Alder, red	Birch, paper	Ash, commercial white
Basswood	Cottonwood	Beech
Butternut	Magnolia	Birch
Chestnut	Sweet gum	Cherry
Yellow poplar	Sycamore	Elm
	Tupelo:	Hackberry
	Black	Hickory:
	Water	True
	Walnut, black	Pecan
		Honey locust
		Locust, black
		Maple
		Oak:
		Commercial red
		Commercial white
SOFTWOODS		
Cedar:	Bald cypress	Douglas fir
Atlantic white	Cedar, eastern red	Larch, western
Incense	Fir:	Pine, southern yellow
Northern white	Balsam	
Port Orford	White	
Western red	Hemlock:	
	Eastern	
	Western	
Pine:		
Eastern white	Pine, lodgepole	
Ponderosa	Redwood	
Sugar	Spruce:	
Western white	Eastern	
	Sitka	

* The groupings in the table are based on the experience of the Forest Products Laboratory and the general reputation of the wood. Direct comparison of species within a group and comparison of hardwoods and softwoods is not intended.

Table 11-5 *Relative Machinability and Bendability of Hardwoods*

Species	Planing	Shaping	Steam Bending
Ash	B	A	B
Basswood	C	D	D
Beech	—	C	B
Birch	C	A	B
Buckeye	—	D	D
Chestnut	B	C	C
Cottonwood	D	D	C
Elm	D	D	B
Hackberry	B	D	A
Hickory	—	C	B
Magnolia	C	C	A
Mahogany	B	A	C
Maple:			
Hard	C	A	C
Soft	C	C	—
Oak:			
Chestnut	—	C	A
Red	A	C	A
White	A	C	A
Pecan	A	B	B
Sweet gum	C	C	B
Sycamore	D	D	C
Tupelo:			
Black	—	C	C
Water	—	A	C
Walnut, black	C	B	B
Willow	C	D	B
Yellow poplar	B	D	C

NOTE: A indicates the highest rating, and D the lowest.

preservatives, its odor is sometimes offensive, it can irritate the skin, and it is flammable. Petroleum oils fortified with chlorinated phenols or with copper naphthene have found increasing use since their introduction after World War II. They can be painted over, are odorless, and are essentially nonflammable. However, they are probably not as effective against insect damage as creosote.

Common water-solution preservatives include various metallic salts such as zinc chloride, acid copper chromate, and chromated copper arsenate. They are used chiefly on wood that will not be in contact with the ground or water and where the treated wood will be painted.

Chemical preservative treatments are applied either by dipping, soaking, or brushing, or by pressure forcing the chemical into the wood. Dipping is the most efficient surface-treatment method. Heating the chemicals and cooling the pieces during dipping often increases penetration. In pressure treatments, generally only moderate penetration below the surface is required to achieve the desired decay resistance. For species that are difficult to penetrate with the chemical, the wood surface is sometimes punctured with tiny holes prior to treatment.

11-13 Impregnated and Compressed Woods

Because wood is naturally hygroscopic, the cyclic presence of moisture causes swelling and shrinking. Dimensional stability can be greatly improved by impregnating the wood's voids with a bulking agent. Such modified wood is called impreg or compreg, depending on the treatment used. The major difference between the two is that compreg wood is compressed before the resin within the wood is cured to produce a higher-density product. In both types, phenolics, ureas, or melamines are the usual impregnants. The final resin content of the treated woods is usually from 20 to 35 percent of the weight of the dry wood. Impreg treatment is limited to relatively thin veneers that are generally made into plywood or laminated-wood products. Compreg treatment is applied to both veneers and such wood products as furniture legs and tool handles.

Both treatments reduce swelling and shrinking to from one-fourth to one-third that of the untreated wood. Some mechanical properties are improved, but others are not. Compressive, flexural, and tensile strength are increased, although generally not in proportion to the density increase. In the case of impreg, a decrease in tensile strength is reported. Both impreg and compreg show considerable increase in hardness, with compreg's being from 10 to 20 times that of unmodified wood. However, the toughness of both is about one-half to three-fourths that of normal wood.

11-14 Reconstituted Woods

There are two major forms of reconstituted woods: wood composition board and wood moldings. Both forms are composed of wood particles, chips, or fibers bonded together either by the wood's natural lignin or by a synthetic resin. Wood composition board is divided into two major types: hardboard and particle board. (There is also a fibrous-felted softboard type, called insulation board, which ranges in density from 10 to 50 lb/ft.^{3} It is used mainly for insulation and low-cost construction.) Because of their basic similarity, hardboard and particle board have many common properties. For example, both are homogeneous over large areas, without the knots or imperfections found in unmodified wood. The absence of grain also gives them equal strength in all directions, and they are less water-absorbent than unmodified wood.

Hardboard. Hardboard is produced by pulverizing wood, forming it into mats, and then pressing the mats into boards. The cellulose provides reinforcing fibers, and the lignin binds the fibers into a solid mass. Binders or other materials may be added during manufacture to improve or obtain certain properties, but the material is primarily bonded together by fiber interfelting. Most hardboard is produced in sheets up to ⅜ in. thick. Thicker, specialty board is also available.

Standard hardboard does not contain additives. It comes in two classes based on density: Class 1 ranges between 60 and 80 lb/ft^3, and class 2 ranges from 50 to 60 lb/ft.^{3} Class 1 has the better strength properties.

Tempered hardboard contains a drying-oil blend of oxidizing resin and is heat-tempered to stabilize the additive. The added chemicals also improve resistance to moisture and increase stiffness. With density ranging between 60 and 80 lb/ft^3 (0.96 to 1.28 specific density), tempered hardboard has tensile strengths as high as 7,800 psi parallel to the surface as compared to 6,000 psi for untempered board. Another premium grade, superhard board, has a tensile strength comparable to that of the regular tempered type, but its modulus of rupture is considerably higher.

Hardboard can be worked by conventional woodworking tools. Also, it can be postformed to single curvatures or moderate double curvatures by heat and pressure. Typical applications include furniture and many other consumer items. Also, it is being used in industry for switch-mounting plates, brush blocks, and panel mountings in chemical plants.

Particle Board. This material is composed of small, discrete pieces of wood bonded together by a synthetic-resin adhesive in the presence of heat and pressure. Particles range in shape and size from fine, fiberlike elements to large flakes. Ureas and phenolics, alone or in combination, are the principal binding materials. Phenolic binders give sufficient heat and moisture resistance for many outdoor uses, and provide somewhat better strength than the urea resins. Besides a low-density insulating grade (largely limited to Europe), there are two types of

Table 11-6 Strength Properties of Some Reconstituted Woods

Type	Density, lb/ft³	Tensile Strength, Parallel to Surface, 1,000 psi	Modulus of Rupture in Bending, 1,000 psi	Modulus of Elasticity in Bending, 10⁶ psi	Compression Strength, Parallel to Surface, 1,000 psi
Hardboard:					
Untempered or standard	50–80	3–6	3–7	0.4–0.8	1.8–6.0
Tempered	60–80	4–7.8	6.5–10	0.8–1.0	4.2–6.0
Super	85–90	7.8	10.0–12.5	1.25	26.5
Particle Board:					
Medium density	26–50	0.5–4	1.5–8	0.15–0.7	1.4–2.8
Hard-pressed	50–80	1–5	3–7.5	0.40–1.0	3.5–4.0
Moldings	55–70	1.5–1.8	2–8	0.7	—

particle board: medium density (26 to 50 lb/ft^3) and hard pressed (50 to 80 lb/ft^3). Also, two different methods of producing particle board provide different characteristics in the end product. Either the blend of particles and resin is extruded and the resin cured, or it is formed into mats, and then hot-pressed and the resin cured in flat-platen presses. Extruded boards have the same type of particle throughout, whereas flat-platen-pressed boards may be either homogeneous or have one type of particle for the core portion and another for the faces.

Particle board has lower strength properties than hardboard. Tensile strength runs up to about 5,000 psi, parallel to the surface, and the modulus of rupture reaches a high of about 8,000 psi. Particle board can be readily shaped and cut with regular woodworking tools, but it is not usually postformed by bending. However, some manufacturers mold this board into curved shapes during the production process. Its major use is for furniture and consumer products. Because of its excellent acoustical characteristics, it is ideal for speaker enclosures and piano components. It is also suitable for some electrical products because of its good dielectric properties.

Wood Moldings. These are similar in composition and structure to composition board. Particles, flakes, or fibers can be used with a resin binder of either urea, melamine, or phenolic. Birch, maple, or pine are the woods usually used. The wood component, with or without binder, is consolidated and pressed in a heated compression mold consisting of a male and female die. Preforms are also used sometimes. Since flow in the mold is limited by the amounts of resin commonly used, there are limitations on the resulting complexity of curvature and shape. However, wood moldings offer greater versatility than composition boards. Density and wall thickness, for example, can be varied in different areas of the part. Slots, holes, and cutouts can be molded, and relatively close tolerance can be held. Overall part sizes can be as large as 4 by 8 ft. However, because the highest densities achieved in moldings are around 60 to 70 lb/ft^3, mechanical properties cannot match those of the higher-density composition boards. But strengths of moldings can be considered similar to those for boards of the same density and the same composition and structure.

Plywoods

Plywood consists of glued wood panels made up of layers, or plies, with the grain of one or more layers at an angle, usually 90 degrees, with the grain of the others. The outside plies are called faces; the

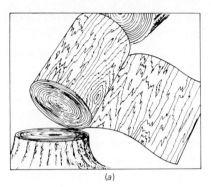 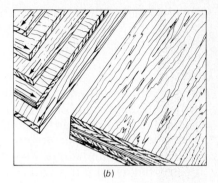

Fig 11-3 (a) Plywood is produced from wood sliced circumferentially from tree trunks. (b) Plywood is built up of plies, with the grain arranged as shown by the black arrows.

center plies are called the core; and the plies immediately below the faces, which are laid at cross angles to them, are called the crossbands. The number of plywood constructions is almost endless, considering the number of woods available, the number of plies used, the placement of the plies, and the types of adhesives.

11-15 Types, Grades, and Sizes

There are two broad classes of plywood: hardwood and softwood. The softwood class is mostly composed of Douglas fir; other woods used are western hemlock, white fir, ponderosa pine, and redwood. The hardwood class contains a great variety of woods.

The grade of a plywood is determined by the quality of the veneer or face, and the type is determined by the moisture resistance. For example, in Douglas fir plywood, the exterior type is expected to retain its form and strength properties when subjected to cyclic wetting and drying, but the interior type is only expected to withstand occasional wet and dry periods. Within each type, several grades are established, depending on the quality of the veneer on the two sides of the panel. The veneer is designated A, B, C, or D, in descending order of quality. For example, grade A-D has grade A veneer on one side and grade D on the other.

Most plywoods are produced to standards established by the plywood industry. There are, however, a number of special-purpose kinds, such as marine, decking, siding, textured, brushed, embossed, and grooved.

Plywood is available in practically any size, but the 4- by 8-ft panel is the standard production size of the industry. Panels with continuous cores and faces can be produced in one piece up to 12 ft long. Oversize panels up to 8 ft wide and of unlimited length can be manufactured by scarf-edge jointing.

11-16 *Properties*
The properties of plywood depend on the characteristics of the wood species used, the number of plies, the thickness of the plies, the grain direction, and so on. It is therefore impossible to test the properties of all possible plywood constructions, and equally impossible to present them in simple tabulations. Plywood strengths are not normally given directly, but as formulas which relate the properties to the particular construction and to the properties of the component plies. They are listed and described fully in *Wood Handbook* (see Bibliography at end of this chapter), and some are given briefly in Table 11-7.

One of plywood's most notable features is its high strength-to-weight ratio. A comparison between birch plywood and other structural materials, for example, shows that plywood's strength-to-weight ratio is 1.52 times that of steel heat-treated to 100,000 psi.

Another important feature is that plywood is virtually split proof because it has no line of cleavage. This is important in fastening because nails can be more closely spaced. In addition, the crisscross arrangement of plies offers exceptional resistance to pull through of nail or screw heads. The absence of a cleavage line also has a pronounced effect on impact resistance. Under extreme impact forces, the side of the panel opposite the impact will rupture along the long-grain fiber, followed by successive shattering of the various plies. Splintering does not occur because pressure is dispersed throughout the panel.

Changes in the moisture content of wood are the primary causes of changes in dimensions. The alternating layers of plies in standard plywood construction restrict expansion and contraction within individual plies. The amount of shrinking and swelling varies in different constructions; however, a softwood plywood panel is said to shrink or swell about 0.2 percent from oven dry to complete saturation. The dimensional change in hardwood plywood is never more than 1 to 2 percent.

Thermal-insulating qualities of plywood are the same as those of the wood of which it is composed, with the important difference that the use of large plywood sheets reduces the number of cracks and joints where leakage could occur. Plywood's ability to absorb sound and reduce its transmission depends on the panel construction and the sound frequency. In general, plywood's acoustical properties are

Table 11-7 *Strength Properties of Some Plywoods*

Species Used	Specific Density, lb/in.^{2}	Ultimate Tensile Strength, Parallel to Surface, 1,000 psi	Modulus of Elasticity in Bending, Parallel to Surface, 10^6 psi	Maximum Crushing Strength, Parallel to Face, 1,000 psi
Douglas fir				
Three ply	0.31	6.5	1.39	3.51
Five ply	0.77	5.7	1.14	3.13
Birch				
Three ply	0.39	9.06	1.74	4.02
Five ply	0.98	7.97	1.42	3.57
Mahogany				
Three ply	0.30	7.02	1.29	3.93
Five ply	0.75	6.17	1.06	3.52
Yellow poplar				
Three ply	0.28	4.82	1.15	2.38
Five ply	0.66	4.65	9.44	2.30
Sweet gum				
Three ply	0.31	6.72	1.26	2.98
Five ply	0.77	5.91	1.03	2.65

NOTE: In three-ply plywood, the face and back are 0.030 in. each, and the core is 0.040 in. In five-ply plywood, the face and back are 0.047 in. each, the core is 0.040 in. and the crossbands are 0.060 in.

good, and this, in conjunction with its beauty and ability to be formed, has led to its extensive use in musical instruments.

One of the chief advantages of plywood is the beautiful decorative effects that can be produced. Colors vary in shade from golden yellow to ebony, from pastels to reds and browns. The use of bleaches, toners, and stains provides even greater variety. Also, there is a great range of possible grain figures, and surfaces can be textured, embossed, and so on.

Plywood can be worked by typical woodworking equipment. It is especially adaptable to portable power-driven saws, drills, and automatic hammers normally used on production or construction jobs.

Plywood can be joined by practically any mechanical fastener and by adhesive bonding. The only precautions required are predrilling some screw holes and selecting proper adhesives for exterior use. Plywood lends itself equally well to butt joints, rabbet joints, dado joints, and frame construction.

Paper

According to the *Dictionary of Paper,* paper is a term used to describe "all kinds of matted or felted sheets of fiber (usually vegetable, but sometimes mineral, animal or synthetic) formed on a fine wire screen from a water suspension." In addition to such materials, paperlike sheets can be produced from platelets or flakes of mica or glass. These are not covered here.

Although there is no distinct line between papers and paperboard, paper is usually considered to be less than 0.006 in. thick. Most fibrous sheets over 0.012 in. thick are considered to be board. In the borderline range of 0.006 to 0.012 in., most are classified as papers.

Properties of papers are controlled by the following variables: (1) type and size of fiber; (2) pulp processing method; (3) web-forming operation; and (4) treatments applied after the paper has been produced.

11-17 Production of Paper

Types of Pulp. Pulps are generally classified as mechanical or chemical wood pulps.

Mechanical wood pulps produced by mechanical processes include:

1. Ground wood, used in a number of papers where absorbency, bulk, opacity, and compressibility are primary requirements, and permanence and strength are secondary.
2. Defibrated pulps, used for insulating board, hardboard, or roofing felts where good felting properties are required.
3. Exploded pulps, used for building and insulation hardboards, or so-called "wood composition materials."

Chemical wood pulps are produced by "cooking" the fibrous material in various chemicals to provide certain characteristics. They include:

1. Sulfite pulps, used in the bleached or unbleached state for papers ranging from very soft or weak to strong grades.
2. Neutral sulfite or monosulfite pulps, used for strong papers for bags, wrappings, and envelopes.
3. Sulfate or kraft pulps, providing high strength, fair cleanliness and, in some instances, high absorbency. Such pulps are used for strong grades of unbleached, semibleached, or bleached paper (called kraft paper) and board.

4. Soda pulps, used principally in combination with bleached sulfite or bleached sulfate pulp for book-printing papers.
5. Semichemical pulps, used for specialty boards, corrugating papers, glassine and grease-proof papers, test liners, and insulating boards and wallboards.
6. Screenings, used principally for coarse grades of paper and board such as millwrapper, and as a substitute for chipboard, corrugating papers, and insulation board.

Stock Preparation. The various operations performed on the pulp prior to the actual making of the paper can have a critical effect on properties. Beating, refining and curlation are three techniques commonly used to modify the fibers of the pulp.

Beating fluffs out the fibers and transfers hydrogen bonds from within fibers to between fibers, permitting them to interlock to a greater degree, thus improving such properties as tensile strength, burst strength, density, folding endurance, stiffness, and transparency. A wide variety of nonfibrous materials can be added at this stage to obtain special properties in the final paper. Common additions are sizes, mineral fillers, starch, silicate of soda, wet-strength resins, and coloring dyes and pigments.

Refining cuts the fibers into shorter lengths, providing better distribution of fibers throughout the paper web.

Curlation curls and crinkles the fibers and breaks up any fiber bundles. Curlation improves such properties as tear strength, bulk, and brightness and provides an even-textured matte finish.

Papermaking. Most papers are made by either the Fourdrinier or the cylinder machine process.

In Fourdrinier papermaking the pulp, mixed to a consistency of 97.5 to 99.5 percent water, is fed continuously to the machine, which consists of an endless belt of fine mesh screen called the "wire." This wire is usually agitated to help the fibers criscross, felt, and mat together to provide a degree of isotropicity in properties.

As the pulp web travels along the wire, water drains from it into suction boxes. As it leaves the wire (at about 83 percent water) it passes through presses, and usually a variety of other types of equipment, such as dryers and calender rolls, depending on the type of paper produced.

In cylinder machine papermaking the web of pulp is formed on a cylindrical mold surface which revolves in a vat of paper stock or pulp. The felting occurs on the face of the cylinder and water drains through the cylinder. A felt conveyor carries the resulting web to the press and dryers.

11-18 *Types of Papers*
There are three basic types of papers: cellulose fiber, inorganic fiber, and synthetic organic fiber papers.

Cellulose Fiber Papers. These papers, made from wood pulp, constitute by far the largest number of papers produced. A great many of the engineering papers are produced from kraft or sulfate pulps. The term "kraft" is used broadly today for all types of sulfate papers, although it is primarily descriptive of the basic grades of unbleached sulfate papers, where strength is the chief factor, and cleanliness and color are secondary. Kraft can be altered by treatments to produce various grades of condenser, insulating, and sheathing papers.

Other types of vegetable fibers used to produce papers include:

1. Rope, used for strong, pliable papers, such as those required in cable insulation, gasketing, bags, abrasive papers, and pattern papers.
2. Jute, used for papers possessing excellent strength and durability.
3. Bagasse, used for paper for wallboard and insulation, usually where strength is not a primary requirement.
4. Esparto, used for high-grade book or printing papers. A number of other types of pulps, also used for these papers, are not discussed here.

Inorganic Fiber Papers. There are three major types of papers made from inorganic fibers:

1. Asbestos papers, the most widely used, are nonflammable, resistant to elevated temperatures, and have good thermal insulating characteristics. They are available with or without binders and can be used for electrical insulation or for high-temperature reinforced plastics.
2. Fibrous glass can be used to produce porous and nonhydrating papers. Such papers are used for filtration and thermal and electrical insulation, and are available with or without binders. High-purity silica glass papers are also available for high-temperature applications.
3. Ceramic fiber (aluminum silicate) papers provide good resistance to high temperatures, low thermal conductivity, and good dielectric properties, and can be produced with good filtering characteristics.

Synthetic Organic Fiber Papers. A great deal of research has been carried out on the use of such synthetic textile fibers as nylon, polyester, and acrylic fibers in papers. Some of the earliest appear highly promising for electrical insulating uses. Others appear promising for chemical or mechanical applications. They are most commonly combined with other fibers in a paper, primarily to add strength.

11-19 Paper Treatments

Papers can be impregnated or saturated, coated, laminated, or mechanically treated. The discussion here covers only the major treatments used, to indicate the extent of treatments available.

Impregnation or Saturation. Impregnation or saturation can be done either at the beater stage in the processing of the pulp, or after the paper web has been formed. Beater saturation permits saturation of nonporous or *ad*sorbent papers, whereas papers saturated after manufacture must be of the *ab*sorbent type to permit complete impregnation by the saturant.

Papers can be saturated or impregnated with almost any known resin or binder. Probably the most commonly used are asphalt for moisture resistance; waxes for moisture vapor and water resistance; phenolic resins for strength and rigidity; melamine and certain ureas for wet strength (not to be confused with moisture resistance); rubber latexes, both natural and synthetic, for resilience, flexibility, strength and moisture resistance; epoxy or silicone resins for dielectric characteristics or dielectric characteristics at elevated temperatures; and ammonium salts or other materials for flameproofing.

A number of proprietary beater saturated papers are available. They are used primarily for gasketing, filtration, simulated leathers, and backing materials. Most of these consist of cellulose or asbestos fibers blended with natural or synthetic rubbers. In some cases cork is added to the blend for increased compressibility. Another type of proprietary beater saturated paper consists of leather fibers blended with rubber latexes.

Coatings. Coating materials, which also impregnate the paper to a greater or lesser degree, include practically every known resin or binder and pigment used in the paint industry. Coatings can be applied in solvent or water solutions, water emulsions, hot melts, and extrusion coatings, or in the form of plastisols or organisols.

The most important properties provided by coatings are (1) gas and water vapor resistance, (2) water, liquid, and grease resistance, (3) flexibility, (4) heat sealability, (5) chemical resistance, (6) scuff resistance, (7) dielectric properties, (8) structural strength, (9) mold resistance, (10) avoidance of fiber contamination, and (11) protection of printing.

Coating materials range from the older asphalts, waxes, starches, casein, shellac, and natural gums, to the newer polyethylenes, vinyl copolymers, acrylics, polystyrenes, alkyds, polyamides, cellulosics, and natural or synthetic rubbers.

Flock coatings can be applied to papers for decoration, sound absorption, vibration cushioning, and surface protection. They consist of extremely short (usually 0.015- to 0.060-in.) fibers, usually of rayon, cotton, hair, or wool, and are applied to adhesive coated paper by spraying, vibrating, or electrostatic techniques.

Laminations. Paper can be laminated to other papers or to other films to provide a variety of composite structures. Probably the most common types of paper laminates are those composed of layers of paper laminated with asphalt to provide moisture resistance and strength. Simple laminations of paper can be so oriented that overall characteristics of the composite are isotropic.

Laminating paper with plastics or other types of films or with metal foils, will, in many cases, combine the desirable properties of the film or foil with those of the paper.

Scrim is a mat of fibers, usually laminated as a "core" material between two faces of paper. It is used to provide strength but can also provide bulk for cushioning, or a degree of "hand" to the composite material.

Mechanical Treatments. Several mechanical treatments can be applied to papers to provide particular special properties.

Crimping, which can be done either on the paper web or on the individual fibers of the paper, essentially adds stretch or extensibility. Crimping the paper web results in crepe paper with improved strength, stretch, bulk, and conformability and texture similar to that of cloth. Typical range of elongation or stretch obtainable is 20 to 300 percent. Cross-creping can provide controllable stretch in directions perpendicular to each other, further improving drapability.

A high degree of stretch, conformability, and flexibility is produced by a patented process which differs from creping in that the individual fibers in the paper web are crimped, rather than the web itself. Amount of stretch is variable, but about 10 percent stretch in the machine direction seems to be optimum for most industrial applications. The major advantages of this type of paper (trade name Clupak) are reported to be a high degree of toughness, combined with a smooth surface, and a high resistance to tearing or punching.

Twisting. Twisting is used to convert paper to twine or yarns. Such yarns have substantially higher strength than the paper from which they are made.

Twisting papers are usually sulfate papers, either bleached or unbleached, with basis weight usually varying from 12 to 60 lb. High tensile strength is required in the machine direction, and the heavier weight papers should usually be soft and pliable. Treatments to impart such characteristics as wear and moisture resistance can be applied during or after the spinning operation.

Embossing and Other Techniques. Decorative papers can be produced by embossing in a variety of patterns. Embossing does not usually improve strength significantly. Embossing or "dimpling" in certain patterns, followed by lamination, can produce composites with added strength as well as bulk and thermal insulation.

Other mechanical methods include: (1) shredding for bulk or padding, (2) pleating, used as a forming aid and for strength in paper cups and plates, (3) die cutting and punching, and (4) molding, which consists of compressing the wet pulp web in a mold to form a finished shape, such as an egg crate.

11-20 *Applications*

As filtration material, paper can be used either as a labyrinth barrier material to guide the fluid or gas to be filtered or, more commonly, as the filtering medium itself. Paper is used for filtering automotive air and oil, and air in room air conditioners, as well as for industrial plant filtration, filtering liquids in tea bags, in addition to machine-cooling oil and domestic hot water filters.

Papers are used for both light- and heavy-duty gaskets, for such applications as high- and low-pressure steam and water, high- and low-temperature oil, aromatic and nonaromatic fuel systems, and for sealing both rough and machined surfaces.

Electrical insulation represents one of the largest engineering uses of papers. For electrical uses, special types include coil papers or layer insulation, cable paper or turn insulation, capacitor papers, and condenser papers. High-temperature, inorganic insulating papers are produced for high-temperature use.

A large and growing use for paper is in structural sandwich materials, where papers are impregnated with a resin such as phenolic and formed in the shape of a honeycomb. The honeycomb is used as a core between facing sheets of a variety of materials including paperboard, reinforced plastics, and aluminum.

Another large-volume application of papers is as backing material for decorative films or other surfacing materials. These provide bulk and depth to the product in simulating leather or fabrics.

Review Questions

1. What is the atomic composition of the basic building block of which wood is composed?
2. What are the two principal ingredients of wood cell walls?
3. What are two differences between heartwood and sapwood?
4. What is the basis of the classification of lumber into softwoods and hardwoods? Which type is mainly used in construction?
5. Which of the three types of lumber shrinks and warps the least in service?
6. What is the general relationship between bulk density and strength properties of wood?
7. Why is seasoned wood generally stronger than green lumber?
8. What effect does grain direction have on wood's tensile and compressive strengths?
9. Name the
    (a) Lightest wood.
    (b) Heaviest wood.
    (c) Stiffest wood.
10. As a class or group, how do American hardwoods compare to American softwoods in
    (a) Toughness?
    (b) Modulus of rupture?
11. Does the decay resistance of wood increase or decrease
    (a) When the moisture content is reduced?
    (b) After dipping in creosote?
    (c) When submerged in water?
    (d) When the air temperature is lowered?
12. When wood is impregnated with a resin, what is the usual effect on
    (a) Swelling and shrinking?   (c) Toughness?
    (b) Hardness?   (d) Strength?
13. Name three characteristics of reconstituted wood that are improvements over unmodified woods.

14. Which of the types of reconstituted wood would you use for
    (a) A part that has different wall thicknesses?
    (b) Highest strength?
15. What is the major reason for the high strength-to-weight ratio of plywood?
16. Why is plywood split proof and high in impact resistance?

Bibliography

Busche, M. G.: "Wood Composition Board," *Materials Engineering*, November 1965.

Forest Products Laboratory: *Wood Handbook*, U.S. Dept. of Agriculture, Washington, D.C., 1955.

Lubars, W.: "A Plywood Primer," *Materials Engineering*, November 1963.

Riley, M. W.: "Paper," *Encyclopedia of Materials, Parts and Finishes* (Clauser, H. R., ed.), Technomic Publishing Co., Westport, Conn., 1975.

Watson, D. A.: "Low-Cost Wood-Particle Moldings," *Materials Engineering*, May 1959.

FIBERS AND TEXTILES

It is quite likely that fibers and textiles were the first man-made material, and they were probably used then for the same reasons they are employed today—for their decorative qualities, their strength and flexibility, and their insulative characteristics. Because of our daily and direct contact with clothing and home furnishings, most of us relate textiles to only these common uses. However, fibers and textiles are one of the most important and versatile engineering materials. Today they are used as a constituent in composites that are among our highest-strength materials. And fibers are a component of many plastics and rubber products.

Fibers and textiles used as materials by themselves are equally versatile. They are used for thermal, vibration, acoustical, and electrical insulation; for wet and dry filtration; for barrier applications such as air seals; for structural coverings; and for a variety of other industrial and consumer-product uses. In this chapter, we will discuss fibrous materials primarily from the standpoint of their use in these broad areas of application. Their use in conjunction with other materials is discussed in other parts of this book.

Polymeric, metallic, and ceramic materials can be produced as fibers. Although metallic and ceramic fibers will be touched on, the major emphasis here will be given to polymeric fibers made from synthetic and natural materials.

Fibers

12-1 Nature and Structure

By definition (ASTM) a fiber has a length at least 100 times its diameter or width, and its length must be at least 5 mm (0.2 in.). Length also determines whether a fiber is classified as staple or filament. Filaments are long and/or continuous fibers. Staple fibers are relatively short, and, in practical applications, range from under 1 to 6 in. long (except for rope, where the fibers can run to several feet). Of the natural fibers, only silk exists in filament form, while synthetics are produced as both staple and filaments.

The internal, microscopic structure of fibers is basically no different from that of other polymeric materials. Each fiber is composed of an aggregate of thousands of polymer molecules. However, in contrast to bulk plastic forms, the polymers in fibers are generally longer and aligned linearly, more or less parallel to the fiber axis. Thus fibers are generally more crystalline than are bulk forms.

Also in contrast to bulk forms, fibers are not used alone, but either in assemblies or aggregates such as yarn or textiles or as a constituent with other materials, such as in composites. Also, compared with other materials, the properties and behavior of both fibers and textile forms are more critically dependent on their geometry (see Sec. 12-2). Hence fibers are sometimes characterized as tiny microscopic beams, and, as such, their structural properties are dependent on such factors as cross-sectional area and shape, and length.

12-2 Geometrical Characteristics

The cross-sectional shape and diameter of fibers varies widely. Glass, nylon, Dynel, and Dacron, for example, are essentially circular. Some other synthetics are oval, while others are irregular and serrated round. Cotton fibers are round tubes, and silk is triangular.

Fiber diameters range from about 10 to 40 microns (0.0004 to 0.0016 in.) in diameter. Because of the irregular cross section of many fibers, it is common practice to specify diameter or cross-sectional area in terms of fineness, which is defined as a weight to length or linear density relationship. One exception is wool, which is graded in microns. The common measure of linear density is the denier, which is the weight in grams of a 9,000-m length of fiber. Another measure is the tex, which is defined as grams per 1 km. A millitex is the number of grams per 1,000 km.

Of course, the linear density, or denier, is also directly related to fiber density. This is expressed as the denier/density value, commonly referred to as denier per unit density, which represents the equivalent denier for a fiber with the same cross-sectional area and a density of 1.

The cross-sectional diameter or area generally has a major influence on fiber and textile properties. It affects, for example, yarn packing, weave tightness, fabric stiffness, fabric thickness and weight, and cost relationships. Similarly, the cross-sectional shape affects yarn packing, stiffness, and twisting characteristics, It also affects the surface area, which in turn determines the fiber contact area, air permeability, and other properties.

12-3 Yarns

Yarns are assemblages or bundles of fibers twisted or laid together to form continuous strands. They are produced with either filaments or staple fibers. Single strands of yarns can be twisted together to form ply or plied yarns, and ply yarns in turn can be twisted together to form cabled yarn or cord. Important yarn characteristics related to behavior are fineness (diameter or linear density) and number of twists per unit length. The measure of fineness is commonly referred to as yarn number. Yarn numbering systems are somewhat complex, and they are different for different types of fibers. Essentially, they provide a measure of fineness in terms of weight per unit or length per unit weight.

Textiles

Textiles, which are assemblages of fibers, or yarns, are of three major types: (1) nonwoven fabrics, (2) woven or knit fabrics, and (3) cordage. The properties and behavior of a textile are dependent on both the fiber used and the geometrical or structural characteristics of the textile.

12-4 Nonwovens

Nonwoven textiles, in the most general sense, are fibrous-sheet materials consisting of fibers mechanically bonded together either by interlocking or entanglement, by fusion, or by an adhesive. They are characterized by the absence of any patterned interlooping or interlacing of the yarns. In the textile trade, the terms nonwovens or bonded fabrics are applied to fabrics composed of a fibrous web held together by a bonding agent, as distinguished from felts, in which the fibers are interlocked mechanically without the use of a bonding agent.

Major Types.   There are three major kinds of nonwovens based on the method of manufacture. Dry-laid nonwovens are produced by textile machines. The web of fibers is formed by mechanical or air-laying techniques, and bonding is accomplished by fusion bonding the fibers or by the use of adhesives or needle punching. Either natural or synthetic fibers, usually 1 to 3 in. in length, are used.

Wet-laid nonwovens are made on modified papermaking equipment. Either synthetic fibers or combinations of synthetic fibers and wood pulp can be used. The fibers are often much shorter than those used in dry-laid fabrics, ranging from ¼ to ½ in. in length. Bonding is usually accomplished by a fibrous binder or an adhesive. Wet-laid nonwovens can also be produced as composites, for example, tissue-paper laminates bonded to a reinforcing substrate of scrim.

Spin-bonded nonwovens are produced by allowing the filaments emerging from the fiber-producing extruder to form into a random web, which is then usually thermally bonded. These nonwovens are limited commercially to thermoplastic synthetics such as nylons, polyesters, and polyolefins. They have exceptional strength because the filaments are continuous and bonded to each other without an auxiliary bonding agent.

Geometrical Characteristics. Fibers in nonwovens can be arranged in a great variety of configurations that are basically variations of three patterns: parallel or unidirectional, crossed, and random (Fig. 12-1). The parallel pattern provides maximum strength in the direction of fiber alignment, but relatively low strength in other directions. Cross-laid patterns (like wovens) have maximum strength in the directions of the fiber alignments and less strength in other directions. Random nonwovens have relatively uniform strength in all directions.

Fiber separation and curvature are two other characteristics that

Fig 12-1 Three basic configurations of fibers in nonwoven textiles: (a) parallel or unidirectional, (b) crossed, and (c) random.

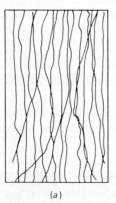
(a)

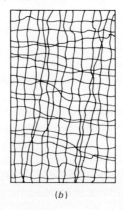
(b)

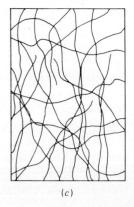
(c)

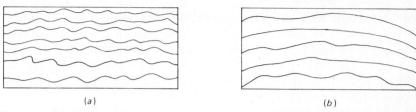

Fig 12-2 Two types of fiber curvature in textiles: (a) crimp and (b) overall curvature.

influence the properties of nonwovens. An evenly spaced fiber pattern distributes applied stresses better than an irregular one, and therefore produces a stronger fabric. Drape, however, is little affected by uniformity of fiber separation.

There are two types of fiber curvature in textiles: overall curvature and crimp. Overall fiber curvature, which is present to some degree in all nonwovens and to the greatest extent in random webs, makes the effectual fiber lengths less than the actual length, which in turn means a lower-strength web than one with perfectly straight fibers. Crimp improves fabric drape but decreases strength as compared to overall curvature. When stress is applied to crimped-fiber textiles, the fibers straighten out unevenly, and thus the stress acts only on some of the fibers.

Bonding. The nature of fiber bonding also significantly influences strength and drapability. Cohesive bonding (between two like fibers) theoretically produces the strongest fabrics because the point at which every fiber crosses over another is a bonding site. Adhesive bonding agents are of three major kinds: liquid dispersions, powdered adhesives, and thermoplastic fibers. A number of different kinds of resin polymers are used to meet specific requirements of strength, flexibility, color stability, and so on. The most common resins used are the vinyls and butadiene styrene. Powdered adhesives, either thermosetting or thermoplastic resins, are sifted into the fiber web while heat (with or without pressure) is applied to produce bonding. Thermoplastic fibers are incorporated as part of the fiber web, and bonding is achieved by heat and pressure or by solvents.

Nonwovens are sometimes bonded in some areas and not in others. This technique, known as pattern bondings, improves drapability with some sacrifice in strength.

12-5 Felts

Felts are nonwoven textiles composed of mechanically interlocking fibers without bonding agents. To produce felts, fibers are first agglomerated and then heat, moisture, and pressure are applied until the fibers shrink to produce interlocking. Felts are made of wool or synthetic fibers. Wool fibers have microscopic surface scales which act as barbs and contribute to interfiber bonding. Because synthetic fibers are usually smooth, barbed needles are inserted into the batt of fibers to promote fiber entanglement (and then are withdrawn).

Felts are produced in thicknesses up to 3 in., with special constructions going up to 12 in. in thickness. Unlike many other textiles, they do not fray or ravel.

Wool felts are produced as sheet stock and in roll form. Standard sheet stock is produced in 36-in. squares, from $1/16$- to 3-in. thick, and is available in densities from 12 to 32 lb (36 by 36 by 1 in.). Roll felts are usually produced in 60- and 72-in. widths and in lengths up to 60 yd. Standard thicknesses range from $1/32$ to 1 in., and densities from 8 to 18 lb/yd^2 for 1-in. thickness.

Synthetic fibers are not as well standardized. However, widths generally range from 52 to 140 in., and thicknesses from $1/32$ to $3/4$ in. Densities can be 8 or 10 lb/yd^2 for a 1-in. thickness.

As a group, felts have low strength compared to other textiles. However, they have good elasticity and resilience and can be readily molded and formed into almost any shape. Other notable properties are mechanical-, thermal-, and acoustic-energy absorption; thermal and chemical stability; high effective surface area per unit volume; and high porosity-to-weight ratio. These characteristics are put to work in such applications as gaskets and seals, wet and dry filtration, thermal and acoustical insulation, vibration isolation, cushioning and packaging, wicking, liquid absorption, and reservoirs.

12-6 Woven and Knit Fabrics

Woven and knit fabrics are composed of webs of fiber yarns. The yarns may be of either filament (continuous) or staple (short) fibers. In knit fabrics, the yarns are fastened to each other by interlocking loops to form the web. In woven fabrics, the yarns are interlaced at right angles to each other to produce the web. The lengthwise yarns are called the warp, and the crosswise ones are the filling (or woof) yarns.

Basic Weaves. The many variations of woven fabrics can be grouped into four basic weaves (Fig. 12-3). In the plain weave, each filling yarn alternates up and under successive warp yarns. With a plain weave, the most yarn interlacings per square inch can be obtained for maxi-

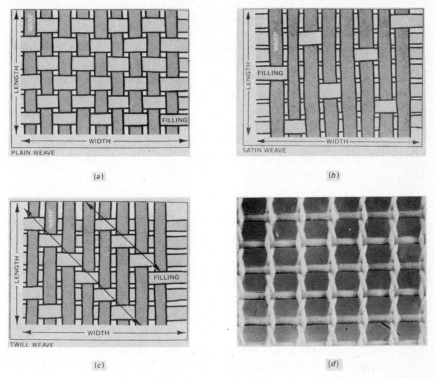

Fig 12-3 Four basic weaves used in woven fabrics: (a) plain (belt duck), (b) satin (sateen), (c) twill, and (d) leno. (J. P. Stevens & Co., Inc.)

mum density, "cover," and impermeability. The tightness or openness of the weave, of course, can be varied to any desired degree.

In twill weave, a sharp diagonal line is produced by the warp yarn crossing over two or more filling yarns.

Satin weave is characterized by regularly spaced interlacings at wide intervals. This weave produces a porous fabric with a smooth surface. Satins woven of cotton are called sateen.

In the leno weave, the warp yarns are twisted and the filling yarns are threaded through the twist openings. This weave is used for meshed fabrics and nets.

Fabric Specifications. Because the variety of woven fabrics is endless, we can only briefly outline here the way woven textiles are characterized or specified. Generally, specifications include the type of weave; the thread count, both in warp and filling; whether the yarn is filament or staple; the crimp, in percent; the twist per inch; and the yarn numbers for warp and fill. Over the years a rather unsystematic fabric-designation system has evolved. For example, some fabrics, such as twills and sateens, are designated by width in inches, number of linear yards per pound, and number of warp and filling threads per inch. Other fabrics are identified by width, ounces per linear yard, and warp and filling count.

While the largest single use of woven fabrics is, of course, wearing apparel, they are used in many other areas: in mechanical applications such as machine and conveyor belting, for filtration, for packaging, and as reinforcement for plastics and rubber.

12-7 Cordage

Cordage, the term applied to all linear forms of textiles, such as thread, twine, rope, and hawser, consists of fibers twisted together, plied, and in many cases cabled to produce continuous lengths (or strands) of fibrous materials. In addition to the fiber type, the properties of cordage are critically dependent on the kind and degree of twist about the axis of the cord. Twist is usually expressed in turns per unit length.

The twist is called S if, when the cord is held vertically, the direction of the spiral slope conforms with the central portion of the letter S. The twist is called Z if the direction of the spiral slope conforms to the central portion of the letter Z. There are two major kinds of twist: (1) the cable twist, in which the twist direction alternates in each successive operation (singles may be S-twisted, plies Z-twisted, and cables S-twisted); and (2) the hawser twist, in which the singles, plies, and cables are twisted SSZ or ZZS. The hawser type of cordage generally provides the highest strength and resilience.

Properties of Fibers and Textiles

The properties and behavior of textiles are dependent on either the fiber or the geometrical or structural characteristics of the textile, or both. In some cases, the properties, for example, resistance to heat and chemicals, are almost solely controlled by the fiber. Textile structure is a major influencing factor of permeability. And both fiber and structural characteristics are often equally important for strength and drape.

12-8 Mechanical

Strength. The ultimate or breaking strength of fibers, referred to as tenacity, is commonly expressed in terms of breaking load per unit of fineness, such as grams per denier. Tenacity can be converted to tensile strength in psi (based on an assumed circular cross section) by multiplying the specific gravity by 12,800. Breaking strength is often obtained under wet and dry conditions. Some fibers, such as cotton, show an increase in strength when wet, while the strength of most synthetic fibers is about the same both wet and dry.

Filaments are generally stronger than staple fibers. The textile structure, such as the direction of the fibers, type of weave, and density, has considerable effect on fabric strength. Also, in bonded fabrics, the type and nature of the bonding affect the strength.

Stiffness. Stiffness or rigidity of fibers and fabrics is the primary factor affecting drape or hand. Because of the viscoelastic behavior of textile fibers, the modulus of elasticity for a fiber is not identical with Young's modulus. The fiber modulus, often called the secant modulus in the literature, is the ratio of tenacity, usually at the breaking point, to the corresponding strain.

Resilience and Toughness. A measure of toughness, or energy absorption, of textile fibers is given by the area under the stress-strain curve. Figure 12-4 gives the typical stress-strain curves for several different fibers. Thus similar toughness can be obtained with a strong, low-extensibility fiber, or with a weak, high-extensibility fiber.

A useful way of ranking fibers for ability to absorb energy is by

Table 12-1 *Mechanical Properties of Typical Fibers*

Fibers	Breaking Point or Tenacity, g/denier	Tensile Strength, 1,000 psi	Secant Modulus, g/denier	Toughness Index, g/denier	Strain Recovery, %
Acetate	1.1–1.4	18–23	4.7	0.17	94
Acrilan	3.5	50	20	0.31	80
Cotton	2.1–6.3	42–125	65	0.14	74
Dacron	4.2–5.0	74–89	18	0.60	97
Dynel	3.0	50	9.7	0.47	97
Fortisan	7.0	138	117	0.21	82
Glass	7.7	250–315	270	0.11	100
Nylon	4.5–7.8	65–114	19–36	0.74–0.66	100
Orlon	4.8–5.8	73–88	32	0.44	97
Polyethylene	1.0–2.5	11–30	4.4	0.35	—
Rayon	1.5–4.0	30–78	9–24	0.14–0.25	82
Saran	1.1–2.9	25–60	7	0.28	—
Silk	2.8–5.2	45–83	18	0.44	92
Wool	1.0–1.7	17–28	3.9	0.24	99

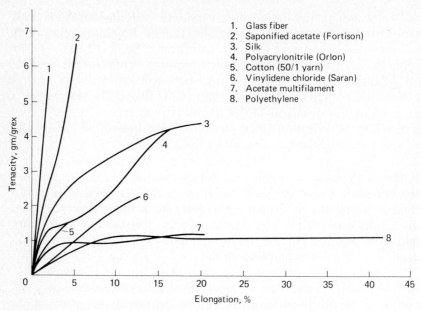

Fig 12-4 Typical stress-strain curves for several fibers. (M. Harris)

means of the toughness index, which is considered a conservative measure. It is defined as the ratio of measured work of rupture to the product of load and extension at break (the area under the secant-modulus line). The higher the index, the greater the ability to absorb energy.

Resilience refers to the ability of a fiber to recover its original dimensions when subjected to a loading and unloading cycle. The recovery from strain for one-time loading varies rather widely among fibers. For example, for nylon, wool, and glass, recovery is virtually 100 percent, whereas for cotton it is only about 75 percent. On repeated loading-unloading cycles, most fibers' resilience is often significantly lower than the one-time value because of the hysteresis effect. In fabrics, this repeated loading behavior, or ability to recover from deformation, is commonly measured in terms of compressional resilience by a test that determines the reduction in thickness resulting from the cyclic loading.

Abrasion Resistance. Both the fiber and the textile form influence abrasion resistance. In general, synthetic fibers have the same abrasion resistance as their bulk form. Nylon generally is considered to have the highest resistance, followed by cotton and wool. Harder, less-resilient fibers such as glass have very low abrasion resistance.

Fabrics can be specifically woven for optimum abrasion resistance. For example, the fabric can be woven so that the abrasion-resistant surface fibers lie predominantly in the direction of the abrading action. And, in general, the smoother the weave, the better the abrasion resistance.

12-9 *Thermal*

Thermal Insulation. The thermal conductivity of the fibers has only a slight effect on the thermal-transmission characteristics of a fabric. The fabric's geometry—the amount and configuration of air spaces—is of primary importance. Therefore, for a given air-permeability value, thermal transmission is a function of textile thickness. However, some physical (geometrical) characteristics of fibers, particularly their bulkiness, influence heat transmission. Certain fiber yarns, such as wool, form bulky or thicker fabrics and provide greater thermal insulation than, for example, silk and some of the less bulky synthetics.

Heat Resistance. Textile fibers can be grouped into three heat-resistant categories. The first includes those few that do not burn in the presence of a flame—the inorganic fibers, including glass, asbestos, graphite, and ceramic fibers, and metal threads. Their maximum continuous use temperatures range from about 600°F for borosilicate glass, 1000°F for asbestos, up to about 2300°F for aluminum-silicate fibers. When protected from oxidation, the strength of graphite fibers actually increases with temperature, up to 4500°F.

The second group are the synthetics. Their heat resistance in fiber form is comparable to that of the bulk form. Except for the fluorocarbons, which can be used in temperatures up to 500°F, most synthetics are not useful above 300 to 400°F. The natural fibers, the third group, have the lowest heat resistance. Silk disintegrates at about 340°F, viscose rayon loses strength at 300°F, and others cannot be used above 250 to 275°F.

12-10 *Chemical Resistance and Weatherability*

Chemical resistance and weatherability of a textile are chiefly a function of its fibers. In general, resistance is comparable to that in bulk form. Briefly, the inorganic fibers are the most inert to both chemicals and weathering. Glass, however, is attacked by strong alkalies. Synthetics as a class are generally more resistant to chemicals and

Table 12-2 *Effects of Sun, Heat, and Flame on Typical Fibers*

Fibers	Effect of Sun	Effect of Heat, °F	Rate of Burning
Acetate	Strength loss	Softens at 400°, melts at 500°	Moderately fast
Acrilan	Slight strength loss	Softens at 465°, 5% shrinking at 490°	Slow
Asbestos	None	Useful to 1490°	Nonflammable
Cotton	Strength loss, yellowing	Useful to 275°	Rapid
Dacron	Strength loss	Sticks at 455°, melts at 480°	Slow
Dynel	Strength loss	Shrinks above 240°	Self-extinguishing
Fortisan	Slight loss	Useful to 275°	Moderately fast
Glass	None	Softens at 1380 to 1550°	Nonflammable
Nylon	Strength loss	Sticks at 455°, melts at 480°	Self-extinguishing
Orlon	Very slight strength loss	Sticks at 455°	Self-extinguishing
Polyethylene	Depends on pigment	Melts at 225°	Slow
Rayon	Strength loss	Decomposes at 350 to 400°	Rapid
Saran	Darkens slightly	Softens at 240 to 280°	Self-extinguishing
Silk	Great strength loss	Disintegrates at 340°	Slow to self-extinguishing
Teflon	None	No effect to 400°, sublime above 550°	Self-extinguishing
Wool	Strength loss less than cotton	Good to 212°	Slow to self-extinguishing

weathering than are the natural fibers, although natural bost (flax, jute, and hemp) and hard fibers (manila and sisal) are essentially unaffected by weather. Rayons and acetates lose strength on prolonged exposure, as do cotton, silk, and wool, with wool being the least affected.

12-11 Gas and Liquid Permeability

Permeability is a function of fiber diameter and textile structure, with structure being the more important because it determines the amount and configuration of the air spaces in the fabric. Permeability of woven fabrics increases with an increase in the following geometrical characteristics: yarns per inch, yarn weight, amount of yarn twist, and overall fabric thickness. In felts, permeability depends primarily on fiber diameter and fabric density. In bonded nonwovens, the binder is an added consideration. For filtration applications, the fiber's cross section can affect its ability to retain particles. Liquid permeability is also affected by the composition of the fiber. Such fibers as glass, vinyl, or acrylics, being hydrophobic, do not affect permeability. However, in cotton or wool, the moisture transmission through the fiber must be considered.

12-12 Tactile

The touch, feel, or sensory characteristics of textiles are of importance chiefly in consumer fabrics. Because tactile properties are subjective, their measurement or description is more or less qualitative. The major tactile characteristics are hand, drape, and luster. Hand refers to the tactile feel of the fabric and is governed by a combination of properties such as the structure of the textile and the stiffness, hardness, and resilience of the fibers (Table 12-3).

Drape is similar to hand, but relates more to the effect of the fabric's weight on folding, bending, or deforming the fabric. Luster is the gloss or sheen resulting from reflectance from the fibers of the fabric. The overall luster of a fabric can be altered by changing the orientation of the fibers and yarns in the fabric.

12-13 Electrical

In most electrical applications, fibers and textiles are used in combination with another material, such as a polymer or coating, in which they serve as the reinforcement. Therefore, although they must be, and are, high in electrical-insulation properties, they are selected primarily on the basis of strength and durability as well as compatability with the coating or impregnant.

Table 12-3 *Terms Relating to Fabric Hand*

Physical Property (Component or Hand)	Description	Terms Describing Property Range
Flexibility	Ease of bending	Pliable (high) to stiff (low)
Compressibility	Ease of squeezing	Soft (high) to hard (low)
Extensibility	Ease of stretching	Stretchy (high) to non-stretchy (low)
Resilience	Ability to recover from deformation. Resilience may be flexural, compressional, extensional, or torsional	Springy (high) to limp (low)
Density	Weight per unit volume (based on measurement of thickness, fabric weight)	Compact (high) to open (low)
Surface contour	Divergence of the surface from planeness	Rough (high) to smooth (low)
Surface friction	Resistance to slipping offered by the surface	Harsh (high) to slippery (low)
Thermal character	Apparent difference in fabric temperature and skin of observer touching it	Cool (high) to warm (low)

SOURCE: ASTM Standards, pt. 10, 1958, p. 59; and M. W. Riley, "Guide to Industrial Textiles," *Materials Engineering*, July 1959.

Review Questions

1. What is the basic difference between fiber staples and filaments?
2. Name the geometrical and dimensional factors influencing the properties of fibers.
3. What is the bonding mechanism(s) characteristic of the following textiles?
   (a) Felts.
   (b) Dry-laid fabrics.
   (c) Wet-laid fabrics.
4. In textiles, which fiber pattern
   (a) Provides the highest strength in one direction?
   (b) Uniform strength in all directions?

5. Which group of textiles is generally lowest in strength but has excellent mechanical- and acoustical-energy-absorption properties?
6. Which type of woven fabric would you use
   (a) For maximum density?
   (b) For smoothness?
7. Which is generally stronger—a textile composed of staple fibers or filament fibers?
8. What is tenacity and how is it expressed?
9. What effect does a wet environment have on the tenacity
   (a) Of cotton?
   (b) Of synthetic textiles?
10. Arrange the following fibers in order of increasing ability to absorb impact energy: glass, nylon, polyethylene, rayon, silk, and wool.
11. What two fibers are nonflammable?
12. Which natural fibers compare with synthetics in resistance to weathering?

Bibliography

Dresher, William H.: "The Age of Fibers," *Journal of Metals,* April 1969.

Hamburger, W. J.: "A Technology for the Analysis, Design, and Use of Textile Structures as Engineering Materials," Edgar Marburg Lecture, American Society for Testing and Materials, Philadelphia, Penna., 1955.

Harris, M.: *Handbook of Textile Fibers,* Interscience Publishers, Inc., New York, 1954.

Riley, Malcolm W.: "Engineer's Guide to Industrial Textiles," *Materials Engineering,* July 1959.

Roden, John J., and Frank Y. Johnson: "The Nonwovens," *Southern Research Institute Bulletin,* Winter 1970.

Smith, H. D.: "Textile Fibers, an Engineering Approach to Their Properties and Utilization," Edgar Marburg Lecture, American Society for Testing and Materials, Philadelphia, Penna., 1944.

CERAMIC MATERIALS

Ceramics, one of the three major materials families, are crystalline compounds of metallic and nonmetallic elements. The ceramic family is large and varied, including such materials as refractories, glass, brick, cement and plaster, abrasives, sanitaryware, dinnerware, artware, porcelain enamel, ferroelectrics, ferrites, and dielectric insulators. In this book, we will be concerned chiefly with industrial ceramics. There are other materials which, strictly speaking, are not ceramics, but which nevertheless are often included in this family (and will be included here). These are carbon and graphite, mica, and asbestos. Also, intermetallic compounds, such as aluminides and beryllides, which are classified as metals, and cermets, which are mixtures of metals and ceramics, are usually thought of as ceramic materials because of similar physical characteristics to certain ceramics.

The principal distinguishing characteristics of ceramics as a class are:

1. They are crystalline materials, like metals, but because of fewer free electrons, they have little or essentially no electrical conductivity at room temperature.
2. They have high stability, and, on average, have higher melting points and greater chemical resistance than metals and organic materials.
3. They are generally the hardest of the engineering materials.
4. They are extremely stiff and rigid. Under mechanical stress, they have little or no yield and exhibit brittle fracture.

Composition and Structure

13-1 *Composition*

A broad range of metallic and nonmetallic elements are the primary ingredients in ceramic materials. Some of the common metals are aluminum, silicon, magnesium, beryllium, titanium, and boron. Nonmetallic elements with which they are commonly combined are oxygen, carbon, or nitrogen. Ceramics can be either simple, one-phase materials composed of one compound, or multiphase, consisting of a combination of two or more compounds. Two of the most common ceramic compounds are single oxides such as alumina (Al_2O_3) and magnesia (MgO), and mixed oxides such as cordierite (magnesia alumina silica) and forsterite (magnesia silica). Other, newer ceramic compounds include borides, nitrides, carbides, and silicides.

13-2 *Structure*

Unlike the relatively simple crystals found in most metals, those of ceramics are quite complex. Because ceramic crystalline phases are compounds, the unit cells often contain three or four different atoms. Also, there are relatively few free electrons in ceramic structures, as compared to metals. Instead, electrons of adjacent atoms are either shared to produce covalent bonds or transferred from one atom to another to produce ionic bonds. These strong bonding mechanisms are what account for many of ceramic's properties, such as high hardness, stiffness, and good high-temperature and chemical resistance.

At the macrostructural level, ceramic materials can have one, two, or three major constituents or components. In classic ceramics, the component termed the body is an aggregate of the crystalline constituents. The other component is a vitreous, or glassy, matrix (or phase) that cements the crystalline particles together. It is often referred to as the ceramic bond. This glassy phase is the weaker of the two components. In most newer refractory, or technical ceramics, it is eliminated and replaced by what is termed crystalline bonding, in which the individual particles of the powder raw material are sintered together in the solid state. That is, the particles are heated just short of complete melting, where the temperature is hot enough to cause the particles to fuse together.

A third component, often but not always present, is a surface glaze. This is a thin, glassy ceramic coating fired onto a ceramic body to make it impervious to moisture or to provide special surface properties. Macrostructurally, then, there are essentially three types of ceramics: (1) crystalline bodies with a glassy matrix, (2) crystalline bodies, sometimes referred to as holocrystalline, and (3) glasses.

Processing

The basic steps in producing ceramic products are (1) preparing the ingredients for forming, (2) shaping or forming the part, (3) drying, and (4) firing or sintering. The drying step depends on the type of ceramic and application, and is sometimes not required.

The raw materials are usually in the form of particles or powder. In the preparation step, the ingredients are weighed, mixed, and blended either wet or dry. In dry processing, the mixture is sometimes heated in order to cause preliminary chemical reactions. In wet processing, the required plasticity for shaping is obtained by grinding and blending plastic clays with finely pulverized nonplastic ingredients and adding alkalies, acids, and salts.

13-3 Forming Methods

Ceramics can be formed by a large number of methods either in a dry, semiliquid, or liquid state, and in either a cold or hot condition.

Slip Casting. This method consists of suspending powdered raw materials in liquid to form a slurry or slip that is poured into porous molds, usually made of gypsum. The mold absorbs the liquid, leaving a layer of solid material on the mold surface. If the part is to be hollow, excess slip is removed after the desired shell thickness has been built up. For solid parts, the slip remains, and more slip is added as shrinkage occurs, until the final solid shape is produced. Large and intricate shapes can be produced by this method. The process is especially economical for short production runs. In general, because slip casting produces parts with green densities up to only about 70 percent of theoretical, a large amount of shrinkage occurs in the firing step.

There are several variations of the basic slip-casting process. In pressure and vacuum casting, the slip is shaped in the mold under pressure or vacuum. In centrifugal casting, the porous mold is rotated, and in thixotropic casting, chemical agents are added to promote curing and to reduce the amount of water required.

Jiggering. This method is limited to circular and oval-shaped pieces. It is extensively used for producing dinnerware and porcelain electric insulators. In this process, which resembles the potter's-wheel technique, a liquid or semiliquid ceramic body is placed on a porous gypsum mold and rotated while a profiling tool forms the surface of the part and cuts away excess material. Once a hand operation, jiggering is now done on automatic machines capable of turning out more than 1,000 pieces per hour.

Pressing. This method can be done with dry, plastic, or wet raw materials. In dry pressing, ceramic mixtures with liquid levels up to 5 percent by weight are pressed into shape under high pressure in a metal die. The method is widely used for manufacturing nonclay refractories, electrical insulators, and electronic ceramic parts, since it produces small uniform parts to close tolerances. Semidry and wet pressing, in which water content is from 5 to 15 percent and 15 to 20 percent, respectively, uses lower pressures and less expensive dies.

In isostatic pressing, dry ceramic powder in a sealed rubber bag, or mold, the approximate size and shape of the finished part is placed in a chamber of hydraulic fluid. The part is formed by hydraulic pressure applied to the rubber bag. Small amounts of binder are used. Sintered densities range from about 80 to 95 percent. The process produces complex, accurate shapes; is widely used with high-grade oxide ceramics; and is the common method of producing spark plug insulators.

Hot pressing, which is a comparatively new method, produces ceramic parts of high density and improved mechanical properties. It combines pressing and firing operations. Isostatic and uniaxial techniques are employed. In the uniaxial method, a plunger compresses the ceramic in either powder or precompacted form in a die that is heated to near the sintering temperature.

Extrusion. Simple cross sections and hollow shapes can be produced by extruding plastic ceramic material through a forming die under pressure, and then cutting to length. Hydraulic extrusion machines are used for producing technical ceramics. Most clay-ceramic products, such as bricks and rain tile, are made with auger extruders. In ram pressing, or plastic pressing, an extruded slug of ceramic is placed between two porous dies, which are moved to the closed position, where they dewater the slug and form the part.

Molding. The process of molding ceramic parts is similar to the injection molding of plastics. The ceramic is mixed with a thermoplastic resin and heated sufficiently to provide the fluidity needed for the mixture to flow into the die cavity. The resin is later burned off in ovens prior to firing. Parts as thin as 0.02 in. and as thick as 0.25 in. can be injection molded. Bulk molding is sometimes used for forming large or irregular shapes or when only a limited number of parts is wanted.

Several coating processes can be used to build up shapes over a mandrel or a mold. In the chemical vapor-deposition process, molecular or atomic particles are deposited on a heated substrate which can be any shape desired. In melt spraying, ceramic particles are simultaneously melted and sprayed onto the mandrel. And in electrophoretic forming, charged ceramic particles in a suitable low dielectric fluid are deposited on an electrode mandrel.

13-4 Drying and Firing

The purpose of drying is to remove any water from the formed plastic-ceramic body before the firing operation. Because excessive drying shrinkage may cause cracking or warping, drying must be performed very carefully. Low-cost ceramic wares are usually dried in the atmosphere under a roof. Quality and technical ceramics often are processed from dry powders, and therefore the amount of moisture present is not an important problem.

The function of firing or sintering is to convert the shaped, dry ceramic part into a permanent product. The firing process and temperatures used depend on the ceramic composition and desired properties. The top temperature to which a ceramic is fired is called the maturing temperature. For traditional ceramic whiteware, maturing temperatures can range from 1700 to 2600°F. In the first stage of firing, any moisture still present after drying is removed. In the next stage, chemical reactions cause the clay to lose its plasticity. In the last state, vitrification of the ceramic begins and continues up to the maturing temperature. During vitrification, a liquid, glassy phase forms and fills the pore spaces. Upon cooling, the liquid solidifies to form a vitreous or glass matrix that bonds the inert, unmelted particles together. Refractory and electronic ceramics are often fired at higher temperatures, sometimes above 3000°F, to obtain the desired vitrification, or ceramic bond. In high-grade refractories, the firing operation produces a crystalline bond instead of the glassy-phase bond resulting from vitrification.

Properties and Characteristics

13-5 Mechanical

As a class, ceramics are low tensile strength, relatively brittle materials. A few have strengths above 25,000 psi, but most have less than that. Ceramics are notable for the wide difference between their tensile and compressive strengths. They are normally much stronger under compressive loading than in tension. It is not unusual for a compressive strength to be 5 to 10 times that of the tensile strength. Tensile strength varies considerably depending on composition and porosity. The stress condition of the outer layers also greatly influences strength.

For example, in glazed ceramic parts, the flexural and tensile strength can be increased or decreased as much as 40 or 50 percent depending on the stress condition in the glaze.

One of the major distinguishing characteristics of ceramics, as compared to metals, is their almost total absence of ductility. Being strictly elastic, ceramics exhibit very little yield or plastic flow under applied loads. This means that when a load is removed, the ceramic returns immediately to its original dimensions. Also, ceramics fail in a brittle fashion. That is, they will stretch or deform only slightly without fracturing. Lack of ductility is also reflected in low impact strength, although impact strength depends to a large extent on the shape of the part. Parts with thin or sharp edges or curves and with notches have considerably lower impact resistance than those with thick edges and gentler curving contours.

As a class, ceramics are the most rigid of all materials. A majority of them are stiffer than most metals, and the modulus of elasticity in tension of a number of types runs as high as 50 to 65 million psi compared with 29 million psi for steel.

Ceramic materials, in general, are considerably harder than most other materials, making them especially useful as wear-resistant parts and for abrasives and cutting tools.

13-6 *Thermal*

Ceramics have the highest known melting points of materials. Hafnium and tantalum carbide, for example, have melting points slightly above 7000°F, compared to 2600°F for tungsten. The more conventional ceramic types, such as alumina, melt at temperatures above 3500°F, which is still considerably higher than the melting point of all commonly used metals.

In general, and as a class, thermal conductivities of ceramic materials fall between those of metals and polymers. However, thermal conductivity varies widely among ceramics. A two-order magnitude of variation is possible between different types, or even between different grades of the same ceramic. The thermal conductivity of refractories depends on their composition, crystal structure, and texture. Simple crystalline structures usually have higher thermal conductivities, as, for example, silicon carbide does. Thermal conductivity versus temperature depends on whether the amorphous (glassy phase) or crystalline constituent predominates. For example, fireclay bricks show an increase with rising temperatures, whereas the more crystalline forsterite and some high aluminas show a decrease with rising temperature.

Compared to metals and plastics, the thermal expansion of ceramics is relatively low, although like thermal conductivity, it varies widely between different types and grades.

Thermal-shock resistance is closely related to thermal conductivity and expansion in brittle materials such as ceramics and glasses. High thermal conductivity and low thermal expansion favors good shock resistance. Also, small differences between tensile and compressive strength lead to good shock resistance. Because the compressive strengths of ceramic materials are 5 to 10 times greater than tensile strength, and because of relatively low heat conductivity, ceramics as a class have fairly low thermal-shock resistance. However, in a number of ceramics, the low thermal expansion coefficient succeeds in counteracting to a considerable degree the effects of thermal conductivity and tensile–compressive-strength differences. This is true in the case of special porcelains, cordierite, and lithium ceramics, and for such glasses as fused silica, Pyrex, and other special compositions.

13-7 Chemical

Practically all ceramic materials have excellent chemical resistance, being relatively inert to all chemicals except hydrofluoric acid and, to some extent, hot caustic solutions. Organic solvents do not affect them. The high surface hardness of ceramics tends to prevent breakdown by abrasion, thereby retarding chemical attack. All technical ceramics will withstand prolonged heating at a minimum of 1830°F. Therefore atmospheres, gases, and chemicals cannot penetrate the material surface and produce internal reactions which normally are accelerated by heat.

13-8 Electrical

Unlike metals, ceramics have relatively few free electrons and therefore are essentially nonconductive and considered to be dielectric. Most porcelains, aluminas, quartz, mica, and glass have volume electrical-resistivity values greater than 10^{15} microhm-cm and dielectric constants up to 12. In general, dielectrical strengths, which range between 200 and 350 V/mil, are lower than those of plastics. Electrical resistivity of many ceramics decreases rather than increases with an increase in impurities, and is markedly affected by temperature.

The diverse types of ceramics for electrical use include everything from low-loss, high-frequency electrical insulation to conductors, semiconductors, ferroelectrics, and ferromagnetics.

13-9 Other
The densities of ceramic materials, which have specific gravities ranging roughly from about 2 to 3, are comparable to those of the light-metal group. Because fired ceramic products show porosity to a variable degree, bulk density is often used to provide a measure of porosity. Bulk density is the ratio of the weight of a material to its bulk volume.

As is well known, glasses have been the major materials used for glazing and optical products for many years. The index of refraction of normal glasses can be varied from 1.458 to 2.008, making them extremely useful for all types of lenses.

Technical and Industrial Ceramics

A wide variety of ceramic materials have been developed over the years for industrial and technical applications—particularly for refractory, and chemical-, electrical-, and abrasion-resistant service. Many of the commercial high-volume-use types are complex bodies composed of high-melting oxides or a combination of oxides of such elements as silicon, aluminum, magnesium, calcium, and zirconium. These are similar in structure to clay products. Other technical ceramics, developed in recent years for service at the very high temperatures encountered in gas turbines, jet engines, nuclear reactors, and high-temperature processes, are relatively simple crystalline bodies composed of very pure metallic oxides, borides, carbides, nitrides, sulfides, and silicides. The major difference between the common and high-grade technical ceramics is that the high-grade types do not have a glassy matrix. Instead, in the sintering process, the fine particles of the ceramic material are bonded together by solid-surface reactions that produce a crystalline bond between the individual particles.

13-10 *Stoneware and Porcelain*
Porcelains and stoneware are highly vitrified ceramics that are widely used in chemical and electrical products. Electrical porcelains, which are basically classical clay-type ceramics, are conventionally divided into low-voltage and high-tension types. The high-tension grades are suitable for voltages of 500 and higher, and are capable of withstanding extremes of climatic conditions.

Chemical porcelains and stoneware are produced from blends of clay, quartz, feldspar, kaolin, and certain other materials. Porcelain is more vitrified than stoneware and is white in color. A hard glaze is generally applied. Stonewares can be classified into two types: a dense, vitrified body for use with corrosive liquids, and a less dense body for use in contact with corrosive fumes.

Both chemical porcelains and stoneware resist all acids except hydrofluoric. Strong, hot, caustic alkalies mildly attack the surface. These ceramics generally show low thermal-shock resistance and tensile strength. Their universal chemical resistance explains their wide use in the chemical and processing industries for tanks, reactor chambers, condensers, pipes, cooling coils, fittings, pumps, ducts, blenders, filters, and so on.

13-11 *Common Refractories*

Common refractory ceramics are produced from clays, and the final product is a glassy matrix binding together the crystalline constituents. The manufacturing methods are designed to produce bodies in which the main ingredient is the glassy matrix, or ceramic bond.

The most widely used common refractories are of the alumina-silica (aluminum oxide and silicon dioxide) type. The compositions range from nearly pure silica through a wide range of alumina silicas to nearly pure alumina. They also contain some impurities, such as basic oxides of iron and magnesium, and smaller amounts of alkaline metal oxides. Refractoriness increases with alumina content.

Other common commercial refractories are silica, forsterite, magnesite, dolomite, silicon carbide, and zircon, some of which are described in the succeeding sections.

13-12 *Oxides*

The oxides can be divided into two groups: single oxides that contain one metallic element, and mixed or complex oxides that contain two or more metallic elements.

From Tables 13-1 and 13-2, which give the properties of these oxide ceramics, it is evident that they differ widely among themselves. As a class, they are low in cost compared to other technical ceramics, except for thoria and beryllia. Each of them can be produced in a variety of compositions, porosities, and microstructures to meet specific property requirements. Thus the data given in the tables are only typical values.

Oxide-ceramic parts are produced by slip casting, pressing, or extrusion, and are then fired at about 3270°F. They are more difficult to fabricate than other types of ceramics because of the usual requirement to obtain a high-density body with minimum distortion and dimen-

Table 13-1 *Properties of Oxide Ceramics*

Property	Alumina	Beryllia	Magnesia	Zirconia	Thoria
Melting point, °F	3700	4620	5070	4710	6000
Modulus of elasticity in tension, 10^6 psi	65	35	40	30	20
Tensile strength, 1,000 psi	38	14	20	21	7.5
Compression strength, 1,000 psi	320	300	120	300	200
Hardness, micro (Knoop)	3,000	1,300	700	1,100	700
Maximum service temperature (Oxidation atmosphere), °F	3540	4350	4350	4530	4890

sional error, except for porous bodies used as thermal insulation. Powder pressing produces bodies with the lowest porosity and highest strength because of the high pressures and the small amount of binder required.

Aluminum Oxide (Alumina). Alumina is the most widely used oxide, chiefly because it is plentiful, relatively low in cost, and equal to or better than most oxides in mechanical properties. Density can be varied over a wide range, as can purity (down to about 90 percent alumina), to meet specific application requirements. Alumina ceramics are the hardest, strongest, and stiffest of the oxides. They are also outstanding in electrical resistivity and dielectric strength, are resistant to a wide variety of chemicals, and are unaffected by air, water vapor, and sulfurous atmospheres. However, with a melting point of only 3700°F, they are relatively low in refractoriness, and, at 2500°F, retain only about 10 percent of room-temperature strength. Besides wide use as electrical insulators, and chemical and aerospace applications, alumina's high hardness and close dimensional tolerance make it suitable for such abrasion-resistant parts as textile guides, pump plungers, chute linings, discharge orifices, dies, and bearings.

Beryllium Oxide (Beryllia). Beryllia is noted for its high thermal conductivity, which is about ten times that of a dense alumina (at 930°F),

Table 13-2 *Properties of Some Mixed-Oxide Ceramics*

Property	Cordierite	Forsterite	Steatite	Zircon
Melting point, °F	2680	3470	2820	2820
Modulus of elasticity in tension, 10^6 psi	7	—	13–16	21
Tensile strength, 1,000 psi	4–8	9	5–10	5–11
Compression strength, 1,000 psi	50–95	80–85	65–90	60–100
Hardness, Mohs	7	7.5	7.5	8
Volume resistance, ohm-cm	$>10^{14}$	$>10^{14}$	$>10^{14}$	$>10^{14}$
Dielectric strength, V/mil	140–230	250	150–280	60–300
Maximum service temperature, °F	1830	1830	1830	2000

three times that of steel, and second only to that of the high-conductivity metals (silver, gold, and copper). It also has high strength and good dielectric properties. However, it is costly and difficult to work with. Above 3000°F it reacts with water to form a volatile hydroxide. Also, because beryllia dust and particles are toxic, special handling

Fig 13-1 Typical parts made of alumina ceramic for wear-, electrical-, and heat-resistant applications. (American Lava Corp.)

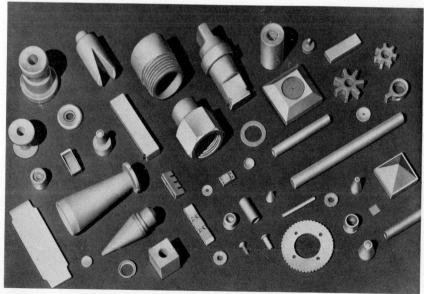

precautions are required. The combination of strength, rigidity, and dimensional stability make beryllia suitable for use in gyroscopes; and because of high thermal conductivity, it is widely used for transistors, resistors, and substrate cooling in electronic equipment.

Magnesium Oxide (Magnesia). Magnesia is not as widely used as alumina and beryllia. It is not as strong, and, because of high thermal expansion, it is susceptible to thermal shock. Although it has better high-temperature oxidation resistance than alumina, it is less stable in contact with most metals at temperatures above 3100°F in reducing atmospheres or in a vacuum.

Zirconium Oxide (Zirconia). There are several types of zirconia: a pure (monoclinic) oxide, a stabilized (cubic) form, and a number of variations such as yttria, magnesia-stabilized zirconia, and nuclear grades. Stabilized zirconia has a high melting point (about 5000°F) and low thermal conductivity, and is generally unaffected by oxidizing and reducing atmospheres and most chemicals. Yttria and magnesia-stabilized zirconias are widely used for equipment and vessels in contact with liquid metals. Monoclinic nuclear zirconia is used for nuclear fuel elements, reactor hardware, and related applications where high purity (99.7 percent) is needed. Zirconia has the distinction of being an electrical insulator at low temperatures, and, as temperatures increase, of gradually becoming a conductor.

Thorium Oxide (Thoria). Thoria, the most chemically stable oxide ceramic, is only attacked by some earth alkali metals under some conditions. It has the highest melting point (6000°F) of the oxide ceramics. But, like beryllia, it is costly. Also, it has high thermal expansion and poor thermal-shock resistance. Because of high cost, its use is limited. It has some application in nuclear reactors.

13-13 Mixed Oxides

Except for zircon, the principal mixed oxides are composed of various combinations of magnesia, alumina, and silica.

Cordierite. Cordierite is most widely used in extruded form for insulators in heating elements and thermocouples. It has low thermal expansion, excellent resistance to thermal shock, and good dielectrical strength. There are three traditional groups of cordierite ceramics: (1) Porous bodies, which have relatively little mechanical strength because of limited crystalline intergrowth and the absence of a ceramic

Ceramic Materials 357

Fig 13-2 Some ceramics are produced in various corrugated and honeycomb forms for use in pollution-control equipment, heat exchangers, gas mixtures, and other products. (American Lava Corp.)

bond. They have thermal endurance and low thermal expansion, and are used for radiant elements in furnaces, resistor tubes, and rheostat parts. (2) Low-porosity bodies, which were developed principally for use as furnace refractory brick. (3) Vitrified bodies, which are used for exposed electrical devices subjected to thermal variations.

Forsterite. This mixed oxide has high thermal-shock resistance, but good electrical properties and mechanical strength. It is somewhat difficult to form and requires grinding to meet close tolerances.

Steatite. Steatites are noted for their excellent electrical properties and low cost. They are easily formed and can be fired at relatively low temperatures. However, compositions containing little or no clay or plastic material have fabricating problems because of their narrow firing range. Steatite parts are vacuum tight, can be readily bonded to other materials, and can be glazed or ground to high-quality surfaces.

Zircon. This mixed oxide provides ceramics with strength, low thermal expansion, and relatively high thermal conductivity and thermal endurance. Its high thermal endurance is used to advantage in various porous ceramics.

13-14 Carbides

There are roughly a dozen different classes of carbides. Silicon carbide, commonly known as Carborundum, is probably the most widely used, followed by tungsten and titanium carbides. The carbide family contains materials with the highest melting points of all engineering materials. Hafnium and tantalum carbides both have melting points of 7100°F. Unfortunately, because of their poor oxidation resistance, they cannot be used unprotected at high temperatures, except for silicon carbide, which is useful at temperatures up to about 3000°F.

Silicon carbide has an attractive combination of high thermal conductivity, low thermal expansion, and low thermal shock. One of the best wear- and abrasion-resistant materials, it is often bonded with other materials. Self-bonded grades are also produced by mixing the silicon-carbide particles with a temporary binder which is replaced after firing by crystalline bonds. This provides a material of high refractoriness, abrasion resistance, density, and strength as well as good chemical resistance.

Boron carbide is best known for its extreme hardness and abrasion resistance. It is also an excellent neutron absorber, and is used for this purpose in atomic-power reactors.

Tungsten carbide, with a specific gravity of about 16, is among the 10 heaviest materials. Its largest single use is for cutting tips and tools. It is also used widely in wear- and abrasion-resistant applications.

13-15 Borides

The major materials in this group of refractory ceramics are borides of hafnium, tantalum, thorium, titanium, uranium, and zirconium. As a class they feature a combination of high strength-to-stiffness ratio, high hardness, and good high-temperature strength retention. However, they have only limited commercial application, being used as rocket nozzles and in molten-metal processing equipment.

13-16 Nitrides

Boron and silicon nitrides are the major commercial materials in this group of refractory ceramics. Like carbides, nitrides are low in oxidation resistance, although silicon nitride is useful up to about 2200°F. Boron nitride is best known as the synthetic-diamond material, Borozon. This is a cubic crystalline form produced under 1 million-psi

Table 13-3 Properties of Other Refractory Ceramics

Property	Carbides	Nitrides	Borides	Silicides	Intermetallics
Melting point, °F	2800–6300	1200–4900	2000–5800	1800–4600	2600–4500
Modulus of elasticity in tension, 10^6 psi	26–65	8–50	32–72	45–55	24–50
Tensile strength, 1,000 psi	—	—	—	24–37	—
Compression strength, 1,000 psi	20–420	40–300	—	100–280	—
Hardness, micro (Knoop)	500–3,200	500–1,900	1,000–3,800	400–2,400	100–1,300

pressure and temperatures above 3000°F. It has a hardness equal to that of diamonds and can withstand temperatures up to 3500°F without appreciable oxidation. Also, it has a thermal conductivity five times that of copper at room temperature, which makes it potentially useful for tiny heat-sink devices. Like commercial graphite, whose structure it resembles, hot-pressed boron nitride is anisotropic. Thermal expansion, for example, parallel to the direction of pressing is 10 times that in the perpendicular direction. Boron nitride differs from graphite in that it has high resistivity and dielectric strength at elevated temperatures.

13-17 *Intermetallics and Silicides*
These materials, although technically metals because they are compounds of metals, are generally classified as ceramics. The three major classes are aluminides and beryllides, and silicides (silicon classed as a metal). They are hard and brittle in their polycrystalline form at room temperature, but they can be deformed plastically, like metals, at elevated temperatures. They are generally considered as having the greatest potential among ceramic materials for achieving low-temperature ductility.

Typical aluminides are nickel-aluminum compounds and titanium-aluminum compounds. Typical beryllides are those in which beryllium is compounded with either columbium, tantalum, or zirconium. Aluminides have the lowest melting point, and beryllides the highest of the three major intermetallics.

Silicides are metalloid compounds of silicon in combination with one or more metallic elements. They are closely related to intermetallic compounds. They have good oxidation resistance and retention of strength up to 2500°F. Of the refractory silicides, the disilicides of molybdebun, tungsten, and tantalum have the greatest potential for high-temperature applications.

13-18 *Electronic Ceramics*
A number of the ceramics already covered are widely used in the electrical field. These include procelains, steatites, zircons, and cordierites. In addition, there are ceramic materials with unusual properties that are of specific use in electronic circuits. They are ferrites and ferroelectric ceramics.

Ferrites are mixed-metal-oxide ceramics, almost completely crystalline, which have a combination of high electrical resistivity and strong magnetic properties. The mineral magnetite is the only naturally occurring ferrite of this type and has been known for centuries as lodestone. In contrast to metallic magnetic materials, ferrites have high-volume resistivity and high permeability. With specific gravities of 4 and 5, they are considerably lighter than iron. They can be divided into soft and hard or permanent-magnet types. Soft ferrites, which have a cubic or spinel crystal structure, can be varied over a wide range for specific uses such as memory cores for computers and cores for radio and television loop antennas. Permanent ceramic magnets have a hexagonal crystal structure. Barium and lead ferrites, which are the best known, have exceptionally high coercive force, and are now widely used in permanent-magnet motors in automobiles, small appliances, and portable electric tools.

Ferroelectric ceramics have the unusual ability to convert electrical signals into mechanical energy, such as sound. They also can change sound, pressure, or motion into electrical signals. Thus they function as transducers. The best-known ferroelectric ceramic is the phonograph pickup "crystal." Barium titanate is the most common of the ferroelectrics. Others are the niobates, tantalates, and zirconates.

13-19 *Glass Ceramics*

Glass ceramics are a family of fine-grained crystalline materials made by a process of controlled crystallization from special glass compositions containing nucleating agents. They are sometimes referred to as devitrified ceramics or vitro ceramics, and some common trademarks are Pyroceram, Cercor, and Cer-Vit. Since they are mixed oxides, different degrees of crystallinity can be produced by varying composition and heat treatment. Some of the types produced are cellular foams, coatings, adhesives, and photosensitive compositions.

Glass ceramics are nonporous, and generally are either opaque white or transparent. Although not ductile, they have much greater impact strength than commercial glasses and ceramics. However, softening temperatures are lower than those for ceramics, and they are generally not useful above 2000°F. Thermal expansion varies from negative to positive values depending on composition. Excellent thermal-shock resistance and good dimensional stability can be obtained if desired. These characteristics are used to advantage in "heat-proof" skillets and range tops. Like chemical glasses, these materials have excellent corrosion and oxidation resistance. They are electrical insulators and are suitable for high-temperature, high-frequency applications in the electronics field.

Glass

Glass, one of the oldest and most extensively used materials, is made from the most abundant of the earth's natural resources—silica sand.

Fig 13-3 Some nose cones for missiles are made from glass-ceramic materials. (Corning Glass Works)

For centuries considered as a decorative, fragile material suitable for only glazing and art objects, today glass is produced in thousands of compositions and grades for a wide range of consumer and industrial applications.

13-20 *Composition and Structure*
The basic ingredient of glasses is silica (silicon dioxide), which is present in various amounts ranging from about 50 to almost 100 percent.

Other common ingredients are oxides of metals such as lead, boron, aluminum, sodium, and potassium.

Unlike most other ceramic materials, glass is noncrystalline. To manufacture it, a mixture of silica and other oxides is melted and then cooled to a rigid condition. Glass does not change from a liquid to a solid at a fixed temperature, but remains in a vitreous, noncrystalline state, and is considered a supercooled liquid. Thus, with the relative positions of the atoms being similar to those in liquids, the structure has a short-range order. However, glass has some distinct differences compared with a supercooled liquid. Glass has a three-dimensional framework and the atoms occupy definite positions. There are covalent bonds present, as in many solids. That is, there is a tendency toward an ordered structure since a continuous network of strongly bonded atoms is present.

13-21 *Production and Processing*
The major steps in producing glass products are (1) melting and refining, (2) forming and shaping, (3) heat-treating, and (4) finishing. The mixed batch of raw materials, along with broken or reclaimed glass, is fed into one end of a continuous-type furnace where it melts and remains molten at around 2730°F. Molten glass is drawn continuously from the furnace and runs in troughs to the working area, where it is drawn off for fabrication at a temperature of about 1830°F. Where small amounts are involved, glass is melted in pots.

Forming Methods. Most glass products are manufactured on automatic high-speed equipment by either blowing, pressing, rolling, drawing, or casting. Pressing, usually the lowest-cost fabrication method, is used to manufacture table- and ovenware, insulators, lenses, and reflectors. In this process, gobs of glass are fed into molds on a rotating press. The molds are moved beneath a plunger which forces the glass into final shape.

Glass blowing is used to produce hollow products such as bottles, jars, and light bulbs. In glass-blowing machines, molten glass in ribbon form sags through holes as air is blown in from above. At the same time, molds move up from below and clamp around the molten glass. More puffs of air force the glass into the mold to form the final shape.

The drawing process, which forms glass into tubing and rods, is performed at up to 40 mi/hr. For tubing, the molten glass passes around a ceramic or metal cone-shaped mandrel as air is blown through the center of the mandrel. The molten glass is fed into a bushing that has a number of orifices through which the glass flows, forming continuous

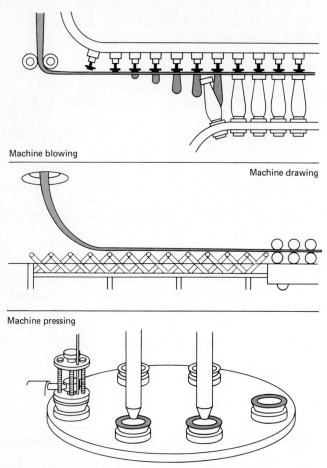

Fig 13-4 Three high-production methods of producing glass products: (a) machine blowing, (b) machine drawing, and (c) machine pressing.

glass fibers, or filaments. The drawn filaments are collected into bundles called strands.

Sheet glass can be manufactured by either drawing, rolling, or floating methods. In drawing, glass is drawn from a molten pool and then passed through or over rollers. In rolling, molten glass passes between cooled rolls. And, in floating, the molten glass is formed into sheet on the surface of a pool of molten tin in a controlled atmosphere.

Stationary casting is perhaps the most difficult method of forming glass, and is usually restricted to large, simple shapes, such as astronomical telescope lenses. Centrifugal casting, in which a gob of glass is spread by centrifugal force over the inside of a rapidly spinning mold, is used to form products such as the funnel portion of TV tubes and missile radomes.

Other fabricating methods include pressing and fusing powdered glass into shapes, and forming molten glass to produce cellular materials.

Heat-Treating. There are two glass heat-treating processes: annealing and tempering. When glass cools from the forming range to room temperature, thermal stresses develop that adversely affect strength properties. Therefore almost all glass products are annealed to eliminate the stresses. The annealing treatment involves heating the glass to its annealing temperature range, holding it there for a period of time, and then cooling it slowly to room temperature.

In contrast, tempering of glass involves heating it to around the softening point and then cooling it rapidly with blasts of air or by quenching it in oil. Tempering produces glass with a rigid surface layer that is in compression and an interior that is in tension. Therefore, in service, compression stresses in the outer skin of the glass resist imposed tensile stresses and thereby greatly enhance overall strength. Heat-tempered glass is three to five times stronger than annealed glass while still retaining its initial clarity, hardness, and expansion coefficient.

Special glass compositions can also be tempered by chemical treatments. The strength achieved is dependent on the specific treatment applied and the glass composition used. Sheets of chemically tempered glass can undergo repeated flexing without failure.

Finishing. Glass surfaces can be treated or finished in a number of ways. Hydrofluoric acid is used for polishing and etching. Glass can also be stained by copper or silver compounds. Metallizing is often used to form a base for sealing, for decoration, or to provide electrical conductivity. Fired-on films are used on such products as instrument windows and electronic components. Mechanical finishing includes grinding to square and smooth edges. Ferric or cerium oxides are used to produce smooth, accurately finished surfaces.

Glass can be sealed to most metallic materials. The principal problem is matching the coefficients of expansion of the two dissimilar materials. The wide range of expansion rates in glass helps solve this problem, as do a number of special metal alloys developed principally for sealing glass.

Fig 13-5 Chemically tempered glass can be bent and flexed without failing. (Corning Glass Works)

13-22 Types
There are a number of general families of glasses, some of which have many hundreds of variations in composition.

Soda-Lime Glasses. The soda-lime family is the oldest, lowest in cost, easiest to work, and most widely used. It accounts for about 90 percent of the glass used in this country. Soda-lime glasses have only fair to moderate corrosion resistance and are useful at temperatures up to about 860°F (annealed) and 480°F (tempered). Thermal expansion is high and thermal-shock resistance is low compared with other glasses. These are the glass of ordinary windows, bottles, and tumblers.

Table 13-4 Typical Properties of Glasses

Property	Soda Lime	Soda Lead	Borosilicate	Aluminosilicate	96% Silica	Fused Silica
Young's modulus, psi	10.0×10^6	9.0×10^6	9.5×10^6	12.8×10^6	9.6×10^6	10.5×10^6
Poisson's ratio	0.24	—	0.20	0.26	0.18	0.17
Specific gravity	2.47	2.85	2.23	2.53	2.18	2.20
Linear coefficient of thermal expansion, cm²/°C	92×10^{-7}	91×10^{-7}	32.5×10^{-7}	46×10^{-7}	8×10^{-7}	5.6×10^{-7}
Service temperature, °C						
Annealed						
Normal	110	110	230	200	800	900
Maximum	460	380	490	650	1100	1200
Tempered						
Normal	220	—	260	400	—	—
Maximum	250	—	290	450	—	—
Volume resistivity, ohm-cm						
At 250°C	2.5×10^6	8×10^8	1.3×10^8	3.2×10^{13}	5×10^9	1.6×10^{12}
At 350°C	1.3×10^5	1×10^7	4×10^6	2×10^{11}	1.3×10^8	2.5×10^{10}
Dielectric constant	7.2	6.7	4.6	6.3	3.8	3.8
Loss factor	0.065	0.01	0.026	0.01	0.0019	0.000038

SOURCE: Corning Glass Works.

Lead Glasses. Lead or lead-alkali glasses are produced with lead contents ranging from low to high. They are relatively inexpensive, and are noted for high electrical resistivity and a high refractory index. Corrosion resistance varies with lead content, but they are all poor in acid resistance compared with other glasses. Thermal properties also vary with lead content. The coefficient of expansion, for example, increases with lead content. High-lead grades are the heaviest of the commercial glasses. As a group, lead glasses are the lowest in rigidity. They are used in many optical components, for neon-sign tubing, and for electric light-bulb stems.

Borosilicate Glasses. Borosilicate glasses are most versatile of the glasses. They are noted for their excellent chemical durability, for resistance to heat and thermal shock, and for low coefficients of thermal expansion. There are six basic kinds. The low-expansion type is best known as Pyrex ovenware. The low-electrical-loss types have a dielectric-loss factor second only to fused silica and some grades of 96 percent silica glass. Sealing types, including the well-known Kovar, are used in glass-to-metal sealing applications. Optical grades, which are referred to as crowns, are characterized by high light transmission and good corrosion resistance. Ultraviolet-transmitting and laboratory-apparatus grades are the two other borosilicate glasses.

Because of this wide range of types and compositions, borosilicate glasses are used in such products as sights and gages, piping, seals to low-expansion metals, telescope mirrors, electronic tubes, laboratory glassware, ovenware, and pump impellers.

Aluminosilicate Glasses. These glasses are roughly three times more costly than the borosilicate types, but are useful at higher temperatures and have greater thermal-shock resistance. Maximum service temperature in the annealed condition is about 1200°F. Corrosion resistance to weathering, water, and chemicals is excellent, although acid resistance is only fair compared with other glasses. Compared with 96%-silica glasses, which they resemble in some respects, they are more easily worked and are lower in cost. They are used for high-performance power tubes, traveling-wave tubes, high-temperature thermometers, combustion tubes, and stove-top cookware.

Fused-Silica Glasses. Fused silica is 100 percent silicon dioxide. If it occurs naturally, the glass is known as fused quartz. There are many types and grades of both glasses, depending on the impurities present and the manufacturing method. Because of its high purity level, fused silica is one of the most transparent glasses. It is also the most heat resistant of all glasses; it can be used at temperatures up to 1650°F in continuous service, and to 2300°F for short-term exposure. In addition, it has outstanding resistance to thermal shock, maximum transmittance to ultraviolet light, and excellent resistance to chemicals. Unlike most glasses, its modulus of elasticity increases with temperature. However, because fused silica is high in cost and difficult to shape, its use is restricted to such specialty applications as laboratory optical systems and instruments and crucibles for crystal growing. Because of a unique ability to transmit ultrasonic elastic waves with little distortion or absorption, fused silica is also used in delay lines in radar installations.

Ninety-six-percent-Silica Glasses. These glasses are similar in many ways to fused silica. Although less expensive than fused silica, they are still more costly than other glasses. Compared to fused silica, they are easier to fabricate, have a slightly higher coefficient of expansion, about 30 percent lower thermal stress resistance, and a lower softening point. They can be used continuously up to 1470°F. Uses include chemical glassware and windows and heat shields for space vehicles.

Fig 13-6 An assortment of glass lighting ware. (Corning Glass Works)

Other Glasses.

1. Nonsilicate glasses include borate glasses, which have very low light dispersion and a high refractive index; phosphate glasses, which are resistant to hydrofluoric acids, but have low resistance to water attack; and calcium-aluminate-germanate and arsenic-trisulfide glasses, which are useful in infrared-transmission systems.
2. Colored glasses are made by adding small amounts of colorants to glass batches. They are used in lamp bulbs, sunglasses, light filters, and signalware.
3. Opal glasses contain small particles dispersed in transparent glass. The particles disperse the light passing through the glass, producing an opalescent appearance.
4. Laminated or safety glass is composed of two or more layers of glass with a layer or layers of transparent plastic, usually vinyl, sandwiched between the glass. Bullet-resisting plate glass is similarly made up of multiple layers of glass bonded together with plastic films.
5. Photosensitive glass is sensitive to ultraviolet light and heat. This property permits reproduction of images on it from a photographic negative. Upon immersion in an acid bath, the image is etched on the surface. The glass can be intricately shaped and patterned without the use of mechanical tools.
6. Cellular or foam glass is made by heating a mixture of pulverized glass and a foaming agent. It is almost as light as cork, and is used as an insulating material.
7. Coated glass has a thin, metallic-oxide surface coating which can conduct electricity. In sheet or panel form, it is used in lighting applications. In rod and tube form, it is used for resistors in electronic devices.

Carbon and Graphite

Carbon and graphite have been used in industry for many years primarily as electrodes, arc carbons, brush carbons, and bearings. In the last decade or so, the development of new types and the emergence of graphite fibers as a promising reinforcement for high-performance composites have significantly increased the versatility of this family of materials.

13-23 *Composition and Structure*
The raw materials for industrial carbon are usually petroleum or anthracite cokes, lampblack, or carbon blacks. These are combined

with carbonaceous binders such as tars, pitches, and resins, and then compacted by molding or extrusion, and baked at temperatures between 1500 to 3000°F to produce what is known as industrial carbon or baked carbon. Conventional industrial graphite is made by mixing mined, natural graphite with carbon to produce, in effect, a carbon-graphite composite; or baked carbon can be heat-treated at about 5400°F, where the carbon graphitizes.

Manufactured or artificial carbon has a two-phase structure consisting of carbon particles (or grains) in a matrix of binder carbon. Both phases consist essentially of disordered, or uncrystallized, carbon surrounding embryonic carbon crystallites. The extent of crystallite development depends on the raw materials and the temperature used in manufacturing. The disordered structure is characteristic of carbons and accounts for their low electrical conductivity and for their abrasiveness as compared with graphite.

Graphites, except for the pyrolytic types, have a two-phase structure similar to that of carbon but, as the result of high-temperature processing, they contain well-developed graphite crystallites in both phases. These multicrystalline graphites exhibit many of the properties of single-crystal graphite, such as high electrical conductivity, lubricity, and anisotropy. Compared with carbon, graphite has higher electrical and heat conductivity and better lubricity, and it is easier to machine. Because of their more favorable properties, graphites have broader application as engineering materials than do carbons. Therefore most of the discussion here will concern graphites.

13-24 Types of Graphite

Industrial and Composite Carbon and Graphite. Conventional industrial carbons and graphites are available in a large number of different grades, sizes, and shapes. For many uses composition is tailored to the application. Because conventionally produced materials are porous, they are often impregnated with synthetic resins, usually phenolics. For bearing-seal and wear applications, carbon and graphite can be blended with various metal powders or plastics, depending on the operating conditions. Industrial carbons and graphites range in size from lamp-filament dimensions to solid cylinders 63 in. in diameter. Custom shapes are readily machined from graphite. However, special alloy or diamond-tipped tools are needed to machine carbon.

Pyrolytic Graphite. This graphite is made by a gas- or vapor-plating process (thermal-decomposition type) in which composition can be closely controlled. It is a highly oriented graphite in which density

Table 13-5 *Typical Properties of Industrial Carbon and Graphite*

		Graphite	
Property	Carbon	General Purpose	Premium
Specific gravity	1.4–1.8	1.4	1.8
Bulk density, lb/in.^{3}	0.05–0.06	0.55–0.064	0.05–0.063
Modulus of elasticity in tension, 10^6 psi	0.6–2.3	0.5–1.8	0.8–1.7
Tensile strength, 1,000 psi	1.0	0.4–1.4	1.1–1.7
Compression strength, 1,000 psi	1.7–9.0	2–6	4–8.5
Electrical resistivity, 10^{-4} ohm-cm	35–80	8–11	8–50

NOTE: Properties are with grain.

(2.25 specific gravity) approaches theoretical limits. Essentially the process consists of passing a heated hydrocarbon gas over a substrate or mandrel in a vacuum furnace. The resulting deposit of highly oriented graphite can either remain as a coating on the substrate, or be stripped from the mandrel to form a self-supporting part. The orientation of the majority of graphite crystals provides a high degree of anisotropy in such properties as strength and electrical and thermal conductivities.

Recrystallized Graphite. This form of graphite is produced by a proprietary hot-working process which yields recrystallized or "densified" graphite with specific gravities in the 1.85 to 2.15 range, as compared with 1.4 to 1.7 for conventional graphites. The material's major attributes are a high degree of quality reproducibility, improved resistance to creep, a grain orientation that can be controlled from highly anisotropic to relatively isotropic, lower permeability than usual, absence of structural macroflaw, and ability to take a fine surface finish.

Graphite fibers. These are produced from organic fibers such as rayon. The fiber or textile form (for example, fabric, yarn, or felt) is graphitized at temperatures up to 5400°F. The resulting fibers are high-purity (99.9 percent) graphite, with extremely high individual fiber strengths (see Sec. 14-7).

PT Graphites. These are graphite fibers impregnated or bonded with an organic resin (such as furfural) and then carbonized. The result is a graphite-reinforced carbonaceous material with a high degree of thermal stability. The composite has a low density (0.93 to 1.2 specific gravity), and what is reported to be the highest strenght-to-weight ratio of any known material at temperatures in the 4000 to 5000°F range.

Colloidal Graphite. This form consists of very fine particles of natural or artificial graphite coated with a protective colloid and dispersed in a liquid. The selection of the liquid (water, oils, or synthetics) is made on the basis of the intended use of the product. In colloidal-graphite dispersions, the graphite particles remain in suspension indefinitely, and the particles "wick"—that is, they are carried by the liquid to most places that the liquid penetrates.

Diamond. Diamond is the cubic crystalline form of carbon (see Sec. 3-1). When pure, diamond is water clear, but impurities add shades of opaqueness, including black. It is the hardest natural material, with a hardness on the Knoop scale ranging from 5,500 to 7,000. It will scratch, and be scratched by, the hardest man-made material, Borozon (Sec. 13-16). It has a specific gravity of 3.5. Its melting point is around 7000°F, at which point it will graphitize and then vaporize. Diamonds are generally electrical insulators and nonmagnetic.

Synthetic diamonds (Borozon) are produced at extremely high pressures (800,000 to 1,800,000 psi) and at temperatures of from 2200 to 4400°F. They are up to 0.01 carat in size, and they are comparable to the quality of industrial diamonds. In powder form, they are used in cutting wheels.

Of all diamonds mined, about 80 percent by weight are used in industry. And roughly 45 percent of the total industrial use is in grinding wheels. However, tests have shown that under many condi-

Table 13-6 *Typical Properties of Some Specialty Graphites*

Property	Recrystallized	Pyrolytic	PT
Specific gravity	1.85–2.15	2.25	0.93–1.2
Bulk density, lb/in.^{3}	0.07	0.07–0.08	0.04–0.06
Modulus of elasticity in tension, 10^6 psi	1.5–2.7	4	0.9–1.8
Tensile strength, 1,000 psi	4–1.2	7–14	1–5
Compression strength, 1,000 psi	5.5–7.2	15	2.5–10
Electrical resistivity, 10^{-4} ohm-cm	—	5	16–80

NOTE: Properties are with grain.

tions synthetic diamonds are better than mined diamonds for this purpose.

13-25 Mechanical Properties

Strength. Compared with metals and most polymers, room-temperature tensile strengths of conventional carbon and graphites are low, ranging from 1,000 to 2,000 psi. Compressive strengths range from 3,000 to 8,500 psi. These are generally with-grain values (that is, specimen length is parallel to the grain). The degree of anisotropy in graphites varies, but cross-grain strengths are usually substantially lower. Tensile strengths of the newer engineering graphites are substantially higher, with that of pyrolytic graphite reaching around 14,000 psi.

The outstanding property of all graphites is their high-temperature strength. Unlike most materials, which rapidly lose strength at elevated temperatures, graphite strength increases continuously from room temperature to a maximum at about 4530°F. For graphites in general, strengths at 4500°F are about double those at room temperature (if protected from oxidation); and their strength at 4500°F exceeds that of any known material at that temperature. Pyrolytic graphite has a fracture strength in excess of 40,000 psi at 5000°F. The strength-to-weight ratio of graphites is of great significance in some applications. Because of its low density and its increase in strength with temperature, graphite is unexcelled among materials at temperatures in the 4500°F range.

Stiffness. The modulus of elasticity in tension for conventionally manufactured graphites is about 0.5 to 1.75 million psi. Recrystallized graphite has a modulus of around 2.5 million psi, and pyrolytic graphite has a modulus as high as 5.5 to 8 million psi along the grain. Higher moduli can be expected in the newer, higher-density materials. Both conventional and recrystallized graphites' moduli increase with increasing temperature. On the other hand, pyrolytic graphite's modulus decreases with increasing temperature (with grain).

Creep. Graphite behaves like a viscoelastic material at high temperatures. The degree of creep depends on many variables, such as density, the particle binder, and the size, type, and orientation of the crystallites and particles. Even at room temperature, graphite is not perfectly elastic and yields plastically to some extent. The greater plasticity of graphite at temperatures above 2900°F greatly increases the rate of creep. Significant improvements in creep resistance are found in both the recrystallized graphites and the pyroelectric materials.

13-26 *Thermal Properties*
Carbon and graphite do not have a true melting point. They sublime at 7590°F. Conventional graphites have exceptionally high thermal conductivity at room temperature, while carbon has only fair conductivity. Conductivity with the grain in graphite is comparable to that of aluminum; across the grain, it is about the same as brass. Conductivity increases with temperature up to about 32°F; it then remains relatively high, but decreases slowly over a broad temperature range before it drops sharply. In pyrolytic graphite, thermal conductivity with the grain approaches that of copper; across the grain, it serves as a thermal insulator and is comparable to that of ceramics.

Thermal expansion of carbon and graphite is quite low (1 to 1.5 × $10^{-6}/°F$)—less than one-third that of many metals. Graphite expansion across the grain increases with increasing density, while along the grain it decreases with increasing density. But expansion increases in both directions with increasing temperatures.

An outstanding feature of graphites is their excellent shock resistance. Unfortunately, no standard test exists to evaluate this property.

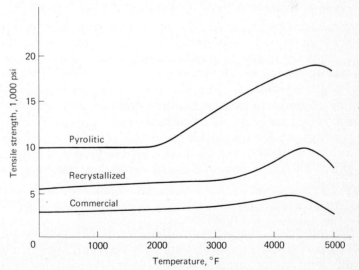

Fig 13-7 Tensile-strength curves of three types of graphite, showing how their strength increases with temperature up to around 4500°F.

However, on a comparative basis, the shock resistance of graphite far surpasses that of many ceramics and metals.

13-27 Chemical and Oxidation Resistance

Graphite is one of the most chemically inert materials. It is subject to only three types of attack: oxidation, formation of lamellar compounds, and reaction with carbide-forming metals at very high temperatures.

Chemical Resistance. Carbon and graphite are highly inert to most acids and alkalies, the exceptions being strong oxidizing media. Hydrofluoric and hydrochloric acid do not attack carbon unless it is the positive electrode in the electrolysis of aqueous hydrofluoric acid. Carbon is slightly attacked by boiling sulfuric acid, the extent depending on the type of carbon. Carbon and graphite are both attacked by concentrated nitric acid, with the resulting formation of melilitic acid, hydrocyanic acid, or carbon dioxide and nitrous oxide, depending on the conditions. Alkali hydroxides do not react with carbon or graphite in solution. But when fused at elevated temperatures, they react to form hydrogen.

Oxidation Resistance. One serious shortcoming of graphite is that it begins to oxidize in air at about 800°F. At low temperatures, however, graphite is actually less reactive than many metals to oxygen. The problem is that, unlike what happens with many metal oxides, the graphite oxide is volatile, and does not form a protective film on the graphite. The oxidation of carbons and graphites varies widely, depending on such variables as the nature of the carbon, the impurities present, the degree of graphitization, and the particle size. No specific data on oxidation rate versus temperature can be given meaningfully.

Much research has been aimed at developing suitable high-temperature coatings to protect graphite from oxidation. One of the most successful has been silicon carbide or siliconized silicon carbide. Such coatings provide reasonable protection for a few hours at temperatures as high as 3000°F. For such uses, special grades of graphite are available with coefficients of thermal expansion matching that of silicon carbide.

13-28 Electrical Properties

The electrical characteristics associated with carbon and graphite's use as electrodes or anodes are relatively well known. They are actually semiconductors; that is, their electrical resistivity, or conductivity, falls between those of common metals and common semiconductors. At temperatures approaching absolute zero, carbon and graphite have few conducting electrons; the number increases with increasing temperature. Thus electrical resistivity decreases with increasing temperature. On the other hand, although increasing electron density tends to reduce resistivity as temperature rises, scattering effects may become dominant at temperatures around 1800°F and thus modify or even reverse this trend. Pyrolytic graphite with its higher density has improved electrical conductivity (along the grain). Furthermore, its high degree of anisotropy results in a high degree of electrical resistivity across the grain.

13-29 *Other Properties*

Surface Qualities. Graphite has excellent lubricity and relatively low surface hardness; carbon has fair lubricity and relatively high surface hardness. Some types of carbon graphitize relatively easily, others do not. Consequently, a wide variety of carbon, carbon-graphite, and graphite materials are available, each designed to provide specific types of surface characteristics for such uses as bearings and seals. Both carbon and graphite can be impregnated with a wide variety of substances, such as synthetic resins, oils, or bearing metals.

Nuclear Grades. The nuclear grades of carbon and graphite are of exceptionally high purity. They have no equals as moderators and reflectors in nuclear reactors because of low thermal neutron absorption, high scattering, high strength at elevated temperatures, and thermal stability in nonoxidizing environments. In general, the properties of carbon and graphite are improved by exposure to nuclear radiation. Hardness and strength increase while thermal and electrical conductivity decrease.

Mica

Mica is a naturally occurring mineral having a wide range of possible chemical compositions and properties. All true mica, however, belongs to one mineral class of silicates having a sheet type of structure, and found in certain areas of the world growing from pegmatite deposits in "book" form. In recent years the increasing demands for high-quality sheet mica have led to the development and commercial production of synthetic mica as an electric furnace product which will be discussed later.

13-30 *Nature and Composition*
Only two types of natural mica are considered to be of commercial importance for most engineering applications. These are muscovite, known generically as "ruby mica," and phlogopite, or "amber mica."

Almost 70 percent of the high-quality natural muscovite mica used in this country comes from overseas sources. Muscovite, which has the composition $KAl_3Si_3O_{10}(OH)_2$, generally comes from India, Brazil, and Argentina, although some lesser deposits are found in the United States, Russia, Rhodesia, and Tanzania.

Ruby mica is generally colorless or has tinges of gray, brown, green, or red. It is the most desirable mica for electrical applications.

Phlogopite mica is a higher-temperature-resistant natural mica having the formula $KMg_3,AlSi_3O_{10}(OH)_2$. It is usually darker in color than muscovite, ranging from brown to greenish-yellow. In general, natural phlogopite is not as desirable as natural muscovite for electrical purposes but finds use in many subcritical applications.

The synthetic mica which is being produced commercially today is a fluor-phlogopite $KMg_3AlSi_3O_{10}F_2$ made by the internal-resistance melting process. This method utilizes graphite electrodes, and melts the mica in its own raw batch so that no crucible is required.

13-31 Properties

The most singular property of mica is its physical structure. As a sheet mineral it can be split into strong, flexible films having good high-temperature resistance and electrical insulating properties.

The natural muscovite and phlogopite micas all contain chemically combined water in the form of hydroxyl $(OH)^-$ ions. Muscovite contains about 4.5 percent water, which at temperatures of 1110°F and above volatilizes, causing breakdown of the mica with resultant blistering delamination. Phlogopite contains about 3 percent water and is stable at temperatures up to 1470°F. Synthetic fluor-phlogopite contains no water. During its formation from the initial oxide and fluoride raw materials the $(OH)^-$ ion positions are filled by fluorine $(F)^-$ as an ionic substitute. The result is that fluor-phlogopite is temperature stable to 1830°F. Above this temperature evolution of fluorine begins and continues until the melting point is reached.

13-32 Glass-bonded Mica and Ceramoplastics

For many years, glass-bonded mica has been used in every type of electrical and electronic system where the insulation requirements are preferably low-dissipation factor at high frequencies, a high-insulation resistance and dielectric-breakdown strength along with extreme dimensional stability.

Glass-bonded micas are made in both machinable grades and precision-moldable grades. Basically, the material consists of natural mica flake bonded with a low-loss electrical glass. This material is heated to the plastic state at temperatures up to 1110°F with subsequent compression or injection molding, depending on the part requirement.

The availability of synthetic mica resulted in the development of so-called ceramoplastics, consisting of high-temperature electrical glass filled with synthetic mica. Ceramoplastics provide an increase in the electrical characteristics over those of natural mica, and, in addition, are more easily molded and have greater thermal stability.

Glass-bonded mica and ceramoplastics have found use in many advanced components such as telemetering commutation plates, molded printed circuitry, high-reliability relay spacers and bobbins, coil forms, transducer housings, miniature-switch cases, and innumerable other component applications.

Asbestos

Asbestos is a general term used to describe six fibrous minerals. The most important of the six is chrysotile, which accounts for over 90 percent of the world's production. The other types of commercial importance are amosite, crocidolite, and anthophyllite. Those of least commercial importance are actinolite and tremolite.

Chrysotile is a hydrated magnesium silicate. Amosite and crocidolite are iron silicates. Anthophyllite is a magnesium silicate with the magnesium isomorphically replaced by varying amounts of iron and aluminum.

The important characteristics of the asbestos minerals that make them unique are their fibrous form; high strength and surface area; resistance to heat, acids, moisture, and weathering; and good bonding characteristics with most binders such as resins and cement.

Review Questions

1. Why do ceramics have very low electrical conductivity?
2. Both metals and ceramics are crystalline materials. Name three microstructural differences.
3. What three macroconstituents compose many ceramics?
4. What is the difference between a ceramic and a crystalline bond?
5. Which ceramic-forming process or processes could be used for (a) High-density parts? (b) Dinnerware? (c) Drain tile? (d) Spark plug insulators? (e) Thin-wall, irregular shapes?
6. What are two differences in the behavior of ceramics and metals under tensile loading?
7. As a class, are metals or ceramics higher in (a) Stiffness? (b) Tensile strength? (c) Hardness? (d) Impact strength? (e) Heat resistance? (f) Chemical resistance?
8. Why do ceramics as a class have relatively low thermal-shock resistance? What property in some ceramics gives them high thermal-shock resistance?

9. Which oxide ceramic is noted for: (a) Low cost and wide use? (b) Best chemical stability? (c) Wide use as heat insulators? (d) Changes from an electrical insulator to an electrical conductor as temperature increases? (e) Extra-high thermal conductivity?
10. What ceramic material has properties similar to diamond?
11. Which ceramic material is widely used for (a) Grinding wheels? (b) Machine cutting tools?
12. Although glass is noncrystalline and is classified as a supercooled liquid, what microstructural characteristics make it behave as a solid material?
13. Why must glass products be annealed after forming operations?
14. What treatment is used to produce "flexible" glass?
15. Which type of glass is noted for: (a) Rigidity increasing with an increase in temperature? (b) Heat-resistant ovenware? (c) Low cost and use for glazing and bottles? (d) Excellent transparency and heat resistance? (e) Automobile windshields?
16. How are intermetallics similar to: (a) Metals? (b) Ceramics?
17. How is the versatility of carbon and graphite extended for bearing and sealing applications?
18. What is the principal industrial use of diamonds?
19. What is unique about graphite's strength at elevated temperatures?
20. What is a major disadvantage in graphite's behavior at temperatures above about 800°F?

Bibliography

Backus, A. S.: "Mica," *Encyclopedia of Materials, Parts and Finishes*, Technomic Publishing Company, Westport, Conn., 1975.

Brady, G. S.: *Materials Handbook*, 10th ed., McGraw-Hill, Inc., New York, 1971.

Chandler, M.: *Ceramics in the Modern World*, Doubleday & Co., Inc., Garden City, 1968.

Corning Glass Works: *This Is Glass*.

"Engineers Guide to Structural Ceramics," *Materials Engineering*, Nov. 1970.

Farrell, E. A.: "Asbestos," *Encyclopedia of Materials, Parts and Finishes*, Technomic Publishing Co., Westport, Conn., 1975.

Hauck, J. E.: "Ceramic Parts," *Materials Engineering*, November 1966.

Henry, E. C.: *Electronic Ceramics*, Doubleday & Co., Inc., Garden City, 1969.

Justrzebski, Z. D.: *Engineering Materials*, John Wiley & Sons, New York, 1959.

Lynch, J. F., C. G. Ruderer, and W. H. Duckworth: *Engineering Properties of Selected Ceramic Materials*, American Ceramic Society, Inc., Columbus, Ohio, 1966.

Materials Selector Issue, *Materials Engineering*, Mid-September 1972.

Riley, M. W.: "Carbon and Graphite," *Materials Engineering*, September 1962.

COMPOSITE MATERIALS

Composites are among the oldest and newest of materials. Paradoxical as this may seem, it is quite true. In primitive times, people discovered empirically that often when two or more different materials are used together as one, the combination does a better job than either of the materials alone. Following this principle, clay bricks were strengthened with straw just as today plastics are often strengthened by embedding glass fibers in them.

Ancient warriors were quite sophisticated in their use of composites. The Mongol bow, for example, was constructed of bull tendons, wood, and silk bonded together with animal glue to give just the right combination of strength and elasticity. And Japanese ceremonial swords were constructed of laminations of different kinds of metal to give the combination of strength, flexibility, and cutting edge that no one material could provide.

However, with some notable exceptions, the potentialities of the composite idea remained unexplored for centuries, as monolithic or homogeneous materials served the major needs of advancing technology. Even in modern times, with the use of reinforced concrete, linoleum, plasterboard, and bimetals, composites have been in an ambiguous position, out of the mainstream of materials development and technology.

Then, in the second quarter of this century, sandwich laminates, powder-metal parts, and glass-fiber-reinforced plastics became commercial realities. These developments began the modern era of composite engineering materials. Since their introduction, the technology of these materials has been steadily developing, and volume use of fiber-reinforced plastics has been increasing at the phenomenal rate of 25 percent annually. But only since 1965 has a distinct discipline and technology of composite materials begun to emerge. That is, 80 percent of all research and development on composites has been done since 1965 when the Air Force launched its all-out development program to make high-performance fiber composites a practical reality.

There are two major reasons for the revived interest in composite materials. One is that the increasing demands for higher performance in many product areas—especially in the aerospace, nuclear energy, and aircraft fields—is taxing to the limit our conventional monolithic materials. The second reason—the most important for the long run—is that the composites concept provides scientists and engineers with a promising approach to designing, rather than selecting, materials to meet the specific requirements of an application.

14-1 *Definition and Nature*

From what we have already learned, we have a general notion of what a composite is and how it differs from homogeneous or monolithic materials. But it is not as simple as it first appears to draw a clear boundary between these two major materials categories, and any definition of composites is arbitrary to some degree.

In the dictionary and in everyday usage, the term composite refers to something made up of various parts or elements. However, when we try to define composite materials in accordance with this general idea, we quickly discover that a useful definition depends on the structural level we are thinking about. At the submicroscopic level—that of simple molecules and crystal cells—all materials composed of two or more different atoms or elements would be regarded as composites. This would include compounds, alloys, plastics, and ceramics. Only the pure elements would be excluded. At the microscopic (or microstructural) level—that of crystals, polymers, and phases—a composite would be a material composed of two or more different crystals, molecular structures, or phases. By this definition most of our traditional materials—which have always been considered monolithic—would be classified as composites. Of all the metals, only single-phase alloys, such as some brasses and bronzes, would be monolithic. Even the steels, which are multiphase alloys of carbon and iron, would be composites by this definition.

It is clear then that this definition is too broad, for it would encompass just about all materials. So we must proceed to the macrostructural level where we deal with constituents such as glass fibers, metal particles, and matrixes. On this basis a useful, but still imperfect, definition of composites is that they are a mixture of macroconstituent phases composed of materials which are in a divided state and which generally differ in form and/or chemical composition.

Note that, contrary to a widely held assumption, this definition does not require that a composite be composed of *chemically* different materials, although this is usually the case. The more important distinguishing characteristics of a composite are its geometrical features and the fact that its performance is the collective behavior of the constituents of which it is composed.

Another important distinction is that in a monolithic or homogeneous material at the microlevel, its composition, structure, and properties are generally considered to be essentially constant from one point to the next inside the material. In comparison, a composite material can vary in composition, structure, and properties from one point to another.

14-2 *Constituent Types*

The major constituents used in structuring composites are fibers, particles, laminas, flakes, filters, and matrixes. The matrix, which can be thought of as the "body" constituent, gives the composite its bulk form. The other four, which can be referred to as structural constituents, determine the character of the composite's internal structure.

Because the constituents are intermixed, there are always regions of contiguity, which can be considered analogous to grain boundaries in metals. The regions may be simply interfaces formed by the contacting surfaces, or they may be composed of a distinct, added phase. Examples of interphases are the coating or coupling agent on glass fibers in reinforced plastics and the adhesive that bonds together the layers of a laminate.

The most familiar composites are those composed of one or more types of constituents dispersed in a matrix. For example, reinforced concrete consists of a stone-and-sand aggregate (particles) and steel rods (fibers) embedded in a matrix of cement. And fiber glass boat hulls are constructed of glass fibers supported by a plastic matrix.

Not all composites have a matrix, however. Some are composed of two or more different material compositions in one of the structural forms. Sandwich materials, such as plasterboard, and laminates such as linoleum and thermostat metal, are composed entirely of layers (lamina). The layers taken together give the composite its form.

14-3 Constituents and Performance

It is evident that composites' properties and behavior are derived from (1) the constituents and (2) the relations and interactions between the constituents.

As is true with monolithic materials, the microstructural nature of the constituents is, of course, critically important. It largely determines the general order or range of properties that the composite will have. But equally important are a number of geometrical characteristics. The shape and size of the constituent elements, and the relative amounts of the different constituent phases, can be varied to provide specific properties or property values. Also, the way the structural constituents are distributed and arranged—whether it is a matrix or nonmatrix type—helps determine a composite's performance. Constituents can be dispersed in a uniform pattern to provide constant property values. Or there can be a nonuniform or nonrepetitive distribution, in which case the nature and properties vary from one region to another over the composite's cross section. And, similarly, the structural constituents can be oriented to provide directional properties. These variables give composites much of their versatility.

14-4 Combination-Dependent Properties

The next important consideration is the collective behavior, or properties, resulting from the combination of the constituents or from their interactions. In systems terms, this collective or composite performance or output can be of one or more different kinds.

Mixture Rule. Perhaps the simplest one is related to the well-known mixture rule. That is, the quantitative value of a given property is the sum of the values of the individual constituents (or the constituent phases). Thus the density (or weight) of a block of concrete is the sum of the separate weights of the sand, stones, and cement of which it is composed.

This summation rule applies at least in an approximate manner to a number of material properties, especially to transport properties such as electrical and heat conductivities. For example, the approximate electrical or heat conductivity of laminar composites can be obtained by adding the conductivities of the different laminas making up the composite. In particulate composites, if metal particles are added to resin encapsulants for electronic parts, the thermal conductivity of the metal is added to that of the resin to dissipate heat more rapidly from the components (heat sink).

Contributory Properties. Another kind of collective output results when each of two or more constituents contributes a particular desired property or properties to the total performance. Many clad materials, for example, are composed of a low-strength surface layer with high corrosion resistance, plus an underlayer of a stronger material that provides structural strength and stiffness. In such composites, the performance of each different constituent, in effect, supplements the others. Many particulate composites have been developed to exploit in this way certain properties of both plastics and metals. For example, powdered iron and iron oxides are blended with plastics to provide magnetic properties. And lead is added to plastic moldings to dampen sounds and vibrations.

Constituent-Interdependent Properties. The third kind of composite output, perhaps the most important of the three, is an expression of the general systems principle that a whole can be something different (other) than the sum of its parts. This means that the behaviors or functions of the constituents are not independent of each other, but rather are *interdependent*, and the composite's performance is therefore the net result of the interrelated functioning of the constituents.

The classic example of this third combination effect is glass-fiber-reinforced plastics. As we learned in Chap. 8, plastics are relatively weak. On the other hand, thin glass fibers are exceedingly strong in tension. But, because glass is brittle, its high strength cannot be realized if the fibers are used alone. The presence of microcracks both inside and on the surface cause load-stress concentrations where the cracks are located. And this causes local overloading that in turn leads to fiber breakage. Also, to be usable, the fibers must be available in bulk form.

Let us now take a detailed look at how a greater share of the potential strength of the glass fibers is achieved by embedding them in a plastic matrix. We will examine what happens inside a piece of a fiber composite when a load is applied to it as shown in Fig. 14-1. To keep the discussion simple, we will assume that all the fibers are arrayed in the same direction as the applied load and that the bonds between the fibers and the matrix act together as a unit. That is, both constituents stretch an equal amount under loading. This means that the strain (ΔL)—the elongation per unit length—of each will be the same.

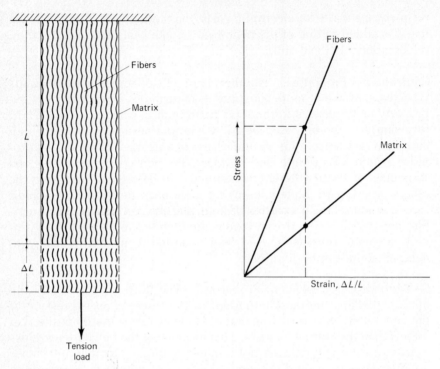

Fig 14-1 The interdependent functions of fiber and matrix. (Left) When tension load is applied, both constituents stretch an equal amount. (Right) Stress-strain diagram shows how stress is transferred from the matrix to the stronger fibers.

When we apply a tension load to the end of the composite, the constituent with greater elasticity (lower modulus)—the plastic—will end up carrying less of the load than the stiffer glass. Thus, in order for the weaker, ductile plastic to stretch no more than the glass (in keeping with our assumption), it must transfer some of its load to the fibers, as shown in the stress-strain chart (Fig. 14-1).

An interaction between the fibers and matrix also occurs if and when the fibers break. When, under a bending load, cracks develop and some fibers break, the stress is distributed to the adjacent fibers, which provide effective barriers against the spread of the cracks (Fig. 14-2).

Hence, the composite concept makes it possible to structure materials in various ways to counteract fracture mechanisms that occur in conventional, monolithic materials, thereby achieving considerably higher strengths than otherwise would be possible.

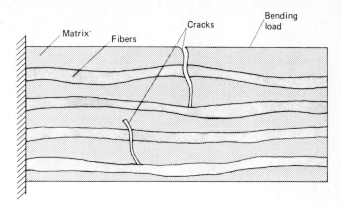

Fig 14-2 Sketch showing how fibers stop propagation of cracks in a fiber composite.

Fiber Composites

It is convenient to classify composites in terms of the major types of structural constituents. On this basis there are five general classes: fiber, particulate or particle, laminar, flake, and filled (Chap. 3 and Fig. 3-10).

Generally, fiber composites are composed of two constituents—fibers and matrix—and an interphase which bonds the individual fibers to the matrix. Perhaps the best known are the load-bearing fiber composites in which the fiber constituent is the principal component both by volume and by weight. However, some fiber composites are composed entirely of fibers, with or without a bonding phase. The polyester-and-wool textile used for men's suits is a familiar example.

14-5 *Property Factors*

As noted, the performance of a fiber composite is determined not only by the chemical nature of the constituents but also by such geometrical characteristics as fiber length, diameter, shape, and orientation. For example, the mechanical properties in any one direction are proportional to the amount of fiber by volume oriented in that direction. As fiber orientation becomes more random, the mechanical properties in any one direction are lower (Fig. 14-3). Also, as the ratio of fiber length to fiber diameter (expressed as the fiber-aspect ratio) increases, the strength of the composite increases (Fig. 14-4).

The individual fibers can be either continuous or discontinuous in the matrix. In general, continuous fibers are easier to handle and to

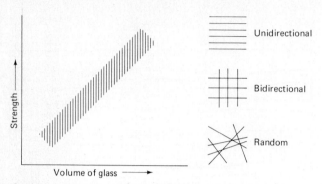

Fig 14-3 Fiber factors influencing composite strength.

orient in given directions than are short fibers, and thus they can develop higher strengths. However, if very short fibers are oriented in one direction, in principle, they could come close to their theoretical strength when embedded in the proper matrix.

Fiber composites can be divided into two broad classes according to their application. One class is composed of composites—chiefly reinforced plastics—used in relatively common structural, industrial, and consumer products. The other class consists of high-performance or advanced fiber composites developed largely for aerospace hardware.

14-6 *Industrial Fiber Composites*

Fiber-Reinforced Plastics. Glass-fiber-reinforced plastics account for over three-quarters of the total fiber-composite production, thus dominating the industrial fiber-composite field. Because we have covered these composites in Chap. 9 (Secs. 9-21 to 9-24), we will give them only cursory treatment here.

Although many types of plastics, both thermosetting and thermoplastic, can be the matrix for glass-reinforced plastics, polyester resins are by far the most widely used. This combination of glass fibers in a polyester matrix provides a good balance of mechanical and electrical properties, corrosion resistance, low cost, and good dimensional stability. In addition, curing can be done at room temperature without pressure.

Other glass-reinforced thermosets include the phenolics and epoxies. Reinforced phenolics are relatively low in cost, and are widely used in electrical products. Epoxies are relatively expensive, and are used primarily in high-performance applications.

Up until a few years ago, glass-reinforced thermoplastics were con-

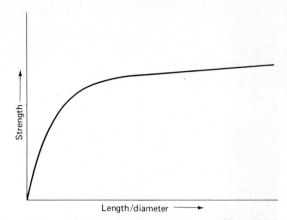

Fig 14-4 Relation of fiber-length-to-diameter (fiber-aspect) ratio to composite strength.

sidered specialty materials. However, today more than 1,000 different grades of reinforced thermoplastics are commercially available. The most widely used are reinforced nylons, polystyrenes, styrene acrylonitrile, polycarbonates, polypropylenes, acetals, polyurethanes, ABS plastics, and polysulfones. As a general rule, reinforcing a thermoplastic with chopped glass fibers at least doubles the plastic's tensile strength. Also, heat-distortion temperatures are usually increased by 100°F, and impact strengths are raised appreciably.

Although glass fibers will continue to dominate the field for many years, other fibers will be more widely used as their technology becomes more highly developed. Metal reinforcements, with properties between glass and such exotics as graphite and boron fibers, will probably find increasing application (see Sec. 14-7).

Fiber-Fiber Composites. Many synthetic fibers, with their good chemical resistance and mechanical and electrical properties, are combined with natural organic fibers to form nonwoven, woven, and knit fabrics that are low in cost and light in weight (Secs. 12-4 and 12-6). Although fibers are not physically joined in most textiles, some bonded fabrics are available. These are bonded either under pressure with an adhesive, or at a temperature that causes the fibers to soften and interlock. The bonding gives these fiber composites greater tensile strength than the unbonded types.

A major limitation of the totally organic-fiber composites is their low strength compared with composites using inorganic fibers. Also, very few organic fibers can be used at temperatures above 300°F.

Thus a number of textile and felt fiber composites are composed of both organic and inorganic fibers. Many of these use metal fibers for strength and toughness and organic fibers for resiliency and chemical resistance. Typical uses are conveyor belts, filters, gaskets, tapes, and seals.

14-7 Advanced Fiber Composites

Boron and Graphite. Boron and graphite are the two fiber constituents that have been under the most intensive development in advanced composites. The matrixes in which they are being used are chiefly epoxy resin and aluminum. Epoxy-resin systems have high bond strength and toughness, but cannot be used for service temperatures higher than around 435°F. The newer polyimide-resin matrixes can be used at temperatures up to over 520°F. Aluminum is the principal metal matrix being used with boron fibers, but others, especially titanium, will find use in the future. The three combinations of these constituents that have been given most attention during the past few years are boron epoxy, boron aluminum, and graphite epoxy. Their combination of strength, stiffness, and light weight exceeds that of any existing commercially available monolithic materials.

Boron composites were developed in the early 1960s in an extensive program by the Air Force. The fibers in boron composites are, themselves, composites. They are produced by vapor deposition of boron on a tungsten substrate. Their specific gravity is about 2.6, and they range from 4 to 6 mils in diameter. They have tensile strengths around 500,000 psi, and a modulus of elasticity of nearly 60 million psi.

Graphite fibers are produced by carbonization of rayon or acrylic fibers. They run about one-third of a mil in diameter, and have a specific gravity of from 1.7 to 2.0. When used in composites, they are generally made into yarn containing some 10,000 fibers. Depending on the precursor fiber, their tensile strength ranges from 200,000 to nearly 500,000 psi, and their modulus of elasticity is from 28 million to 75 million psi.

While the properties of boron and graphite fibers are impressive, only a fraction of these values are attained when they are used as structural constituents in composites. Even so, the composite's specific strengths and moduli are still far above those attainable in bulk-material form or in monolithic structural materials such as aluminum, steel, and titanium. Unidirectional boron-aluminum composites and graphite-epoxy composites (55 percent fiber content) have tensile strengths ranging from 110,000 to over 200,000 psi, specific strengths in inches of about 2.0 to 2.5 million, and specific moduli of 350 to 400 million. For comparison, specific strength for high-strength aluminum is about 0.75 million, and its specific modulus is about 100 million. In actual applications, properties such as these can be translated into significant weight savings.

Fabrication Methods. In addition to superior mechanical properties, much of the utility of fiber composites comes from their inherent design and fabricating versatility. One of the oldest methods of fabrication is filament winding, which involves continuously winding fibers on a mandril of the desired shape (Sec. 9-23). Multilayer lay-up methods are the most widely used. Each layer is made of composite tapes of graphite or boron fibers in a resin or aluminum-foil matrix. After the desired shape is formed, heat and pressure are applied to complete the manufacturing operation. The end product is a laminar composite, in which the laminas are fiber composites. This fabricating method permits the designer to design a material structure tailored to the stress pattern of the application.

Another fabricating method uses conventional metal forms. It involves bonding a unidirectional composite tape to the surfaces of structural shapes or inserting composite rods into them. This method has greatly increased the stiffness of aluminum shapes, and has achieved weight savings of between 25 and 60 percent.

Three-dimensional weaving techniques also have been developed to produce isotropic or orthotropic composite structures. Various cylindrical and block shapes are possible, and gears, bearings, and other highly stressed components can be fabricated by machining methods.

Other Fiber Composites. While it is likely that boron and graphite will dominate the high-performance composites field for some years to come, a variety of other fiber composites probably will find application in the future. Silicon carbide deposited over boron, because of its superior compatibility with metals, shows promise in metal-matrix composites operating in the high-temperature region of 1500 to 1800°F. Boron nitride, another candidate for metal-matrix composites, has the advantage of low precursor-fiber cost, low density, and good high-temperature stability. However, both its tensile strength (150 to 200,000 psi) and its modulus of elasticity (20 million psi) are relatively low. Alumina, or sapphire, fibers drawn continuously from a melt are also being considered for use with metal and ceramic matrices because of their high-temperature capabilities and chemical compatibility.

A recent addition to the high-performance fiber field is an organic polymeric fiber (developed by DuPont) known as Kevlar. With a spe-

Fig 14-5 (left)   Three-dimensional composite cone. Fig 14-6 (right)   Three-dimensional composite ring. (Avco Corp.)

cific gravity of only 1.45 and a tensile strength of 525,000 psi, the fiber has a higher specific strength than glass, boron, and graphite. Its specific modulus is five times that of glass, and is as high as some graphite fibers.

Whisker Composites. The ultimate in fiber constituents is generally considered to be materials in the form of very fine, single crystals. Known as whiskers, they range from 3 to 10 microns in diameter and have length-to-diameter ratios of from 50 to 10,000. Since they are single crystals, their strengths approach the calculated theoretical strengths of the materials (Fig. 14-7). Alumina whiskers, which have received most attention, have tensile strengths up to 3 million psi and a modulus of elasticity of 62 million psi. Other potential whisker materials are silicon carbide, silicon nitride, boron carbide, and beryllia.

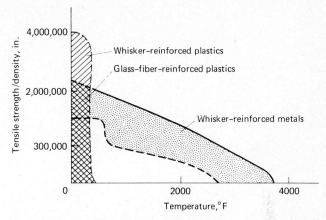

Fig 14-7 High-temperature strength of whisker-reinforced metals compared to other materials.

Hybrid Composites. And, finally, hybrid composites (composed of two or more different fiber types) will be developed as composite technology advances. Preliminary work in combining graphite and boron fibers in an epoxy matrix has shown that increases of up to 30 percent in the modulus of elasticity and strength over conventional boron-epoxy composites may be possible.

Particulate Composites

14-8 *Property Factors*

The particles in particulate composites are discrete, and, by definition, they do not combine chemically with the matrix. Actually, however, in some cases, such as cermets, there is a small amount of solubility at the interface between the matrix and the particles.

As mentioned earlier, the geometrical characteristics of the particles and their structural arrangement in the matrix largely determine the properties of the composite. These characteristics include particle size and shape, the spacing between the particles, the amount or volume fraction of particles, and the manner in which the particles are distributed in the matrix. As a general rule, for a given particle shape, any two of the three variables of size, spacing, and volume fraction determines the third (Fig. 14-8).

Because of the wide range of particle characteristics and compositions, a large variety of composites have been developed. Two major groups of particulate composites are cermets and dispersion-hardened alloys (Fig. 14-9).

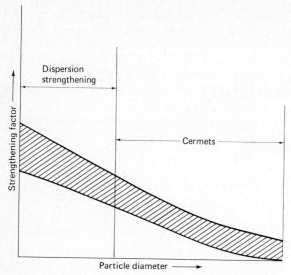

Fig 14-8 Relation of particle diameter and composite strength.

14-9 Cermets

Cermets are composed of ceramic particles (or grains) dispersed in a metal matrix. Particle size is greater than 1 micron, and the volume fraction is over 25 percent and can go as high as 90 percent.

Bonding between the constituents results from a small amount of mutual or partial solubility. Some systems, however, such as the metal oxides, exhibit poor bonding between phases and require additions to serve as bonding agents.

Cermets are produced by powder metallurgy (P/M) techniques. They have a wide range of properties, depending on the composition and relative volumes of the metal and ceramic constituents. All P/M tech-

Fig 14-9 Difference in particle characteristics between cermets and dispersion-hardened alloys: (a) cermet and (b) dispersion-hardened matrix.

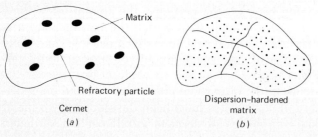

niques are applicable, including cold and hot pressing and slip casting. Cermets can also be produced in most of the shapes normally feasible by P/M methods. Some cermets are also produced by impregnating a porous ceramic structure with a metallic matrix binder. Cermets can also be used in powder form as coatings. The powdered mixture is sprayed through an acetylene flame, and it fuses to the base material.

Although a great variety of cermets have been produced on a small scale, only a few types have significant commercial use. These fall into two main groups: oxide-base and carbide-base cermets.

Oxide-base Cermets. The most common type of these cermets contains aluminum-oxide ceramic particles (ranging from 30 to 70 percent volume fraction) and a chromium or chromium-alloy matrix. In general, oxide-base cermets have specific gravities between 4.5 and 9.0, and tensile strengths ranging from 21,000 to 39,000 psi. Their modulus of elasticity runs between 37 and 50 million psi, and their hardness range is A70 to 90 on the Rockwell scale. The oxide-base cermets are used as a tool material for high-speed cutting of difficult-to-machine materials. Other uses include thermocouple-protection tubes, molten-metal-processing equipment parts, and mechanical seals.

Carbide-base Cermets. There are three major groups of carbide-base cermets: tungsten, chromium, and titanium. And each of these groups is made up of a variety of compositional types or grades.

Tungsten-carbide cermets contain up to about 30 percent cobalt as the matrix binder. They are the heaviest type of cermet (specific gravity is 11 to 15). Their outstanding properties include high rigidity, compressive strength, hardness, and abrasion resistance. Their modulus of elasticity ranges between 65 and 95 million psi, and they have a Rockwell hardness of about A90. They are used for gages and valve parts.

Most titanium-carbide cermets have nickel or nickel alloys as the metallic matrix, which results in high-temperature resistance. They have relatively low density combined with high stiffness and strength at high temperatures (above 2200°F). Typical properties are specific gravity, 5.5 to 7.3; tensile strength, 75,000 to 155,000 psi; modulus of elasticity, 36 to 55 million psi; and Rockwell hardness, A70 to 90. Typical uses are gas-turbine nozzle vanes, torch tips, hot-mill-roll guides, valves, and valve seats.

Chromium-carbide cermets contain from 80 to 90 percent chromium carbide, with the balance being either nickel or nickel alloys. Their tensile strength runs about 35,000 psi, and they have a tensile modulus of from about 50 to 56 million psi. Their Rockwell hardness is about

A88. They have superior resistance to oxidation, excellent corrosion resistance, and relatively low density (specific gravity is 7.0). Their high rigidity and abrasion resistance makes them suitable for gages, valve liners, spray nozzles, bearing seal rings, bearings, and pump rotors.

Other Cermets. Barium-carbonate-nickel cermets are used in higher-power pulse magnetrons. Some proprietary compositions are used as friction materials. In brake applications, they combine the thermal conductivity and toughness of metals with the hardness and refractory properties of ceramics.

Uranium-dioxide cermets have been developed for use in nuclear reactors. Other cermets developed for use in nuclear equipment include chromium-alumina, nickel-magnesia, and iron-zirconium-carbide cermets.

14-10 *Dispersion-Hardened Alloys*

Like cermets, dispersion-hardened alloys are composed of a hard particle constituent in a softer metal matrix. However, unlike cermets, the particles are less than 1 micron in size, and the particle volume fraction ranges from only 2 to 15 percent. Also in contrast to cermets, the matrix is the primary load bearer while the particles serve to block dislocation movement and cracking in the matrix. Therefore, for a given matrix material, the principal factors that affect mechanical properties are the particle size, the interparticle spacing, and the volume fraction of the particle phase. In general, strength, especially at high temperatures, improves as interparticle spacing decreases.

Depending on the materials involved, dispersion-hardened alloys are produced by either powder metallurgy, liquid metal, or colloidal techniques. They differ from precipitation-hardened alloys (see Sec. 5-6) in that the particle is usually added to the matrix by nonchemical means. Precipitation-hardened alloys derive their properties from compounds that are precipitated from the matrix through heat treatment.

There are a rather wide range of dispersion-hardened-alloy systems. Those of aluminum, nickel, and tungsten, in particular, have proved commercially significant. Tungsten thoria, a lamp-filament material, has been in use for more than 30 years. And dispersion-hardened aluminum alloys are in wide commercial use today. Known as SAP (aluminum–aluminum-oxide) alloys, they have a unique combination of good oxidation and corrosion resistance plus high-temperature stability and strength considerably greater than that of conventional high-strength aluminum alloys (Fig. 14-10).

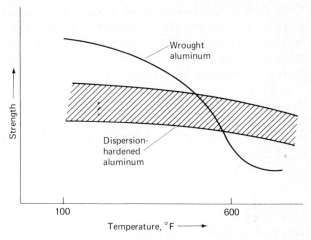

Fig 14-10 High-temperature strength of SAP alloys compared to wrought aluminum.

Another dispersion-hardened alloy is TD nickel, a dispersion of thoria in a nickel matrix. The alloys are three to four times stronger than pure nickel at 1600 to 2400°F. Other metals that have been dispersion strengthened include copper, lead, zinc, titanium, iron, and tungsten alloys.

14-11 *Metals in Plastics*
A number of useful particulate composites consist of metal particles in a plastic matrix. Such filled plastics may contain metal particles up to 90 percent of volume.

Aluminum, as a filler, has applications ranging from a decorative finish to the improvement of thermal conductivity. Aluminum powder added to metal-bonding adhesives delays degradation and improves retention of strength at 400 to 500°F. And cold plastic solder is composed of aluminum powder in a vinyl or epoxy binder. Aluminum powder also improves thermal and electrical conductivity in castable thermosetting plastics. At 30 percent by volume, aluminum improves thermal conductivity 300 percent (compared with an unfilled epoxy resin) and somewhat lowers electrical resistivity.

Unlike aluminum particles, iron and steel—especially when added to liquid resin polymers—tend to settle because of their higher density relative to the resin. Yet, steel-filled plastics have been successfully used for many years on small-lot production tooling. These parts range in weight from a few ounces to several hundred pounds.

Copper particles are used in plastics to provide electrical conductivity and as a coloring agent. Generally, the alloys are used in conjunction with thermoplastic solutions, such as cellulose nitrate or vinyl lacquers. Epoxies predominate as the binder in thermosetting plastic systems because copper inhibits vinyl-type polymerization and the curing of thermosetting polyesters.

Three properties of lead are used in lead-particle–plastic composites: its ability to dampen sound vibration, its effectiveness as a barrier against gamma radiation, and its high density. In spite of their high density, it has been found that lead powders blend more easily than other metals with liquid plastics, so that a greater volume percentage can be introduced.

Laminar Composites

Laminar, or layered, composites consist of two or more different layers bonded together. The layers can differ in material, form, and orientation. For example, clad metals are made up of two different materials. In sandwich materials, such as honeycombs, the core layer may or may not differ in form from that of the facings. In plywoods, although the layers are often of the same type of wood, the orientation of the layers differs.

14-12 *Major Characteristics and Types*

Because of the many possible combinations, it is difficult to generalize about laminar composites. However, for the most part: (1) properties of laminar composites tend to be anisotropic, (2) properties may vary from one side of the composite to the other, and (3) each layer may perform a separate and distinct function. Another distinction is that laminar composites can incorporate the advantages of other composites. That is, the improvement in properties obtained from combining fibers, particles, or flakes in a matrix may be utilized in combination with other materials in layered constructions.

Laminar composites can be divided into two major classes: laminates and sandwiches (Fig. 14-11). Laminates consist of two or more superimposed layers bonded together. Sandwiches, which might be considered a special case of laminates, consist of a thick, low-density core (such as a honeycomb or foamed material) between thin faces of comparatively higher density. In sandwich composites, a primary objective is improved structural performance, or, more specifically, high strength-to-weight ratios. Laminates, on the other hand, are most often designed to provide characteristics other than superior strength. There are, of course, exceptions, the most notable being plastics laminates,

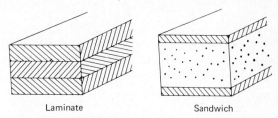

Fig 14-11 Two major types of laminar composites.

which consist of layers of resin-impregnated fabric, paper, or glass cloth, and which possess high strength-to-weight ratios. In general, laminates are designed to protect against corrosion and high-temperature oxidation, to provide impermeability, to facilitate fabrication, to cut costs, to improve appearance, to reduce thickness, to modify electrical and other properties, or to overcome size limitations.

14-13 *Sandwich Composites*

Structural sandwiches can be compared to I beams (Fig. 14-12). The facings correspond to the flanges, the object being to place a high-density, high-strength material as far from the neutral axis as possible, thus increasing the section modulus. The core which supports the facings is comparable to the I-beam web, which supports the flanges of the beam and allows them to act as a unit. The core, like the web, carries the shear stresses. However, the sandwich core differs from an I-beam web in that it maintains continuous support for the facings, allowing the facings to be worked up to or above their yield strength without crimping or buckling.

Sandwich structures are actually more efficient than I beams. The combination of high-density facings and low-density cores provides a much higher section modulus per unit density than any other known construction. Thus, for an equivalent rigidity factor, the weight of an aluminum-faced honeycomb-sandwich-structure beam is only about one-fifth that of solid aluminum.

Facings. One of the advantages of sandwich construction is the wide choice of facings, as well as the opportunity to use thin sheet materials. The facings carry the major applied loads and therefore determine the stiffness, stability, and, to a large extent, the strength of the sandwich. Theoretically, any thin, bondable material with a high tensile- or compressive-strength–weight ratio is a potential facing material for sandwich panels. The materials most commonly used are aluminum, stainless steel, glass-reinforced plastics, wood, paper, and vinyl and

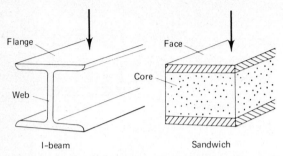

Fig 14-12 Sandwich composites function like an I beam.

acrylic plastics, although magnesium, titanium, beryllium, molybdenum, and ceramics have also been used.

Cores. The bulk of a sandwich is the core. Therefore it is usually lightweight for high strength-to-weight and stiffness-to-weight ratios. However, it must also be strong enough to withstand normal shear and compressive loadings, and it must be rigid enough to resist bending or flexure.

Core materials can be divided into three broad groups: cellular, solid, and foam. Paper, reinforced plastics, impregnated cotton fabrics, and metals are used in cellular form. Balsa wood, plywood, fiberboard, gypsum, cement-asbestos board, and calcium silicate are used as solid cores. Plastic foam materials—especially polystyrene, urethane, cellulose acetate, phenolic, epoxy, and silicone—are finding increasing use for thermal-insulating and architectural applications. Foamed inorganics such as glass, ceramics, and concrete also find some use. Solid cores, although important, obviously cannot provide the strength-to-weight ratio inherent in cellular construction. Foam cores are particularly useful where the special properties of foams are desired, such as insulation. And the ability to foam in place is an added advantage in some applications, particular in hard-to-get-at areas.

Of all the core types, however, the best for structural applications are the rigid cellular cores. The primary advantages of the cellular core are that (1) it provides the highest possible strength-to-weight ratio, and (2) nearly any material can be used, thereby satisfying virtually any service condition.

There are, essentially, three types of cellular cores: honeycomb, corrugated, and waffle (Fig. 14-13). Other variations include small tubes or cones and mushroom shapes. All these configurations have certain advantages and limitations. Honeycomb, for example, can be isotropic,

Composite Materials

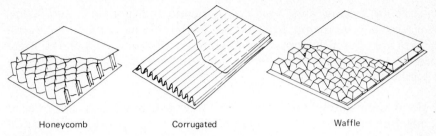

Fig 14-13 Some types of cellular cores.

and it has a high strength-to-weight ratio, good thermal and acoustical properties, and excellent fatigue resistance (Table 14-1). Corrugated-core sandwich is anisotropic and does not have as wide a range of application as honeycomb, but it is often more practical than honeycomb for high production and fabrication into panels.

Theoretically, any metal that can be made into a foil and then welded, brazed, or adhesive bonded can be made into a cellular core. A number of materials are used, including aluminum, glass-reinforced plastics, and paper. In addition, stainless steel, titanium, ceramic, and some superalloy cores have been developed for special environments.

14-14 Laminates

In principle, there are as many different types of laminates as there are possible combinations of two or more sheet materials. But, for our purposes here, if we divide all materials broadly into metals and nonmetallics, and if we divide nonmetallics into organic or inorganic, then there are six possible combinations in which laminates can be produced: (1) metal and metal, (2) metal and organic, (3) metal and inor-

Table 14-1 *Weight Comparison of Structures to Carry 3,600-lb Beam Load**

Material and Structure	Weight, lb
Honeycomb sandwich	7.8
Nested I beams	10.9
Steel angles	25.9
Magnesium plate	26.0
Aluminum plate	34.2
Steel plate	68.6
Glass-reinforced plastic	83.4

* Simple beam loading. Beam supported on 2-in. centers with load applied at midpoint. Deflection = 0.058 in.

ganic, (4) organic and organic, (5) organic and inorganic, and (6) inorganic and inorganic. In laminates containing more than two layers, there are obviously considerably more possibilities. And one or more of the layers may be a composite, making the combinations even more variable and complex.

Metal-Metal Laminates. There are three basic functional categories of metal-metal laminates: (1) laminates whose face is primarily decorative; (2) laminates whose face provides one or more important surface properties (other than appearance) and whose base makes the laminate cheaper and/or stronger than the equivalent face material alone; and (3) laminates that provide special bulk properties, or properties resulting from a reaction between the face and base.

Within this framework, there are two major classes, based on the methods of producing them: precoated metals and clad metals. In precoated metals, the face is formed by building up the second constituent on a substrate to form a thin, essentially continuous film. This is usually done by electroplating or hot dipping, although chemical plating is also used. In clad metals, the face is a solid, wrought, and sometimes cast material. Clads are more suitable than precoated metals when the environment is more severe, calling for a thicker face.

The face of a clad metal is usually considerably thicker than that of a precoated metal. A range of 5 to 20 percent of the thickness of the combination is most common, although the proportion may be as high as 90 percent in some special cases. Actual face-metal thicknesses may range from almost one coat for lead-clad steel to as little as 1 mil in some special applications. A great variety of common metals are used; recently, many exotic metals have been used, including titanium and beryllium.

Metal-Organic Laminates. There are several types of metal-organic laminates. As in the case of metal-metal laminates, they can be conveniently divided into those in which a thin face is built up on a base and those in which the face is considerably thicker and usually formed by bonding on a solid layer.

The best-known metal-organic laminate is prefinished or prepainted metal, whose primary advantage is the elimination of final finishing by the user. Almost any organic coating can be used. Some of the most popular coatings are alkyds, acrylics, vinyls, epoxies, and epoxy phenolics. Face thicknesses range from 0.1 to 2 mils, but they are usually ½ to 1 mil. The most common base metals are cold-rolled steel, tinplate, tin-milled blackplate, hot-dipped and electrogalvanized steel, and standard aluminum alloys. Although there are many possible combinations of solid organic films and metals, plastic-metal combinations are probably most familiar. Vinyl-metal laminates probably account for most of the plastic-metal laminates now used.

Metal-Inorganic Laminates. Although they are normally not considered as composites, porcelain-enameled and ceramic-coated metals are as much composites as are clad and prepainted metals. Glass-lined steel, now widely used for tanks and farm silos, is also in this group. The advantages of a glass-steel combination are almost self-evident. Glass is smooth, inert, and corrosion resistant, but fragile. Steel is strong, durable, and formable, but is attacked by the weather and corrosives. However, when they are put together, a composite results that is strong and corrosion resistant.

Ceramic-lined metals are used for severe environments, as they provide excellent resistance to impact, abrasion, and thermal shock up to 1400°F. Typical applications are chemical reactors and immersion heaters.

Organic-Organic Laminates. Laminated wood is probably the oldest composite in this group. Glued-wood laminates are made by bonding together parallel layers of boards, joists, or planks. Their advantages over conventional lumber are their (1) more uniform distribution of properties, (2) high strength where needed, (3) elimination of checking, splitting, warping, and twisting, (4) uniform moisture content, and (5) ability to overcome the size limitations of lumber. (See also Secs. 11-14 and 11-15.)

The most important of the organic-organic composites is the high-pressure, thermosetting-plastic laminate composed of resin-impregnated paper, cotton, or plastic fabric. These laminates are used for a variety of mechanical and electrical products. High-performance laminates of this type generally have an inorganic constituent—glass fiber or cloth reinforcement, as discussed below. Other examples of organic-organic laminates are plastic-faced wood, laminated paper, rubber-fabric belting, and synthetic felts bonded into layers for a variety of uses.

Organic-Inorganic Laminates. High-pressure, thermosetting-plastic laminates (with glass or asbestos reinforcement) and glass-plastic laminates are the two most common forms of organic-inorganic laminates. High-pressure laminates are produced by impregnating a reinforcing material with a thermosetting resin, laminating the material into multiple layers, and curing with heat and high pressure to form a dense, hard solid with good mechanical strength. Reinforcements used include several organic materials as well as inorganic materials such as glass and asbestos (Sec. 14-6). Resins used include phenolic, melamine, silicone, epoxy, and polyester. Outstanding characteristics of glass- and asbestos-reinforced-plastic laminates are their excellent electrical insulation, strength-to-weight ratios, and corrosion resistance. By varying the proportion and composition of the base materials and resin systems, an extremely broad range of characteristics can be obtained.

Glass-plastic laminates usually consist of two or more layers of glass sheet and one or more layers of plastic. The most common laminated glass is safety glass. This material usually consists of two outer ⅛-in. sheets of glass bonded to a slightly thinner layer of polyvinyl-butyral resin.

Inorganic-Inorganic Laminates. There are not many all-inorganic laminates. Most of them involve glass. Structural glass-glass laminates are used for partitions and for construction of table tops, counter tops, and signs. They are also used for special lenses.

Flake Composites

Flake composites could be considered as being in the particulate class, since flakes are a form of particle. However, they are sufficiently different from the cermets and dispersion-hardened composites to warrant a separate class. A flake composite consists of flakes held together by an interface binder or incorporated into a matrix. Depending on the material's end use, the flakes can be present in a small amount or can comprise almost the entire composite.

14-15 *Property Factors*

The special properties that can be obtained with flakes are, of course, due in large part to their shape. Being flat, they can be tightly packed to provide a high percentage of reinforcing material for a given cross-sectional area. Because of the considerable amount of overlap between flakes in a composite, they naturally form a series of barriers that can stop or reduce the passage of liquid and vapors, as well as reduce the danger of mechanical damage by penetration. Overlapping and touching metal flakes provide electrical conductivity through the composite. Or, with nonconductive flakes, such as glass or mica, it is possible to obtain good dielectric properties as well as resistance to heat. And, by controlling flake shape and orientation, it is possible to obtain special decorative effects. Flake aluminum, for example, is used in automobile paints and in molded plastics to provide decorative color effects and various degrees of transparency.

14-16 *Types Available*
A limited number of materials are used in flake composites. Most metal flakes are aluminum, although silver is used to a minor extent. The other important flake materials are mica and glass. In almost all cases, flakes can be used with a wide variety of organic or inorganic binders or matrixes, provided that the material has chemical, mechanical, and processing compatibility with the flakes.

Glass-flake-reinforced plastics and laminates have an excellent combination of strength and electrical properties. They are used for printed circuits, molded insulators, polarized lighting panels, and electrical potting mixes. The impermeable barrier provided by glass flakes is also put to use in heavy-duty coating systems. Because of their good electrical insulation and dielectric strength, glass flakes are used in insulating paper. For example, paper containing 90 percent glass-flake and 10 percent glassine-grade pulp has a dielectric strength at least six times higher than kraft pulp.

Mica-flake composites are the most familiar and widely used flake composites. Mica flakes, natural or synthetic, are available in a large range of shapes and sizes. Synthetic flakes in a plastic matrix, such as potting compounds, provide good dielectric properties and heat resistance. And, combined with glass, they produce a patented specialty composite known as Ceramoplastic. This material provides a good hermetic seal with excellent high-temperature properties and vacuum tightness.

Natural-mica composites, being older, have a considerably wider range of uses, compared to the synthetic type. Natural-mica flakes are laminated in several layers and bonded with about 5 percent shellac, epoxy, alkyd, or other resin for segment insulations in electric motors. Up to 15 percent of these resins is used to produce a composite that softens when heated so that it can be molded into various shapes. In some cases, mica flakes are bonded with organic or inorganic resins that volatilize when the end product is put into use. This principle is used in some domestic appliances.

Metal-flake composites are confined principally to aluminum and silver. Probably the biggest deterrent to a wider variety of metal flakes is that they are difficult to produce. Aluminum flakes are extremely thin—only 0.012 to 0.042 mil thick. In plastic moldings, they provide a silvery, metallic luster. Silver flakes are preferred for composites where electrical conductivity is needed.

Filled Composites

14-17 *Nature and Structure*
In its simplest form, a filled composite consists of a continuous, three-dimensional structural matrix infiltrated or impregnated with a second-phase filler material. The filler also has a three-dimensional shape, determined by the voids in the matrix. The matrix itself may be an ordered honeycomb, a group of cells, or a random, spongelike network of open pores.

In effect, both the matrix and the filler exist as two separate constituents that do not alloy or, except for a bonding action, chemically combine to a significant extent. The matrix is always continuous, but the filler may be either continuous, as in an impregnated casting, or discontinuous, as in a filled honeycomb. In most filled composites, the matrix provides the framework and the filler provides the desired engineering or functional properties. To obtain the optimum properties in filled composites, the two materials must be compatible so they do not react in ways that would adversely affect or destroy their inherent properties.

14-18 *Open and Filled Types*
The types of skeletal structures that are best for impregnation are an open honeycomb and a spongelike network of open pores. In general, the open-pore structure presents more processing problems than the honeycomb because it is randomly oriented, and it is not usually possible to introduce filler materials in the solid state. In general, the bond between the structure and filler is not as strong in honeycomb or cellular structures as it is in spongelike structures. The random orientation of the spongelike structure helps keep the filler material in place even if there is little or no adhesive action between it and the matrix. Because honeycomb cells have flat, unbroken areas, the bond strength between structure and filler is not likely to be high.

Most filled composites in use today consist of a matrix formed from a random network of open passages or pores. Unlike the matrix in a filled honeycomb, which is specifically designed to a given shape, the open matrix in a sponge or pore composite is formed naturally during processing. Typical materials that have this kind of structure, and that lend themselves to filling, are metal castings, powder-metal parts, ceramics, carbides, graphites, and foams.

The open network of a spongelike structure can be filled with a wide range of materials including metals, plastics, and lubricants, depending on the desired end properties. Metal impregnants, for example, can be used to improve the strength of a matrix or to provide better bearing properties. Plastics can be impregnated into metals to make them pressure tight or—like lubricants—to provide special bearing properties. They can also be incorporated into porous ceramics and graphite as a structural binder.

Filled honeycombs have been developed chiefly for high-temperature applications. A number of combinations are available. A common one is a composite consisting of a metal-honeycomb structure filled with a ceramic. It is useful where high-heat fluxes are encountered for short periods.

Review Questions

1. Define composite materials.
2. Name five common composite materials.
3. Describe two major differences between monolithic or homogeneous materials and composites.
4. Name two kinds of composites that do not have a matrix.
5. Name three geometric variables of structural constituents that determine composite properties.
6. State the mixture rule that applies to some composite properties. Name two composite properties that are determined by using the mixture rule.
7. How does a composite property that is the net result of interacting constituents differ from a mixture-rule property?
8. When a tension load is applied to a fiber-reinforced plastic, why do the fibers carry more of the stress than the matrix?
9. Name the five major classes of composites.
10. If two fiber composites are identical, except for the orientation of the fibers, explain the differences in strength between the one whose fibers are randomly oriented and the one whose fibers are all oriented in the same direction.
11. Name a fiber composite that does not have a matrix.
12. What are two outstanding properties of boron and graphite composites for structural applications?
13. Name four forms in which fiber composites are produced.
14. Which kind or type of fiber constituent has the highest potential tensile strength?
15. What is the major structural difference between cermets and dispersion-hardened alloys?
16. Name three uses of cermets.
17. Give two advantages that dispersion-hardened aluminum (SAP) alloys have over conventional aluminum alloys.

18. What are two reasons for mixing metal particles in plastics?
19. What is the major structural difference between laminates and sandwich materials?
20. Give the similarities and differences between a honeycomb sandwich and a solid I beam.
21. Name three broad groups or types of sandwich cores.
22. What are three reasons for using laminates?
23. Give an example of a laminar composite in which the following constituent is present:
    (a) Wood.
    (b) Paper.
    (c) Glass.
    (d) Plastics.
    (e) Metal.
24. Name two materials used as the flake constituent in flake composites.
25. What are three uses of filled composites?

Bibliography

Broutman, L. J., and R. H. Krock (eds.): *Modern Composite Materials*, Addison-Wesley Publishing Co., Reading, Mass., 1967.

Campbell, J. B., H. R. Clauser, R. J. Fabian, J. E. Hauck, W. Lubars, and W. Peckner: "The Promise of Composites," *Materials in Design Engineering*, September 1963.

Clauser, H. R.: "Advanced Composites," *Scientific American*, July 1973.

———: "A Profile of Composite Materials," Conference Preprint, American Institute of Chemical Engineers, April 1968.

Dietz, A. G. H.: "Fibrous Composite Materials," *International Science and Technology*, August 1964.

Grimes, D. L.: "Trends and Applications of Structural Composites," Whittaker Corp., Los Angeles.

Holliday, L. (ed.): *Composite Materials*, Elsevier Publishing Co., New York, 1966.

Krock, R. H.: "Whisker-Strengthened Materials," *Science and Technology*, November 1966.

FINISHES AND COATINGS

In this book, the terms finish and coating will be used interchangeably to refer to a surface layer on a material body. This layer differs in composition and/or structure from that of the base material. In this sense, a material with a finish layer is a laminar composite as discussed in Chap. 14. A finish layer can be either an additive coating or a conversion finish. In additive coatings, as the name implies, a layer of material is applied to the surface of the material body. Common examples are paints and plated coatings. In conversion finishes, the surface of the material body itself is changed by the application of either a reactive substance (such as a chemical), or by thermal, electrochemical, or mechanical action. Two examples are anodized and diffusion coatings.

15-1 *Functions and Finishes*
Finishes have many different uses and functions:

1. They protect the base material against deterioration by corrosion, oxidation, heat or cold, and mechanical wear and/or deformation.
2. They improve or provide functional properties not attainable with the base material alone. These properties may be electrical conductivity or insulation, light reflectivity or absorbability, thermal insulation and bearing surfaces.
3. They provide a different surface appearance from that of the untreated base material, either by color, polish, or decoration.

4. They reduce costs. Lower costs often result, because of reduced deterioration of the base materials, and because functional performance is improved. At times, a finish permits use of a lower-cost material than otherwise possible. For example, in the common "tin" can, the thin tin coating allows use of low-cost carbon steel, which otherwise would be unsatisfactory because of rusting. The use of expensive metals as a finish, rather than as the total body or part, is another cost reduction. For example, precious metals are frequently applied as a coating or plating on a less costly material.
5. They improve materials processing or fabrication. Coatings applied to a material body can change surface characteristics and thus enhance formability, machinability, or joinability. For example, oils or compounds on metal surfaces facilitate deep drawing and machinability, thin nickel coatings promote bonding of porcelain enamel to steel, and low-melting metal coatings aid soldering.

While all the above coating functions are important, the two widest uses of finishes are to protect a base material from deterioration and for appearance or decorative purposes. Therefore, we will mainly focus on these two functions, with greater emphasis on protective properties.

15-2 *Coating Performance Factors*

Adhesion.   Regardless of its function, a finish must adhere to the base material. (However, because conversion finishes are an integral part of the base material, their adhesion is usually not as critical as additive coatings.) Coatings adhere differently to different base materials. For example, a coating suitable for steel may flake or peel when used on aluminum.

Adhesion is also critically dependent on the condition of the surface being coated. Both surface texture and cleanliness are important considerations. Highly polished surfaces cause more adhesion difficulties than rough surfaces. On most surfaces, foreign matter of one sort or another—rust, dirt, or grease—is present, often as thin and invisible films. For optimum adhesion, these contaminants must be removed. Cleaning and preparing surfaces for coating application is a large and important subject beyond the scope of this book. We will just emphasize, in passing, that surface preparation is as critical to the performance of a finish as is the finish itself.

Appearance Characteristics.   The appearance or decorative qualities of finishes include color, brightness, mass tone, gloss or reflectivity, color retention, and hiding power or opacity. (Some of these characteristics are briefly defined in Chap. 4.)

From a performance standpoint, color retention is an important consideration since it indicates how long a finish will remain attractive under normal service conditions. Some colors are more susceptible to fading or discoloring than others. With white finishes, nonyellowing properties are especially critical.

Opacity, or the coating's ability to obscure the base-material surface or previous finishes, is important because it influences thickness and/or number of coats or layers required.

Protection Characteristics. There are two ways in which finishes can protect a base material against deterioration by corrosion and/or oxidation. First, practically all finishes function as a physical barrier that isolates the base material from the environment. How well a finish functions this way depends mainly on its inertness to the environment and on its thickness. Second, a number of finishes, particularly some types of metallic coatings, also provide protection by a sacrificial, electrochemical action. That is, when a break occurs in the coating, an electric (galvanic) cell is formed between the coating and the exposed base metal. If the coating material is anodic with respect to the base material, it will dissolve anodically, while the base material, which is the cathode in the galvanic cell, will not be attacked. For example, since the metals zinc, cadmium, and aluminum are above iron in the galvanic series, they function as anodic coatings on steel (see Table 4-4). This type of protection is often referred to as cathodic protection.

Organic Coatings

Organic coatings are mostly additive type finishes, and are used on almost all types of materials. They can be monolithic (one layer or coat), or they can be composed of two or more layers. The total thickness of coating systems varies widely. Some are less than 1 mil thick. Others are 10 to 15 mils thick. Generally, by definition, coatings that are more than 10 mils thick are called linings, films, or mastics. Although organic coatings are mostly applied for their decorative and protective qualities, they are also used for the other purposes listed in Sec. 15-1.

Organic coatings depend principally on their chemical inertness and impermeability to provide a protective barrier against corrosion and oxidation. In addition, however, some coatings contain inhibiting pigments that render surfaces, especially metal ones, less reactive chemically. Some coatings also contain metallic pigments that give electrochemical protection to metals.

15-3 *Coating Application and Drying*

Organic coatings are commonly applied by almost all production methods. Application by brushing, however, is usually done by hand, and it is the slowest method.

Organic coatings dry, or cure, by one or more of the following mechanisms: (1) evaporation or loss of solvent, (2) oxidation, and (3) polymerization. After the coating is applied, the volatile ingredients, which are almost always present in at least small amounts, evaporate. Some finishes, such as lacquers, dry completely by evaporation of solvents. Other coatings, although they dry by evaporation, are still in a semifluid state after evaporation. They depend on oxidation or polymerization, or a combination of both, to convert to their final form.

Drying by oxidation, which is usually done at room temperature, is the slowest of the three methods. Polymerization, which involves the polymer chain-forming mechanism (Secs. 3-2 and 8-2), can be done at normal or at elevated temperatures. Polymerization time can be shortened by heat. In recent years, radiation curing, which involves using an electron beam to polymerize the coating in a few seconds, has found increasing use.

15-4 *Coating Types and Systems*

Coating Composition. An organic coating is made up of two principal components: a vehicle and a pigment. The vehicle (see Sec. 15-5) is always there. It contains the film-forming ingredients that enable the coating to convert from a mobile liquid to a solid film. It also acts as a carrier and suspending agent for the pigment. Pigments (see Sec. 15-6), which may or may not be present, are the coloring agents, and, in addition, contribute a number of other important properties.

Organic coatings are commonly divided into about a half-dozen broad categories, based on the types and combinations of vehicle and pigment used in their formulation. These classifications are paints, enamels, varnishes, lacquers, dispersion coatings, emulsion coatings, and latex coatings. However, the complexity of modern formulations makes distinguishing between these various types often difficult.

The various layers of organic finishes are commonly classified as primers, intermediate coats, and finish coats.

Primers. These are the first coatings placed on the surface, (except for fillers, in some cases). If chemical pretreatments have been used, primer coats may often be unnecessary. Primers for industrial or production finishing are of two types: air dry and baking. The air-dry types have drying-oil vehicle bases, and are usually called paints. They may be modified with resins. They are not used as extensively as the baking-type primers, which have resin or varnish vehicle bases. These dry chiefly by polymerization. Some primers, known as flash primers, are applied by spraying, and dry by solvent evaporation within 10 min.

In practically all primers, the pigments impart most of the anticorrosion properties to the primer, and, along with the vehicle, determine its compatibility and adherence with the base metal.

Intermediate Coats. These are fillers, surfacers, and sealers. They can be applied either before or after the primer, but they are more often applied after the primer, and sometimes after the surfacer coat. Their function is to fill in local imperfections or large irregularities in the surface. A variety of materials is used, producing puttylike substances. Their chief characteristics are that they must (1) harden with a minimum of shrinkage, (2) have good adhesion, (3) have good sanding properties, and (4) work smoothly and easily.

Surfacers are often similar to primers. They usually have the same composition as the priming coat, except that more pigment is present. They are applied over the priming coat to cover all minor irregularities in the surface.

Sealers, as a rule, are used over either the fillers or surfacers. Their chief function is to fill up the pores of the undercoat to avoid "striking in" of the finish coat. This filling-in quality of the porous intermediate coats also tends to strengthen the entire coating system. If sealers are used over surfacers, they are usually formulated with the same type of pigment and vehicle as the final coat.

Finish Coats. Finish or top coats are usually the decorative and/or functional part of a paint system. However, they often have a protective function as well. The primer coats may require protection against the service conditions because, although the pigments used in primers are satisfactory for corrosion protection of the metal, they are frequently not satisfactory as top coats. Their color retention when weathered and their physical durability may be poor.

There are also one-coat applications where finish coats are applied directly to the base-material surface, and therefore provide the sole protective medium.

15-5 *Vehicles*
Vehicles are composed of film-forming materials and various other ingredients, such as thinners (volatile solvents), which control vis-

cosity, flow, and film thickness; and driers, which facilitate application and improve drying qualities. We will be concerned here chiefly with the film-forming part of the vehicle, because it determines, to a large extent, the quality and character of an organic finish. It determines the possible ways in which the finish can be applied, and how the "wet" finish will dry to a hard film; it provides for adhesion to the metal surface; and it usually influences the finish's durability.

Vehicles can be divided into three main types: oil, resin, and varnish. The simplest and one of the oldest vehicles is the drying-oil type. Resins, as a class, can serve as vehicles in their own right, or they can be used with drying oils to make varnish-type vehicles. Varnish vehicles are composed of resins, either drying or nondrying oils, and required amounts of thinners and driers. Varnishes are often considered full-fledged organic finishes (see Sec. 15-9).

Drying Oils. Vehicles consisting of oil alone have limited use in industrial finishes. Linseed oil is probably the most widely used of the oils. There are a number of different kinds of oils that differ in their rate of drying, and in such properties as water resistance, color, and hardness.

Tung oil or China wood oil, when properly treated, excels the other drying oils in speed of drying, hardening, and water resistance. Oiticica oil is similar to tung oil in many of its properties. Dehydrated castor oil dries better than linseed oil, but slower than tung oil. Some of its advantages are good color, good color retention, and flexibility. The oils from some fish are also used as drying oils. If processed properly, they dry reasonably well and have little odor. They are often used in combination with other oils. Perilla oil is quite similar in properties to fast-drying linseed oil. Its use is largely dependent on price and availability. Soybean oil is the slowest drying oil, and is usually used in combination with some faster-drying oil.

Resins. Although both natural and synthetic resins can serve as organic coating vehicles, today the plastic resins have largely replaced the natural types. Nearly all the plastic resins—both thermosets and thermoplastics as well as many elastomers—can be used as film formers, and frequently two or more kinds are combined to give the set of properties desired. Typical thermoplastics used are acrylics, acetates, butyrates, and vinyls. Commonly used thermosets include phenolics, alkyds, melamines, ureas, and epoxies. The properties of these and other plastics as coatings are similar to those of the bulk form as discussed in Chaps. 9 and 10.

15-6 Pigments

Pigments are the second of the two principal components of most organic finishes. They contribute a number of important characteristics to a coating. Of course, they serve a decorative function. The choice of color and shade is practically unlimited. Closely associated with color is the pigment's hiding power, or its ability to obscure the surface of the material being finished. In many primers, the principal function of the pigment is to prevent corrosion of the base metal. In other cases, they are added to counteract the destructive action of ultraviolet light rays. Pigments also help give body and good flow characteristics to the finish. And, finally, some pigments give to organic coatings what is termed package stability; that is, they keep the coating material in usable condition in the container.

Pigments can be conveniently divided into three classes: white (commonly called white-hiding), colored, and extender or inert pigments. White pigments are used not only in white paints and enamels, but also in making white bases for the tinted and light shades.

Colored pigments give both opacity and color to the finish. They may be used by themselves to form solid colors, or combined with whites to produce tints. They often inhibit rust as well. For example, red lead, certain lead chromates, zinc chromates, and blue lead are used in iron and steel primers as rust inhibitors. There are two general classes of colored pigments: earth colors, which are very stable and are not readily affected by acids and alkalies, heat, light, and moisture; and chemical colors, which are produced under controlled conditions by chemical reaction. The metallic pigments, of which aluminum powder is perhaps the best known, are also included in the chemical class.

The chief functions of extender pigments are to help control consistency, gloss, smoothness and filling qualities, and leveling and check resistance. Thus particle size and shape, oil absorption, and flatting power are important selection considerations. For the most part, extender pigments are chemically inactive. They also usually have little or no hiding power.

15-7 Enamels

By definition, enamels are an intimate dispersion of pigments in either a varnish or a resin vehicle, or in a combination of both.

They may dry by oxidation at room temperature and/or by polymerization at room or elevated temperatures. They vary widely in composition, in color and appearance, and in properties. Although they generally give a high-gloss finish, some give a semigloss or eggshell finish, and still others give a flat finish. Enamels as a class are hard and tough and offer good mar- and abrasion-resistance. They can be formulated to resist attack by the most commonly encountered chemical agents and corrosive atmospheres.

Because of their wide range of useful properties, enamels are probably the most widely used organic coating in industry. One of their largest areas of use is as coatings for household appliances—washing machines, stoves, kitchen cabinets, and the like. A large proportion of refrigerators, for example, are finished with synthetic baked enamels. These appliance enamels are usually white, and therefore must have a high degree of color and gloss retention when subjected to light and heat. Other products finished with enamels include automotive products; railway, office, sports, and industrial equipment; toys; and novelties.

15-8 *Lacquers*

The word lacquer comes from lac resin, which is the base of common shellac. Lac resin dissolved in alcohol has been used for many centuries. Nowadays, shellac is called spirit lacquer. It is only one of several different kinds of lacquers; these, except for spirit lacquer, are named after the chief film-forming ingredient. The most common ones are cellulose acetate, cellulose acetate butyrate, ethyl cellulose, vinyl, and nitrocellulose.

A distinguishing characteristic of lacquers is that they dry by evaporation of the solvents or thinners in which the vehicle is dissolved. This is in contrast to oils, varnishes, or resin-base finishes, which are converted to a hard film chiefly through oxidation and/or polymerization.

Because many modern lacquers have a high resin content, the gap between lacquer and synthetic-type varnishes diminishes until finally we have what might be called modified-synthetic, air-drying varnishes. They may dry by oxidation and/or polymerization.

Lacquers normally dry hard and dust-free in a very few minutes at room temperature. In production-line work, forced drying is often used. That makes it possible to do a multicoat job without having to lose time between coats. Because of their drying speed, and the fact that they are permanently soluble in the solvents used for application, lacquers are usually not applied by brush. They are applied by spraying or dipping.

Lacquers can be either clear and transparent or pigmented, and their color range is practically unlimited. Lacquers in themselves have good color retention, but sometimes the added pigments, modifying resins, and plasticizers may adversely affect this property. They lack good adhesion to metal, but modern lacquer formulations have greatly improved their adhesion properties.

Lacquers are hard and mar-resistant. They can be made resistant to a large variety of chemicals, including water and moisture; alcohol; gasoline; vegetable, animal, and mineral oils; mild acids; and alkalies. But, because of the volatile solvents, lacquers are inflammable in storage and during application, and this sometimes limits their use.

15-9 Varnishes

Varnishes consist of thermosetting resins and either drying or nondrying oils. They are clear and unpigmented, and they can be used alone as a coating. However, their major use in industrial finishing is as a vehicle to which pigments are added, thus forming other types of organic coatings.

All varnishes follow the same general pattern of drying. First, any volatile solvents present evaporate; then, drying by oxidation and/or polymerization takes place, depending on the nature of the resin and oil. At high temperatures, of course, there is more tendency to polymerize. So varnishes can be formulated for either air or bake drying. Varnishes may be applied by brushing or by any of the production methods.

Because of the large variety of raw materials to choose from and the unlimited number of combinations possible, varnishes have an extensive range of properties and characteristics. Their color range is from almost clear white to a deep gold, and they are transparent, lacking any appreciable amount of opacity. Japan, a hard-baked, black-looking varnish, is an exception. It is opaque since carbon and carbonaceous material are present.

There are some distinctions in properties between oil-modified alkyd varnishes and the other types. In general, oil-modified alkyds have better gloss and color retention and better resistance to weathering. They form a harder, tougher, more durable film, and they dry faster. On the other hand, they have less alkali resistance than the other varnishes. There are no distinctive differences in such properties as adhesion and rust inhibition.

15-10 Paints

The word paints is sometimes used broadly to refer to all types of organic coatings. However, by definition, a paint is a dispersion of a pigment or pigments in a drying-oil vehicle. Paints are seldom used today as industrial finishes. Their principal use is as primers. Paints dry by oxidation at room temperature. Compared to enamels and lacquers, their drying rate is slow. They are also relatively soft, and tend to chalk with age.

15-11 Other Organic Coatings

Dispersion and Emulsion Coatings. These are known as water-base paints or coatings because many of them consist essentially of finely divided ingredients, including plastic resins, fillers, and pigments, suspended in water. An organic medium may also be involved. There are three types of water-base coatings: emulsions or latexes, dispersion coatings, and water-soluble coatings. Emulsions, or latexes, are aqueous dispersions of high-molecular-weight resins. Strictly speaking, latex coatings are dispersions of resins in water, whereas emulsion coatings are suspensions of an oil phase in water.

Emulsion and latex coatings are clear to milky in appearance, have low gloss, excellent resistance to weathering, and good impact resistance. Their chemical and stain resistance varies with composition. Dispersion coatings consist of ultrafine, insoluble resin particles present as a colloidal dispersion in an aqueous medium. They are clear or nearly clear. Their weathering properties, toughness, and gloss are roughly equal to those of conventional solvent paints.

Water-soluble types, which contain low-molecular-weight resins, are clear finishes, and they can be formulated to have high gloss, fair to good chemical and weathering resistance, and high toughness. Of the three types, they handle and flow most like conventional solvent coatings.

Plastic-Powder Coatings. Several different methods have been developed to apply these coatings. In the most popular process—fluidized bed—parts are preheated and then immersed in a tank of finely divided plastic powders, which are held in a suspended state by a rising current of air. When the powder particles contact the heated part, they fuse and adhere to the surface, forming a continuous, uniform coating.

Another process, electrostatic spraying, works on the principle that oppositely charged materials attract each other. Powder is fed through a gun, which applies an electrostatic charge opposite to that applied to the part to be coated. When the charged particles leave the gun, they are attracted to the part where they cling until fused together as a plastic coating. Other powder-application methods include flock and flow coating, flame and plasma spraying, and a cloud-chamber technique.

Although many different plastic powders can be applied by the above techniques, vinyl, epoxy, and nylon are most often used. Vinyl and epoxy provide good corrosion and weather resistance as well as good electrical insulation. Nylon is used chiefly for its outstanding wear and abrasion resistance. Other plastics frequently used in powder coating include chlorinated polyethers, polycarbonates, acetals, cellulosics, acrylics, and fluorocarbons.

Hot-Melt Coatings. These consist of molten thermoplastic materials that solidify on the metal surface. The plastic is either applied in molten form by spraying or flow coating, or it is applied in solid form and then melted and flowed over the surface. Since no solvent is involved, thick single coats are possible. Bituminous coatings are also commonly applied by the hot-melt process.

Metallic Coatings

Metallic coatings are additive-type finishes that are mostly used on metals, although in recent years they have been applied to nonmetallics, especially plastics. They are either monolithic or multilayer, and are classified as anodic or cathodic, depending on the protection mechanism employed by the coating (Sec. 15-2). When a coating is above the base metal in the galvanic series (Table 4-4), it is anodic with respect to the base metal. Therefore, anodic coatings, besides functioning as a physical barrier, also provide sacrificial, or cathodic, protection. Typical anodic coatings are zinc, cadmium, and aluminum. They are applied to steel. Cathodic coatings are lower (less noble) than the base metal in the galvanic series, and therefore can provide protection only as a physical barrier. Thus cathodic coatings must be continuous and nonporous to be effective. In fact, because the base metal is anodic with respect to the coating, if the cathodic coating is porous, accelerated corrosion of the base metal can result. Typical cathodic coatings for steel are tin, nickel, and chromium.

15-12 *Electroplates*
Electroplated coatings are applied by an electrodeposition process. In this process, a direct electric current passing through a plating bath deposits the metal coating on the base metal, which serves as the cathode of the electrolytic cell. The anode is either the metal to be deposited or another electrically conductive material, such as graphite. The plating bath, or electrolyte, contains a salt solution of the plating metal. Factors influencing plated-coating properties include the com-

position of the plating bath, the current density, agitation, and the solution pH and temperature.

In general, electroplates are more uniform in thickness, less porous, and of higher purity than other types of metallic coatings. Common electroplated metals are chromium, nickel, cadmium, copper, tin, zinc, lead, silver, and gold. Thicknesses of electroplates range from a few millionths of an inch to more than 100 mils.

Nickel. Nickel plating is one of the most widely used electroplates. It has excellent corrosion resistance. Proprietary plating baths are available for depositing nickel with a fully bright, semibright, or satin-finished surface. Thicknesses up to 60 mils or more are used on equipment subject to severe corrosive environments, while lower thicknesses—0.2 to 3 mils—are used for protection and appearance on steel, copper, brass, zinc, aluminum, and magnesium. Nickel deposits up to ¼ in. thick are used in building up worn parts and for producing parts by the electroforming process.

Chromium. Decorative chromium-plating systems, consisting of a top layer of chromium applied over layers of copper, nickel, or copper plus nickel, have an attractive blue-white appearance. In addition, chromium plates can be produced in blue, black, and gray. The underlayers of copper and nickel provide a nonporous undercoat for the relatively brittle and porous chromium layer. Besides colored and bright chromium plates, no-gloss finishes can be obtained by using satin-nickel undercoats.

Hard chromium, also known as industrial chromium, consists of thick, hard layers adding up to coats as thick as 20 mils, and, in some cases, 100 mils or more. Hard chromium is usually applied to steel, zinc, and aluminum to provide a combination of hardness, corrosion resistance, and low coefficient of friction.

Porous chromium plates are produced by special techniques, and are used on piston rings where oil-retaining surfaces are desired.

Zinc and Cadmium. These plates have good resistance to many atmospheres. They are often used as anodic coatings on steel and as a corrosion-resistant paint base. Zinc is lower in cost than cadmium. Lacquered or chromate-treated zinc plates are sometimes used to simulate chromium plate. Cadmium is sometimes used as a base for zinc plating. It is seldom used for decorative purposes.

Tin. Tin is a relatively low-cost plate with good resistance to corrosion and tarnish. Because its corrosion products are not toxic or objectionable in taste, it is used on "tin" cans and copper kitchenware. Thick tin plates are used to resist special chemical environments.

Tin-copper plates are known as speculum. They contain about 45 percent tin, and resemble polished silver when buffed to a high luster. Tin-zinc (about 80 percent tin) plates, under certain conditions, offer better outdoor protection to steel than zinc or cadmium plates. Tin-nickel plates are bright, tarnish-resistant coatings that are sometimes competitive with chromium plate.

Copper and Brass. Copper is used primarily as an undercoating for deposits of other plating materials, such as nickel and chromium. Lacquered, bright copper plates sometimes serve as a relatively inexpensive decorative finish for steel. Because brass plates have low corrosion resistance, they are used chiefly for decorative purposes. They are applied as thin coatings (from 0.07 to 0.03 mil), and are usually protected with a clear organic coating.

Lead. Lead plates provide good protection for steels exposed to industrial atmospheres. They also provide a good paint base which can be severely deformed without being stripped off the base metal. Lead-tin plates are used on bearings and as a base for soldering. Plates with about 5 to 6 percent tin have excellent corrosion resistance and are competitive with terne plate (hot dipped).

Precious Metals. Most precious metals can be plated. However, since they are expensive, they are used only where their high cost can be justified, and then in thin layers only. Silver plate is most common It has a pleasing appearance, high chemical resistance, good resistance to high-temperature oxidation, high electrical conductivity, and good bearing qualities. Because gold plates are fine grained and dense in structure, they can be used in extremely thin layers—for example, as thin as 0.0001 in. on brass. Their hardness is considerably increased by alloying them with cobalt and other metals. Such plates have hardnesses of over 300 Vickers.

15-13 *Chemical-Deposition Coatings*
These coatings are often referred to as electroless, or immersion, or displacement coatings. They are similar to electroplates, but they are produced without using plating anodes and electric current. The base material is immersed in an aqueous salt solution of the plating metal, and the plate is deposited by one of two mechanisms. Immersion or displacement coatings are produced by simple displacement, where the coating metal has a lower solution potential than the base metal.

These coatings are quite thin, since deposition continues only as long as the base metal is exposed to the solution. Thicker coatings can be obtained by a chemical-reduction process or by keeping the base metal in contact with a metal less noble than the coating metal.

Nickel displacement coatings, about 0.05 mil thick, are used on steel as a base for porcelain enamel. Hard, dense, chemical-reduction plates (electroless nickel) range from about 0.5 to 3 mils thick, and contain about 5 to 7 percent phosphorus. They are used on steel, aluminum, and copper for corrosion and wear resistance, especially where a uniform electroplate is difficult to achieve.

Tin coatings, about 0.02 to 0.5 mil thick, provide a bright, decorative finish for brass, iron, and steel. Other chemical-deposition coatings include gold and silver for decoration, tarnish resistance, and reflectivity; and zinc applied to aluminum and magnesium as a base for nickel plating.

15-14 *Hot-Dip Coatings*

Hot-dip coatings of zinc, lead, tin, or aluminum are applied to a metal surface by immersing the base metal in a bath of the molten coating metal. The base metals are limited to metals with high melting points, such as cast iron, steel, and copper. Most hot-dip coating systems consist of at least two distinct layers. The first is an alloy layer, usually a brittle intermetallic compound, formed by diffusion of the coating metal into the base metal. The top layer is usually composed of relatively pure coating metal.

Zinc, or galvanized, coatings, which are usually 0.0008 to 0.002 in. thick, are widely used as a low-cost corrosion-resistant coating on steel sheet and strip used for roofing, wire, nails, and guard rails.

Hot-dip tin coatings are best known for their use in "tin" cans. On fabricated steel products, tin coatings run from 0.0003- to 0.0005-in. thick. The coatings have very good resistance to tarnishing and staining in contact with foods. Tin-coated sheet can be extremely formed without suffering coating damage.

Lead-tin coatings, known as lead coatings, contain about 3 to 8 percent tin. Those containing 12 to 20 percent tin are known as terne. Lead coatings have high resistance to atmospheric corrosion and chemicals, but are poor in wear and abrasion resistance. Terne coatings are less corrosion resistant than lead coatings, and they are quite similar to tin coatings.

Aluminum, or aluminized, coatings consist of a 0.002-in. pure-aluminum overlayer plus a refractory, intermetallic layer, usually about 0.002 in. thick. Aluminum coatings of this kind are often used to protect steel from oxidation at temperatures up to 1000°F.

15-15 Sprayed-Metal Coatings

Sprayed-metal coatings are produced by impinging molten particles against a base-metal surface where they are flattened into flakes that interlock with surface irregularities. The most common application method involves drawing the coating metal wire through a nozzle where it is melted by a gas flame and atomized by a blast of compressed air, which also carries the particles to the base-metal surface. Coatings can also be sprayed in powder form against the base metal by an explosion process, called flame or detonation spraying. A plasma-spraying process has also been developed, in which a high-velocity inert gas stream heated to 20,000°F melts the metal powder and blows it against the base-metal surface.

Most sprayed-metal coatings are fairly hard, and they usually have good wear resistance. Because of their relative high porosity, they are usually sealed or overcoated for corrosion-resistant applications. Almost any metal or alloy can be sprayed, but the most commonly sprayed metals are zinc and aluminum, which are primarily used to protect steel exposed to water or outdoor atmospheres. These coatings range in thickness from 0.001 to 0.015 in., depending on requirements.

15-16 Vapor-Deposited Coatings

These are thin single or multilayer coatings applied to base surfaces by deposition of the coating metal from its vapor phase. Most metals and even some nonmetals, such as silicon oxide, can be vapor-deposited. Vacuum-evaporated (or vacuum-metallized) films of aluminum are most common. They are applied by vaporizing aluminum in a high vacuum and then allowing it to condense on the object to be coated. Vacuum-metallized films are extremely thin, from 0.002 to 0.1 mil.

In addition to vacuum evaporation, vapor-deposited films can be produced by ion sputtering, chemical-vapor plating, and a glow-discharge process. In ion sputtering, a high voltage applied to a target of the coating material in an ionized gas media causes target atoms (ions) to be dislodged and then to condense as a coating on the base material. In chemical-vapor plating, a film is deposited when a metal-bearing gas thermally decomposes on contact with the heated surface of the base material. And in the glow-discharge process, applicable only to polymer films, a gas discharge deposits and polymerizes the plastic film on the base material.

15-17 Diffusion Coatings

Diffusion coatings, also called cementation coatings, involve impregnating the surface of a metal with fine particles of another material, usually a metal also. The base metal is heated in the presence of the coating material to a temperature that promotes the diffusion process. The resulting coating system can consist of several layers. The layer formed adjacent to the base metal may be an intermetallic compound, a solid solution, or just particles diffused in the grain boundaries. Subsequent layers are richer in the coating material. The principal reasons for using diffusion coatings are to increase hardness, and to improve resistance to wear, corrosion, and oxidation.

Diffusion-coating materials are limited to those capable of alloying with iron or steel. There are over a half-dozen types in common use.

Calorized (aluminum) coatings are applied either by treating the metal in a powdered-aluminum compound or in aluminum-chloride vapor, or by spraying the aluminum and then heat-treating the coated part. The coatings range in depth from 0.005 to 0.040 in., and make parts serviceable in temperatures up to 1400°F.

Carburized-steel surfaces are obtained by introducing carbon into a steel surface. The metal is heated above the critical (transformation) temperature in contact with a carbonaceous material, which can be a solid, gas, or liquid. In general, carburizing is limited to steels with less than 0.45 percent carbon.

Cyaniding and carbonitriding cause carbon and nitrogen to diffuse into the base-metal surface. A liquid bath is used in cyaniding, and a gas atmosphere is used in carbonitriding. For extra-high hardness, parts are quenched after the diffusion operation.

Nitriding consists of exposing steel parts to gaseous ammonia at about 1000°F to form metallic nitrides at the surface. The hardest coatings are obtained with aluminum-bearing steels. Nitriding of stainless steel is known as Malcomizing.

Chromizing consists of diffusing chromium into ferrous materials by packing parts in a proprietary powdered-chromium compound and heating it at 1500 to 1900°F. High-chromium—iron alloy coatings, about 0.003 in. thick, are produced on low-carbon-content alloys. On high-carbon materials, a chromium-carbide coating with exceptional hardness is obtained.

Sheradizing involves applying zinc coatings to ferrous and nonferrous metals by heating the parts in zinc powder at 650 to 700°F for 3 to 12 hr. And siliconizing involves impregnating steel and iron parts with silicon to form a case containing about 14 percent silicon. The depths of the cases can vary from 5 to 10 mils.

Conversion Finishes

Conversion coatings are inorganic barrier films produced by chemical reaction with the surface of the base steel. (In anodic coatings, the process involves electrochemical action.) Conversion finishes differ from organic and metallic coatings in that they are an integral part of the base metal. They are particularly useful as a base for organic coatings. They also can serve as a decorative finish, corrosion-resistant barrier, adherent base for lubricants, wear-resistant surface, electrically resistant surface, and an aid in cold forming. The main types are phosphate, chromate, chemical oxide, and anodic.

15-18 Phosphate Coatings

Phosphate coatings are formed by chemically reacting a metal surface with an aqueous solution of a soluble metal phosphate, phosphoric acid, and accelerators. They vary in color from iridescent green to light green, depending on thickness. They are widely used on iron, steel, and zinc, and, to a lesser extent, on aluminum, cadmium, and tin. Paint-base phosphates for iron and steel consist of zinc-iron or manganese-iron phosphates. Heavy phosphate coatings are normally used in conjunction with an oil or wax for corrosion resistance. The combination has a synergistic effect, giving much greater protection than the sum of two taken separately.

15-19 Chromate Coatings

Chromate coatings are essentially insoluble chromium compounds, generally less than 0.02 mil thick. Unlike phosphate coatings, they are amorphous, nonporous, and tend to be self-healing. They are used on zinc die castings and electroplates, on cadmium electroplates, and on copper, aluminum, and magnesium metals.

Chromate films have better corrosion resistance than phosphate coatings, but they are low in abrasion and electrical resistance. Their color varies from iridescent gold to brown, depending on thickness. These coatings are generally used as a paint base.

15-20 Chemical Oxide Coatings

Oxide coatings vary widely in composition, but most are produced by exposing the base metal to hot, oxidizing solutions or gases. They are usually quite thin (0.02 to 0.2 mil), have various colors, and are used for decoration, corrosion protection, and as a paint base.

Black oxide coatings are produced on steels by immersion in a hot (300°F), strongly alkaline solution. They are usually sealed with oil or wax. Stainless steels are coated by immersion in fused dichromate. Oxide films of various colors, ranging from straw yellow to light blue, are produced by heating steel in air at temperatures from 400 to 640°F for specific time periods. A black gun-metal finish that can be oiled is produced by heating steel with carbon and oil at 650 to 750°F.

15-21 Anodic Coatings

Oxide coatings can also be produced by an electrochemical process in which the base metal is immersed in an electrolyte solution, becoming the anode. The resulting electrolytic action converts the surface to an oxide layer. Sulfuric acid is the most commonly used electrolyte. The thickness and porosity of the coating depend on electrolyte composition and operating conditions, and can be controlled over a considerable range. Coatings range from 0.001 to 0.003 in. thick, and they can be clear or colored.

The anodic-coating process is mostly used on aluminum alloys for both decoration and protection. Electrolytes include sulfuric, chromic, and oxalic acids. Hard, dense films, about 0.002 to 0.004 in. thick, are used where normal corrosion and wear resistance are specified. Colored coatings are obtained either by immersion in warm dye solutions or by chemical precipitation of mineral pigments. The film is then sealed by immersion in hot water or dilute nickel acetate.

Integrally colored anodic finishes can be obtained as a result of the chemical constituents in certain alloys reacting with the electrolyte. They have greater stability than impregnated dyes. Typical colors possible are brown, gold, yellow, bluish white, and gray.

Exceptionally thick (up to 0.01 in.) anodic coatings are produced where good wear, electrical, and/or heat resistance are required. Pre-anodized aluminum sheet is manufactured for a wide variety of uses, such as auto components, appliance parts, sporting goods, and furniture.

Anodizing is also used on magnesium alloys, zinc die castings, wrought zinc, and galvanized steel. Anodized coatings on magnesium are much softer, less dense, and not as corrosion resistant as those on aluminum. Ordinarily, anodized coatings on magnesium are covered with an organic finish or an inorganic sealant.

The anodizing process applied to zinc converts the surface to a complex, fritted compound that is harder, thicker, and more corrosion resistant than other conversion coatings on zinc. The principal colors possible are green, gray, and brown.

Ceramic Coatings

The terminology for coatings mainly composed of ceramic ingredients is not very precise. For our purposes here, they will be divided into two groups: (1) porcelain enamels and glass linings, often referred to as vitreous enamels, and (2) ceramic coatings.

15-22 Vitreous and Porcelain Enamels

Vitreous enamels are glasslike coatings that are fused to base metals, giving a tightly adhering, hard, and durable finish. They are composed of a glassy matrix in which crystalline opacifiers and pigments are suspended. The matrix is usually borosilicate, and the opacifiers are oxides of titanium, antimony, zirconium, lead, and tin. The basic material from which vitreous enamels are made is referred to as frit.

Vitreous enamels are sometimes divided into porcelain enamels and glass linings. The distinction is on the basis of coating thickness and application. Generally, glass linings have thicknesses of 0.025 in. or greater, and are used chiefly on industrial products, particularly cast-iron and steel equipment such as piping, tanks and vessels, tank cars, and chemical-processing equipment. Porcelain enamel thicknesses are generally under 0.025 in., and they are formulated for use on consumer products, such as stoves, refrigerators, and plumbing fixtures.

Not all metals can be satisfactorily vitreous-enameled. Steel and iron are most amenable to vitreous coatings. An exceptionally pure iron, known as enameling iron, is specifically designed and processed for porcelain enameling. Also, special steels, called enameling steels, have a low metalloid content and are specially processed to accept a porcelain-enamel cover coat without the need of a ground coat. One of these steels is titanium bearing, and the other is extremely low in carbon. In the nonferrous family, application of vitreous enamels is limited mainly to copper and aluminum. In aluminum, the frits are formulated to permit firing at or below 1000°F.

Coating Systems. Vitreous-enamel systems consist of one or more layers. The first coat is called the ground coat; it is analogous to a primer in organic coating systems. It is about 0.003 to 0.004 in. thick, and contains cobalt oxide and, often, oxides of nickel and manganese, to promote good adhesion. While ground coats are generally intended as a base for cover coats, they sometimes serve as the final finish when their dark color is not objectionable.

One or two cover coats, 0.003 to 0.005 in. thick, are normally applied over the ground coat in porcelain-enamel systems. Cover coats in glass-lining systems are thicker. There are hundreds of different cover-coat formulations to meet various decorative and protection requirements. They are usually classified either according to the type of opacifier used or according to their acid resistance. There are two general types of opacifier: opaque white and nonopaque.

One-coat procelain enamels are also available. They consist of a titanium-opacified cover coat, 0.003 to 0.006 in. thick, fired directly onto low-carbon enameling steels. They can be produced in white and some solid colors.

Properties. White porcelain enamels are the most widely used, but vitreous enamels come in practically any color or shade. In general, colored vitreous enamels are highly durable and will retain their color for a ling time. Porcelain enamels can be produced in a variety of finishes, from an extremely high gloss to a dull, full matte, as well as in marblelike designs and other specialty patterns.

Their chemical resistance varies over a wide range. Acid resistance is frequently rated in accordance with a standard test developed by the Porcelain Enamel Institute. The grades run from class AA to class D, with class AA being the highest. Porcelain enamels are reasonably resistant to alkalies, but they are attacked by those containing free caustic.

The softening temperature of most vitreous enamels is around the firing temperature. However, their useful service temperature usually ranges between about 600 to 1000°F. Some special high-temperature enamels will withstand temperatures up to 1700°F for a short time. Thermal-shock resistance is usually good. Most types withstand rapid changes in temperature of at least 200°F without damage. For best thermal-shock resistance, the difference in thermal expansion between the enamel and the base metal should be as low as possible.

Vitreous enamels are hard and dense. Their hardness is approximately 4 to 6 on Mohs' scale, compared with around 2 or 3 for organic finishes. The high hardness contributes greatly to their good abrasion and scratch resistance. Another related property is high smoothness, which provides low friction characteristics. This, combined with high hardness, gives an abrasion-resistant coating that is ideal for material handling and conveying equipment.

Although vitreous enamels are susceptible to chipping, this disadvantage can be minimized by proper product design and careful selection of coating thickness. In general, the thinner the coating, the less the tendency to chip.

15-23 *Ceramic Coatings*

Ceramic coatings are similar in most respects to vitreous enamels, except that they have considerably better high-temperature resistance. This higher refractoriness is obtained by adding high-refractory compounds, such as alumina, chromium oxide, cobalt oxide, and beryllium oxide, to the frit. The highest temperature-resistant ceramic coatings consist of pure metal oxides such as alumina and zirconium oxide. They are applied by flame or plasma spraying. The molten ceramic particles impinge on the base material, which is heated to a temperature of 600 to 800°F.

Many materials can be coated in this way, for example, ferrous metals, aluminum, graphite, glass, and even some plastics. Flame-sprayed ceramic coatings have a porosity of about 8 to 15 percent, and are therefore generally lower in corrosion resistance than vitreous coatings.

Review Questions

1. Explain the difference between the protective mechanism of a barrier-type coating and an anodic metal coating.
2. Name three ways in which organic coatings change from a liquid or semiliquid to a solid finish.
3. Vehicles and pigments are the two principal components in many organic formulations. State their functions.
4. Name the three principal kinds of layers that often make up an organic coating system, and state one function of each.
5. Identify the following by type of vehicle (oil, resin, or varnish):
   (a) Japan.
   (b) Phenolic.
   (c) Linseed oil.
   (d) Alkyd and drying oil.
6. Describe two distingishing features in the composition of so-called water-base coatings.
7. How is adhesion obtained with plastic powder coatings?
8. Name the electroplate metal that best fits each of the following descriptions:
   (a) Resembles polished silver.
   (b) Provides high hardness and low coefficient of friction.
   (c) Used primarily as undercoat.
9. Why is nickel used as an undercoat for chromium electroplates?
10. Describe the two layers of hot dip coatings.
11. State the essential difference between the structural nature of sprayed metal coatings and electroplates.
12. Name the "impregnant," or substance, that is used to produce diffusion coatings with the following processes:
    (a) Carburizing
    (b) Nitridizing
    (c) Chromizing
    (d) Calorizing
    (e) Sheradizing

13. Name two mechanisms by which conversion finishes are produced on metals.
14. Name the conversion finish that best fits each of the following descriptions:
    (a) Oxide coating on aluminum.
    (b) Paint base on irons and steels.
    (c) Produced by exposing metal to hot oxidizing solution.
15. How do porcelain enamels, as a class, compare with organic coatings in the following properties?
    (a) Resistance to temperatures in the range from 600 to 1000°F.
    (b) Hardness and wear resistance.
    (c) Resistance to impact.

Bibliography

Clauser, H. R.: "Organic Finishes for Metals," *Materials Engineering*, December 1947.

———: "Porcelain Enamels," *Materials Engineering*, February 1950.

Dreger, Donald R.: "Paints that Don't Pollute," *Machine Design*, February 21, 1974.

"Finishes for Metal Products," *Materials Engineering*, September 1955.

Jastrzebski, Z. D.: *Engineering Materials*, John Wiley & Sons, Inc., New York, 1959.

Mock, John A.: "Plastic Powder Coatings, Cut Costs, Up Performance," *Materials Engineering*, March 1969.

INDEX

Abrasion resistance, 277, 338–339
ABS plastics, 242–243, 249
Acetal, 239–241
Acrylic, 234–235, 250
Additives in plastics, 208–210
Adhesion, 410
Age hardening (see Precipitation hardening)
Air-melted steels, 113
Alkyd, 254, 255, 269
Alloys:
  definition of, 81
  multi-phase, 82
  single-phase, 82
Allotrophy, 83, 104
Alloying in ferrous metals, 110–112
Allylics, 256
Alumina (see Aluminum oxide)
Aluminide, 360
Aluminum, 155–166
  anodic coating of, 426
  applications of, 161, 164–166
  cast alloys, 164–166
  coatings, 424
  composite matrix, 390
  designation systems, 157, 160
  flakes, 405
  heat-treatable alloys, 162–164
  microstructure, 156–163
  nonheat-treatable alloys, 160–161
  processing of, 156–157, 164
  properties of, 156, 161, 163, 165
  SAP alloy, 396
  temper designations, 158–159
Aluminum brass, 180
Aluminum bronze, 180–181
Aluminum oxide (alumina), 354
Aluminum powders, 166, 396, 397
Amber mica, 377–378
Amino, 258–259
Amosite asbestos, 379

Annealing of metals, 92–93
Anodic coating, 426
Antimony, 188
Appearance characteristics, 410–411, 415
Application analysis, 7–13
Arc resistance, 68
Aromatic polyester, 246–248
Asbestos, 379
  fibers, 262
  paper, 322
Atomic bonding, 24
Atoms:
  in diffusion, 83–84
  interstitial, 80
  packing in crystals, 76–77
  in solid solutions, 82
  substitutional, 80
  structure of, 24
Austempering, 90
Austenite, 88, 108
Austenitic manganese steel, 136

Bainite, 110
Balsa wood, 304
Barium-carbonate-nickel cermets, 396
Beryllide, 360
Beryllium, 173
Beryllium copper, 182–183
Beryllium oxide (beryllia), 354–355
Bismuth, 188
Black oxide coating, 425
Blow molding, 213–214
Blowing glass, 363
Body-centered unit cell, 29, 76–77
Bonds:
  atomic, 24
  in ceramics, 346, 349, 353
  in composites, 33–34
  covalent, 26, 346, 363

431

Bonds:
　in crystals, 76, 346, 349
　in elastomers, 274
　in glass, 363
　intermolecular, 204, 206, 250
　ionic, 26, 346
　metallic, 26, 27
　in polymers, 29, 204, 206, 250, 274
　secondary, 27
　van der Waals, 27
Borate glass, 370
Boride, 358
Boron carbide, 358
Boron fiber composites, 390, 391
Boron nitride, 358–359, 391
Borosilicate glass, 368
Borozon, 358, 373
Branched polymers, 205
Brass, 175, 177–180
Brass electroplates, 422
Brass powder, 183
Brazing, 99–100
Brinell hardness, 57
Brittle fracture, 41
Bronze, 180–183
Bulk modulus, 47
Butyl elastomer, 290–291

Cadmium, 189
Cadmium electroplate, 420
Calendering, 214
Calorized coating, 425
Capped steel, 113
Carbide, 358
Carbide cermets, 395–396
Carbon:
　composition of, 370
　effects on properties, 106
　in ferrous metals, 104–105
　microstructure, 24–25, 371
Carbon black, 279, 288
Carbon steel (see Plain carbon steel)
Carbonitriding, 424
Carburizing, 424
Cast iron:
　composition of, 139–142
　microstructure, 139–142
Cast steels (see specific types)
Casting glass, 364–365
Casting metals, 94–95

Casting plastics, 214
Catalysts, 209
Cathodic protection, 65, 411
Cellular glass, 370
Cellulose acetate, 236
Cellulose acetate butyrate, 237
Cellulose fiber paper, 322
Cellulosics, 235–237
Cementite (see Iron carbide)
Centrifugal casting, reinforced
　　plastics, 266
Ceramic bond, 346, 353
Ceramic coatings, 426–429
Ceramic fibrous paper, 322
Ceramics:
　chemical properties of, 351
　composition of, 346
　electrical properties of, 351–352
　macrostructure, 346
　major characteristics of, 345–346
　mechanical properties of, 349–350
　microstructure, 346
　processing, 347–349
　thermal properties of, 350–351
Ceramoplastic, 378–379
Cermets, 394–396
Charpy impact test, 52
Chemical corrosion, 63
Chemical-deposition coating, 421–422
Chemical oxide coating, 425
Chemical resistance:
　of fibers, 339, 341
　of wood, 309–310
Chlorinated polyether, 245
Chloroprene elastomer, 289–290
Chlorosulfonyl polyethylene elastomer, 293–294
Chlorotrifluoroethylene, 245
Chromate coatings, 425
Chromium, 194
　in stainless steels, 127
Chromium electroplates, 420
Chromizing, 425
Chrysotile asbestos, 379
Clad-metal composites, 402
Coated glass, 370
Coating of plastics, 215
Coatings (see Finishes)
Cobalt, 192–193

Cold heading of metals, 96
Cold working, 84–88
Colloidal graphite, 373
Color, 70, 411, 415
Colorants, 209
Colored glass, 370
Columbium, 196
Composites:
  constituents in, 383–384
  contributory properties of, 385
  definition of, 382–383
  filled, 406–407
  flake, 404–405
  history of, 381–384
  interdependent properties of, 385–386
  laminar, 398–404
  macrostructure, 33–35
  mixture rule, 384
  particulate, 393–398
Compressed wood, 313
Compression molding, 211
Compression-set values in elastomers, 276
Compression stress, 38
Compressive strength, 45
Constituents in composites, 32
Constraints in selection, 11–12
Conversion coatings, 424–426
Copolymer, 205
  definition of, 29
Copper, 155, 174–177
  properties of, 174–177
  temper designations of, 175
  types of, 175–177
Copper electroplates, 421
Copper powders, 184, 397–398
Cordage, 336
Cordierite, 356–357
Corrosion, 63–66, 411
Corrosion-resistant cast steels, 132–134
Creep, 48–51
  in elastomers, 276
Creep modulus (apparent), 51
Creep strength, 50
Crimping of paper, 324
Crocidolite asbestos, 379
Crystalline bonding:
  in ceramics, 346, 349, 353

Crystalline bonding:
  in elastomers, 274
  in polymers, 207
Crystals:
  ceramic, 346
  dislocations, 81
  grain size, 78–79, 88, 92
  imperfections, 31, 79–81
  lattice, 29
  line defects, 81
  multi-phase, 31
  point defects, 79
  polycrystals, 77–79
  single-phase, 31
  structures, 29–31
  unit cells, 76–77
  work hardening, 85
Cupronickel, 182
Cyaniding, 424
Cyclic stresses, 54

Decay resistance of wood, 309
Deformation, 38, 39
Denier (density) of paper, 300–331
Density of polymers, 207, 208
Deterioration resistance of elastomers, 277
Dezincification, 65
Diallyl phthalate, 254, 256
Diamond, 24–25, 358–359, 373
Die casting, 95
Dielectric constant, 68
Dielectric strength, 67–68
Diffusion in metals, 83–84
Diffusion coatings, 424
Diffusion heat treatment, 91
Dislocations, 31, 81
Dispersion coating, 418
Dispersion-hardened alloys, 396–397
Drape, textiles, 341
Drawing of glass, 363–364
Drawing of metals, 96–97
Drying oil, 414
Ductile fracture, 41
Ductile iron, 146–148
Ductility, 47–48, 51, 77, 80, 87, 92, 350
Durometer hardness, 59
Durometer (Shore) test, 276

Ebony wood, 304
Economic factors, 12–13
Elastic modulus (see Modulus of
    elasticity)
Elastic strain, 40, 43
Elastomers:
    classification system, 278–279
    definition of, 273
    major characteristics of, 273–274
    microstructure, 29, 274
    types of, 278
    uses of, 266–288
Electrical conductivity, 67, 310
Electrical insulation, 67–68, 341
Electrical resistivity, 67
Electrical steel, 120
Electrochemical corrosion, 64–66
Electroforming, 95
Electroless coatings, 421
Electrons, 24, 26
Electroplated coatings, 419–421
Electrosprayed coatings, 418
Elongation, 47
Embossed paper, 325
Emissivity, 62
Emulsion coating, 418
Enamel, 415–416
Enameling steel, 427
Endurance limit, 55
Epichlorohydrin elastomer, 294
Epoxy, 256–258, 269, 390
Ethyl cellulose, 237
Ethylene copolymer, 230–231
Ethylene ethyl acrylate, 230
Ethylene propylene elastomer, 293
Ethylene vinyl acetate, 230
Extrusion:
    of ceramics, 348
    of metals, 96
    of plastics, 217

Face-centered unit cell, 29, 77
Factory lumber, 301
Failure, definition of, 42
Failure analysis, 18–20
Fatigue, 52–56
Fatigue resistance, 308
Fatigue strength, 53, 54
Felts, 333–334
Ferrite in steel, 107

Ferrites, 360–361
Ferroelectrics, 361
Ferrous-base powder, 136–138
Ferrous metal, definition of, 103–104
Fiber-aspect ratio, 387
Fiber composites, 33, 387–393
    advanced types of, 390–393
    fabrication of, 391
    hybrids, 393
    industrial types of, 388–390
    property factors, 387–388
Fiber-fiber composites, 389–390
Fiber-reinforced plastics, 260–270,
    389
    processing methods, 264–266
    resins, 262–264
Fibers:
    definition of, 330
    geometrical characteristics of,
        330–331
    graphite, 372
    Kevlar, 391–392
    microstructure, 330
    in plastics, 262
Fibrous glass paper, 322
Filament winding, 265–266, 391
Filled composites, 33, 406–407
    structure of, 406
    types of, 406–407
Filters, 209
Finish coat, 413
Finishes:
    definition of, 409
    functions of, 409–410
    performance factors of, 410–411
Finishing:
    of glass, 365
    of plastics, 218
Fire retardants, 210
Firing of ceramics, 349
Flake composites, 33, 404–405
Flame hardening, 91
Flexural strength, 46
Fluidized bed coating, 418
Fluorocarbon, 244
Fluorocarbon elastomer, 294–295
Fluoroplastic, 243–245
Fluorosilicone elastomer, 295
Foam glass, 370
Forging, 96

Forsterite, 357
Fracture, 41–42
Fused-silica glass, 368
Fusion welding of metals, 99

Galvanic series, 64–65
Gas permeability of textiles, 341
Glass:
  composition of, 362–363
  microstructure, 363
  processing, 363–365
  types of, 366–370
Glass-bonded mica, 378–379
Glass ceramic, 361
Glass-fiber-reinforced plastics:
  performance principle, 385–386
  processing methods, 264–266
  properties of, 266–270
  thermoplastics, 263–264
  thermosets, 262–263
Glass fibers, 262
Glass flakes, 412
Glass lining, 427
Glass transition temperature, 208
Gold, 198
Gold electroplates, 421
Grain boundaries, 31, 79
Grains (see Crystals)
Graphite, 24–25, 105–106, 371–377
Graphite fiber composites, 390, 391
Graphite fibers, 372
Gray iron, 142–146
Green lumber, 304
Growth rings in wood, 298

Hand, textiles, 341–342
Hand lay-up molding, 264–265
Hardboard, 314
Hardenability of metals, 88–90
Hardening of metals, 88–91
  (See also specific methods)
Hardness, 56–60, 80, 87, 276, 308–309
Hardwood, 299–301, 304
Heartwood, 298
Heat, 60–61
Heat capacity, 61
Heat distortion point, 61
Heat resistance, 61–62, 339, 340
Heat-resistant cast steel, 132–133

Heat treatment, 88–93, 162, 365
Hexagonal unit cell, 29, 77
High-alloy quenched-and-tempered steel, 127
High-copper alloy, 182
High-energy radiation effects, 69–70
High-pressure laminating, 215–216
High-strength low-alloy steel, 123–125
Homopolymer, 29, 205
Honeycomb composites, 400, 401
Honeycomb filled composites, 406
Hooke's law, 43
Hot-dip coating, 422
Hot-melt coating, 419
Hot-rolled steel, 113
Hot working, 84, 85
Hypalon, 293–294
Hysteresis in elastomers, 275–276

Impact strength, 52
Impregnated wood, 313
Indium, 189
Induction hardening, 91
Industrial carbon, 371
Ingot iron, 104
Injection molding, 210–211, 264
Inorganic fiber paper, 322
Interface, 33
Intergranular corrosion, 66
Intermediate coat, 413
Intermetallics, 360
International Annealed Copper Standard (IACS), 67
Interphase, 33
Interstitialcy, 80, 82
Investment casting, 95
Ionomer, 229, 230
Iridium, 199
Iron:
  allotropic forms, 104
  cast, 139–152
  microstructure, 104–105, 139, 142
  wrought, 139
Iron-base superalloys, 134–135
Iron carbide, 105, 107
Isobutylene-isoprene elastomer, 290–291
Isoprene elastomer, 291
Izod impact test, 52

435

Jiggering, 347
Joining:
 metals, 98–100
 plastics, 219
Jominy test, 90

Kevlar, 391–392
Killed steel, 113
Kiln-dried lumber, 304
Knit fabrics, 334–336
Knoop hardness, 59
Kraft paper, 322

Lacquer, 416–417
Laminar composites, 33, 398–404
Laminates, 401–404
 glass, 370
 inorganic-inorganic, 404
 metal-inorganic, 403
 metal-metal, 402
 metal-organic, 402–403
 organic-inorganic, 403–404
 organic-organic, 403
 paper, 324
Laminating, continuous, 266
Latex, 278
 coating, 418
Lauan (tanguile), 304
Lead, 187
Lead electroplates, 422
Lead glass, 368
Lead powder, 398
Leaded brass, 178–179
Ledeburite, 110
Loss factor, 68
Low-alloy carbon steel:
 alloying elements, 118
 applications of, 121
 grades, 118–122
 properties of, 120
 wrought, 118–122
Low-alloy steel, cast, 122–123
Low-melting alloys, 155, 186–189
Low-temperature low-alloy steel, 120

Machining:
 of metals, 97–98
 of plastics, 218
Macrostructure, 32–35, 383

Magnesium, 166–169
 applications of, 168
 cast alloys, 169
 properties of, 166–169
 wrought, 168–169
Magnesium oxide (magnesia), 356
Magnetic properties, 68–69
Mahogany wood, 304
Malleable iron, 146–151
 alloyed, 150–151
 ferritic, 150
 microstructure, 148
 pearlitic, 150
 properties of, 148–149
Manganese brass, 180
Maraging steel, 126
Martempering, 89
Martensite, 91, 108
Matched die molding, 264
Materials:
 application of, 2
 application analysis, 7–13
 classification of, 4–5
 definition of, 4
 history of, 1–2
 number of, 2
 production of, 2
Matrix in composites, 32, 33, 383, 387
Maximum service temperature, 61
Melamine, 258, 259, 269
Melting point, 61
Metal powders:
 aluminum, 166
 in composites, 393–398
 copper and alloys, 183–184
 ferrous, 136–138
 spray coating, 423
Metal processing, 93–100
Metallic coatings, 419–421
Metals:
 definition of, 75
 major characteristics of, 85
 number of, 75
 (See also specific types)
Mica:
 composition of, 378
 flakes, 405
 properties of, 378
 types of, 377–378

Microstructure:
　amorphous, 27, 31–32, 363
　bonding, 24
　crystal, 27
　glass, 27, 363
　of metals, 76–81
　of polymers, 27
Modulus of elasticity, 46–47
Modulus of toughness, 51
Mohs hardness, 56–57
Moisture in wood, 298, 304, 305, 308, 309, 318
Molding:
　ceramics, 348
　fiber-reinforced plastics, 264–265
　plastics, 210–214
Molecular structure, 27–29
Molecular weight, 205
Molybdenum, 194
Monomer, 27
Muscovite mica, 377–378

Natural rubber, 278, 288–289
Neoprene, 289–290
Nickel, 189–192
　applications of, 190–192
　electroless coating, 421
　microstructure, 189
　processing, 189–190
　properties of, 189–191
　TD nickel, 397
Nickel alloys, 190–192
Nickel electroplates, 420
Nickel-silver, 182
Nitride, 358
Nitriding, 425
Nitriding steel, 122
Nitrile butadiene elastomer, 291
Nonferrous metal:
　definition of, 155
　types of, 155, 156
　(See also specific types)
Nonwoven textiles, 331–333, 389–390
Normalizing, 92
Notch sensitivity, 52
Nylon (see Polyamide)

Oil resistance of elastomers, 277–278

Olefin copolymers, 229–231
Opacity, 70, 411, 415
Opal glass, 370
Organic coatings, 411–419
　coating application, 412
　coating systems, 412
　definition of, 411
　thickness of, 411
Osmium, 200
Oxidation, 66, 411
Oxide-base cermet, 395
Oxides, 353–358
　single, 353–356
　mixed, 356–358
Oxygen-free copper, 176

Paint, 417–418
Palladium, 199
Paper:
　coatings, 323–324
　definition of, 320
　production of, 320–322
　treatments, 323–325
　types of, 322–323
　types of pulp, 320
Particle board, 314, 316
Particulate composites, 33, 393–398
Pearlite, 108
Permanent mold casting, 94
Phenolic, 251–254, 269
　in particle board, 314
Phlogopite mica, 377–378
Phosphate coating, 425
Phosphor bronze, 181–182
Phosphorus-deoxidized copper, 176
Photosensitive glass, 370
Pigments, 415
Plain carbon steel:
　cast, 122–123
　wrought: applications of, 118
　　composition of, 114
　　grades of, 114–116
　　properties of, 106, 114, 116
Plain-sawed lumber, 299
Plain weave, 334
Plaster mold casting, 94
Plastic alloys, 248–250
Plastic deformation, 40, 41, 44, 47, 48, 51, 77, 85
Plastic flow (see Plastic deformation)

Plastic-powder coating, 418–419
Plasticizers, 209
Plastic strain, 40, 47
Plastics:
   definition of, 204
   major characteristics of, 204
   microstructure, 204, 205, 207
   processing methods, 210–219
Platinum group metals, 198–200
Plywood:
   grades and sizes of, 317–318
   properties of, 318–319
Polyacrylate elastomer, 291
Polyallomer, 229, 230
Polyamide, 238–239
Polybutadiene elastomer, 292
Polycarbonate, 240, 241, 249
Polyester, 254, 255, 262–263, 267
Polyethylene, 225–227
Polyimide, 246–248
Polymer:
   definition of, 27
   microstructure, 27, 205–208
   rubber, 274
   types of, 205–206
   wood, 297–298
Polymerization, 27
Polyolefins, 225–231
Polyphenylene oxide, 245, 250
Polyphenylene sulfide, 246, 248
Polypropylene, 227–229
Polystyrene, 231–233
Polysulfide elastomer, 293
Polysulfone, 247, 248
Polyvinyl chloride, 233–234, 248–250
Polyvinyl dichloride, 234
Porcelain, 352–353
Porcelain enamel, 427–428
Porosity, 70
Powder metallurgy (P/M) process, 97, 394–395
Powder metals (see Metal powders)
Precious metals, 156, 197–200
Precipitation hardening, 90–91, 162
Precoated metals, 402
Prefinished steel, 139–141
Pressing:
   ceramics, 348
   glass, 363

Pressing:
   metals, 96–97
Pressure welding, 99
Primers, 412–413
Process annealing, 92
Properties, 10–11
Property-structure relationships, 23, 24–25, 77, 205–207, 266, 274, 298, 330, 346
Proportional limit, 43
PT graphite, 372–373
Pulp molding, 213
Pultrusion, 266
Pyrolytic graphite, 371–372

Quarter-sawed lumber, 299
Quenched-and-tempered low-alloy steel, 125–126

Radiation effects, 69–70
Reconstituted wood, 313–315
Recrystallization, 84
Recrystallized graphite, 372
Reduction of area, 47–48
Refractories, 353
Refractory metals, 155, 193–196
Reinforcements in plastics, 210
Reliability, 8
Resilience:
   in elastomers, 275–276
   in fibers and textiles, 337–338
Resin vehicles (coating), 414
Rhodium, 199
Rimmed steel, 113
Rockwell hardness, 57–58
Rolling of metals, 95
Rotational molding, 212–213
Rosewood, 304
Rubber (see Elastomers)
Ruby mica, 377–378
Rupture strength, 44, 51
Ruthenium, 199

Safety, 8–9
Safety glass, 370
Sand casting, 94
Sandwich composites, 399–401
   cores, 400–401
   facings, 399–400
SAP alloy, 396

Sapwood, 298
Satin weave, 335
Scleroscope hardness, 59
Selection methodology, 11–20
Semikilled steel, 113
Shear modulus, 47
Shear strength, 45–46
Shear stress, 38
Sheet glass, 364
Sheradizing, 424
Shop lumber, 301
Silica glass, ninety-six percent, 369
Silicide, 360
Silicon brass, 180
Silicon bronze, 182
Silicon carbide, 358, 391
Silicon nitride, 358
Silicones:
   elastomer, 295
   plastics, 259–260, 269
Siliconizing, 424
Silver, 197–198
Silver-bearing copper, 176
Silver electroplate, 421
Silver flakes, 405
Slip casting, 347
Soda-lime glass, 366
Softwood, 299–300
   grades of, 300–301
Soldering, 99–100
Solid solutions, 82, 84
Solution anneal, 90
Solution heat treatment, 162
Solvent molding, 213
Sound absorption coefficient, 70
Specular gloss, 70
Spherodizing, 92
Spray-up molding, 265
Sprayed-metal coating, 423
Stabilizers, 209
Stainless steel, 127–132
   composition of, 127
   corrosion resistance of, 127–129
   oxide coating of, 425
   properties of, 129–130
   types of (grades), 130–132
   ultrahigh-strength, 126
Stamping:
   of metals, 96–97
   of plastics, 218

Steatite, 357
Steel, 103
   (See also specific types)
Stiffness:
   of fibers and textiles, 337
   of wood, 305, 308
   (See also Modulus of elasticity)
Stoneware, 352–353
Straight brass, 178
Strain, 38–41
Strength, 80, 87
   of fibers and textiles, 336–337
   of wood, 305–307
Strengthening heat treatment (see
   Hardening)
Stress, 38–40
Stress relieving, 92
Structural lumber, 301
Structural transformations, 82–84
Styrene acrylonitrile, 232–233
Styrene butadiene elastomer, 289
Sulfur-bearing copper, 177
Superalloys, 134–135
Superpolymers, 246–248
Surface hardening, 91
Swelling in elastomers, 277–278
Synthetic textile fibrous paper, 323
Systems approach, 2–13, 37, 38

Tactile properties, textiles, 341
Tantalum, 196
TD nickel, 397
Tear resistance of elastomers, 276
Tellurium-bearing copper, 177
Temperature, 60–61
Tempered martensite, 108
Tempering of steel, 91
Tensile strength, 43–44
Tensile stress, 38
Terpolymer, 29, 205
Thermal conductivity, 62, 309, 339
Thermal expansion, 62–63, 309
Thermal stresses, 62–63
Thermoforming, 216
Thermoplastics:
   definition of, 204
   microstructure, 29, 204
   raw material forms, 221
Thermosetting plastics:
   bonding, 250

Thermosetting plastics:
  definition of, 204
  microstructure, 29, 204
  processing cycle, 251
  raw materials, 250–251
Thiokol, 293
Thorium oxide (thoria), 356
Tin, 187
  electroless coating, 422
  electroplate, 420–421
Tin brass, 180
Titanium, 169–173
  applications of, 171
  castings, 171
  commercially pure, 171
  microstructure, 170
  processing, 171
  properties of, 169, 170–172
Titanium alloys, 171–172
Titanium carbide, 358
Tool and die steels, 135–136
Torsion strength, 45–46
Tough-pitch copper, 176, 177
Toughness, 51–52
  of fibers and textiles, 337–338
  of wood, 308
Transfer molding, 211–212
Transformations in metals, 82–84
Transmittance, 70
Tungsten, 195
Tungsten carbide, 358
Tungsten thoria, 396
Twill weave, 335
Twisting of paper, 325

Ultimate strength (see Tensile strength)
Ultrahigh-strength steels, 126–127
Unit cells, 29, 76–77
Uranium-dioxide cermet, 396
Urea, 258, 259, 314
Urethane, 260, 293

Vacancies, 31, 80
Vacuum-degassed steels, 113
Vacuum-melted steel, 113
Vacuum-metallized coating, 423

Vapor-deposited coating, 423
Varnish, 414, 417
Vehicles (coating), 413–414
Vickers hardness, 58, 59
Vinyl, 233–234
Vinylidene chloride, 234
Viscoelasticity, 40, 205
Vitreous enamel, 427–428
Vulcanization, 274

Water absorption, 67
  (See also Moisture)
Water-base coating, 418
Wear resistance, 56, 60
Weatherability, 339–341
Weighted property indices, 17–18
Welding, 99
Whiskers, 392
White irons, 151–152
Wood:
  chemical composition of, 297
  consumption of, 297
  imported, 304
  kinds and grades of, 299–304
  laminated, 403
  macrostructure, 298–299
  microstructure, 297–298
  moldings, 316
  processing, 310–312
  treatments, 310, 313
Work hardening, 85–88
Woven fabrics, 334–336, 389–390
Wrought iron, 139

Yard lumber, 300–303
Yarn, 331
Yield strength, 44
Young's modulus, 47
  (See also Modulus of elasticity)

Zinc, 184–186
  applications of, 185, 186
  casting alloys, 184–186
  electroplates, 421
  properties of, 185
  wrought alloys, 186
Zircon, 357–358
Zirconium oxide (zirconia), 356